Finite Element Method for Hemivariational Inequalities

Nonconvex Optimization and Its Applications

Volume 35

The titles published in this series are listed at the end of this volume.

Finite Element Method for Hemivariational Inequalities
Theory, Methods and Applications

by

Jaroslav Haslinger
Charles University,
Czech Republic

Markku Miettinen
University of Jyväskylä,
Finland

and

Panagiotis D. Panagiotopoulos
Aristotle University,
Greece

KLUWER ACADEMIC PUBLISHERS
DORDRECHT / BOSTON / LONDON

A C.I.P. Catalogue record for this book is available from the Library of Congress.

ISBN 978-1-4419-4815-1

Published by Kluwer Academic Publishers,
P.O. Box 17, 3300 AA Dordrecht, The Netherlands.

Sold and distributed in North, Central and South America
by Kluwer Academic Publishers,
101 Philip Drive, Norwell, MA 02061, U.S.A.

In all other countries, sold and distributed
by Kluwer Academic Publishers,
P.O. Box 322, 3300 AH Dordrecht, The Netherlands.

Printed on acid-free paper

This book is dedicated
to the memory of
Professor P.D. Panagiotopoulos

Contents

The mathematical formulation of many problems in physics leads to the following inclusion type problem:

$$\begin{cases} \text{Find } u \in V \text{ such that} \\ f - Au \in B(u) \quad \text{in } V^*, \end{cases} \qquad (0.1)$$

where V^* is the dual space to a Banach space V, $f \in V^*$ is given, $A : V \to V^*$ is a single-valued mapping while $B : V \to V^*$ is generally a multivalued one. Firstly inclusions involving maximal monotone operators were analysed. One of the most investigated cases is this one in which B corresponds to the subgradient of a convex functional Φ. Then (0.1) is equivalent to the variational inequality:

$$\text{Find } u \in V : \langle Au, v - u \rangle + \Phi(v) - \Phi(u) \geq \langle f, v - u \rangle \quad \forall v \in V. \qquad (0.2)$$

If A is a potential mapping, i.e., there exists a functional $\Psi : V \to \mathbb{R}$ such that $A = \text{grad}\, \Psi$ then (0.2) is equivalent to the following minimization problem:

$$\text{Find } u \in V : \mathcal{L}(u) \leq \mathcal{L}(v) \quad \forall v \in V, \qquad (0.3)$$

where $\mathcal{L} = \Psi + \Phi - \langle f, \rangle$. The fact that the subgradient of a convex function is maximal monotone is very important from the theoretical point of view (the existence and the uniqueness of solutions to (0.1)) as well as from the computational point of view (reliable and efficient mathematical programming methods for the resolution of (0.3) are available). The monotonicity assumption on B is however very restrictive. There are many problems in mechanics in which constitutive laws defining the mapping B are no longer monotone. If such a situation occurs, the problem cannot be treated in the frame of convex analysis and more general tools have to be employed. The progress in nonconvex and nonsmooth analysis allowed to involve nonmonotone multivalued mappings into (0.1). P.D. Panagiotopoulos was one of the first who recognized the importance of this new tool for the mathematical formulation of problems in mechanics of solids. Since eighties he paid his attention to differential inclusions with nonmonotone multivalued mappings. He termed them *hemivariational inequalities*.

Recently the mathematical analysis of hemivariational inequalities is well established.

In parallel with the mathematical analysis of hemivariational inequalities a natural demand appears, namely how to solve them numerically. In other words, how to discretize (0.1) in order to get a model with a finite number degrees of freedom which is close to the original (continuous) setting and whose solutions can be found by using appropriate numerical methods.

The convergence analysis for hemivariational inequalities is the main goal of this monograph. We shall restrict our presentation to a class of hemivariational inequalities in which the multivalued operator B is given either by the generalized gradient $\bar{\partial}\Phi$ of a locally lipschitz function $\Phi : V \to \mathbf{R}$ (unconstrained hemivariational inequalities) or by the sum $\bar{\partial}\Phi + N_K$, where N_K is the normal cone of a closed convex subset K of V (constrained hemivariational inequalities). The approximation of hemivariational inequalities will be based on a Galerkin type procedure which will be applied to the following equivalent form of (0.1):

$$\begin{cases} \text{Find } (u, \Xi) \in V \times Y \text{ such that} \\ Au + \Xi = f \quad \text{in } V^* \\ \Xi \in B(u) \quad \text{in } V^*, \end{cases} \qquad (0.4)$$

where $Y \subset V^*$ is an appropriate function space. We shall approximate both components u and Ξ separately on their own finite dimensional spaces V_h and Y_h. The approximation of (0.4) reads as follows:

$$\begin{cases} \text{Find } (u_h, \Xi_h) \in V_h \times Y_h \text{ such that} \\ A_h u_h + P_h \Xi_h = f_h \quad \text{in } V_h^* \\ \Xi_h \in B_h(u_h), \end{cases} \qquad (0.5)$$

where A_h, B_h, f_h, V_h, Y_h are the respective approximations of A, B, f, V, Y and P_h is a linear mapping from V_h into Y_h. We shall formulate sufficient conditions on approximated data under which (0.4) and (0.5) are close. Just the fact that not only u but also Ξ is approximated is important. First of all Ξ represents a mechanical quantity (a reaction or a friction force on contact surfaces, e.g), whose knowledge is useful. There is yet another reason. Besides (0.5) we shall also consider a substationary type problem:

$$\begin{cases} \text{Find } \bar{u}_h \in V_h \text{ such that} \\ 0 \in \bar{\partial}\mathcal{L}_h(\bar{u}_h), \end{cases} \qquad (0.6)$$

where $\mathcal{L}_h : V_h \to \mathbf{R}$ is the so called *superpotential* of the problem, whose form depends on A_h, B_h, P_h and f_h. This new problem is defined in such a way that knowing $\bar{u}_h$ one can find $\Xi_h \in Y_h$ such that the pair $(\bar{u}_h, \Xi_h)$ solves (0.5). Instead of (0.5) we shall solve (0.6). The reason why (0.5) is replaced by (0.6) is the absence of mathematically justified numerical methods for the realization of (0.5) on the one hand and the existence of a well developed class of nonsmooth optimization methods for the realization of (0.6) on the other hand. As usually, minimization methods do not discover $\bar{u}_h$ exactly but with a certain accuracy. Let u_h^* be numerically computed $\bar{u}_h$. The question is if u_h^* is good enough. Let $P_h : V_h \to Y_h$ be such that the equation

$$P_h \Xi_h = g_h \in V_h^* \tag{0.7}$$

has a unique solution for any $g_h \in V_h^*$. Then inserting $g_h := f_h - A_h u_h^*$ into the right hand side of (0.7) we obtain its unique solution Ξ_h^* . If u_h^* is "a good approximation" of $\bar{u}_h$ then Ξ_h^* should belong to $B(u_h^*)$ as follows from the last inclusion in (0.5). If not, u_h^* has to be found with a higher accuracy. Thus the knowledge of Ξ_h is important for taking a decision on reliability of the numerical result.

Our aim was to write a self-contained book. For this reason, after an introductory part presenting a very simple hemivariational inequality demonstrating all difficulties, Chapter 1 containing mathematical preliminaries follows. It collects basic results from the theory of Lebesgue, Sobolev and Bochner spaces as well as elements of convex and nonconvex analysis which will be used in subsequent parts. This chapter is also completed by classical results on the approximation of elliptic and parabolic equations and inequalities with monotone operators. As we have already mentioned, hemivariational inequalities represent a mathematical tool enabling us to model nonmonotone phenomena.

In Chapter 2, the motivation for the study of such a type of problems in mechanics of solids is discussed. Problems with nonmonotone reaction-displacement relations on a contact boundary and nonmonotone strain energy density functions, relating the stress and strain tensors are shown. The rest of the book is devoted to the approximation of hemivariational inequalities.

Approximation of elliptic hemivariational inequalities is studied in Chapter 3. We start with the simplest case, i.e., with unconstrained hemivariational inequalities of scalar type. Sufficient conditions for the existence of solutions to the discretized problem are formulated. Then we present sufficient conditions for systems of finite dimensional spaces $\{V_h\}$, $\{Y_h\}$ and systems of mappings $\{A_h\}$, $\{B_h\}$, $\{f_h\}$, $\{P_h\}$ being the approximations of the original data under which solutions to (0.5) are close on subsequences to solutions of (0.4). Then we describe the construction of V_h and Y_h by using the finite element method and verify all the assumptions guaranteeing the convergence. A special section is devoted to the construction of the superpotential $\mathcal{L}_h$ and to the analysis of the mutual relation between solutions of (0.5) and (0.6). In the remaining part of this chapter convergence results are extended to constrained and vector-valued hemivariational inequalities.

Chapter 4 is devoted to the approximation theory of scalar parabolic hemivariational inequalities. We introduce a fully discrete Galerkin type approximation scheme in which both the space and the time variables are discretized. As in the elliptic case we firstly analyse the solvability of the discretized problem. Under appropriate assumptions on approximated problem data we prove the convergence of fully discrete schemes to a solution of the original problem. Then we show that the discretized problem can be solved as a sequence of substationary point problems of type (0.6). At the end of this chapter extensions to the constrained problems are discussed.

As we have mentioned our goal was not only to establish a convergence analysis but also to propose numerical methods for solving (0.6). A traditional way of finding solutions to (0.6) is to use heuristic methods whose convergence is not guaranteed. Another possibility is to impose additional restrictions on the nonconvex perturbation Φ, for instance that Φ is the difference of two convex functions. If it is so then one can find and use rigorous minimization methods for the realization of (0.6). Our intention was different: we wanted to use mathematically justified minimization methods with minimum assumptions on minimized functions. For this reason we decided to apply nonsmooth minimization methods enabling the minimization of locally lipschitz functions. In Chapter 5 the main representatives of this class of methods are briefly described. A special attention is paid to the bundle type methods of the first and the second order which are used in the next chapter.

In Chapter 6 we illustrate how the theoretical results can be used for the numerical realization of several model examples from nonsmooth nonconvex mechanics: an elastic structure supported by a foundation having nonmonotone multivalued responses on the contact part corresponding to nonmonotone friction and contact conditions and a laminated composite structure under loading when the binding material between the laminae obeys a nonmonotone multivalued law. These examples are solved by using the bundle type methods mentioned above.

This project started in spring 1997. During this and the next year we visited each other several times having discussions on the book. Last time we all met in Thessaloniki in July, 1998. Besides of scientific work we enjoyed a friendship of Panos (how Prof. Panagiotopoulos was called by his friends) combined with the traditional greek's hospitality. When we were saying good bye to him we did not know that we see him for the last time. A few weeks later Prof. Panagiotopoulos died. Since we did not feel to be competent to make a good presentation of "the mechanical" part of the book we asked Prof. Stavroulakis, the collaborator of Prof. Panagiotopoulos to help us. We are deeply indebted to him for the preparation of Chapter 2.

We would also like to acknowledge the great assistance we received from Dr. M.M. Mäkelä from University of Jyväskylä who wrote Chapter 5 and provided us with his proximal bundle optimization code for the numerical calculations. We also express our appreciation to Dr. L. Lukšan and Dr. J. Vlček from the Czech Academy of Sciences for providing us with their proximal bundle and bundle-Newton code. Many thanks are also due to Zuzana Moravkova for programming most of the numerical applications of Chapter 6 and to Prof. R. Mäkinen and Dr. J. Toivanen from University of Jyväskylä for providing us with the finite element method solver which was used in Example 4 of Chapter 6.

The research of the first author was supported by the grant A1075707 of the Grant Agency of the Czech Academy of Sciences and the grant 101/98/0535 of the Grant Agency of the Czech Republic. His visits at University of Jyväskylä were supported by the grants #34063 and #41933 of the Academy of Finland

and his visit at Aristotle University in Thessaloniki was realized in the frame of NATO fellowship. The second author was supported by the grants #32572 and #38962 of the Academy of Finland. Further, this project has been supported by the grant #8583 of the Academy of Finland.

JAROSLAV HASLINGER, MARKKU MIETTINEN

PRAGUE-JYVÄSKYLÄ, JUNE 1999

List of Notations

Sets

$\mathbf{R}$	the real line
$\tilde{\mathbf{R}} = \mathbf{R} \cup \{+\infty\}$	the extended real line
$\mathbf{R}_+$	the set of nonnegative reals
$\mathbf{R}_-$	the set of nonpositive reals
$[a, b]$	a closed interval in $\mathbf{R}$
(a, b)	an open interval in $\mathbf{R}$
$(a, b], [a, b)$	semiopen intervals in $\mathbf{R}$
$\mathbf{N}$	the set of all positive integers
$\mathbf{N}_0$	the set of all nonnegative integers
X^n	$X \times \ldots X$ (n-times), where $X \subseteq \mathbf{R}$
Ω	a bounded domain in $\mathbf{R}^n$
$\overline{\Omega}$	the closure of Ω
$\partial\Omega$	the boundary of Ω
$\mathrm{int}_B A$	the interior of A in B
B_X	the unit ball in a normed space X
$\mathrm{conv}\, S$	convex hull of a set S

Functional spaces

$Y(\Omega)$	a set of functions $v : \Omega \to \mathbf{R}$
$Y(\Omega; \mathbf{R}^d)$	a set of functions $v = (v_1, ..., v_d) : \Omega \to \mathbf{R}^d$
$C(\Omega)$	the space of functions, continuous in Ω
$C^k(\Omega)$	the space of functions whose derivatives up to the order $k \in \mathbf{N}$ belong to $C(\Omega)$
$C^\infty(\Omega)$	$\cap_{k=0}^\infty C^k(\Omega)$
$C_0^\infty(\Omega)$	the space of infinitely differentiable functions with a compact support in Ω
$C^{0,1}(\overline{\Omega})$	the space of Lipschitz functions in $\overline{\Omega}$
$L^p(\Omega)$	the space of measurable functions in Ω such that $\int_\Omega |v|^p \, dx < +\infty$, $p \in [1, \infty)$

$L^\infty(\Omega)$ the space of measurable functions in Ω such that $\displaystyle\inf_{\substack{\text{meas } M=0, \\ M\subset\Omega}}\ \sup_{x\in\Omega\backslash M}|v(x)| < +\infty$

$\|\cdot\|_{L^p(\Omega)}$ the norm in $L^p(\Omega)$, $p \in [1,\infty]$

$W^{k,p}(\Omega)$ the set of measurable functions whose generalized derivatives up to the order k belong to $L^p(\Omega)$, $p \in [1,\infty]$, $k \in \mathbf{N}$

$W_0^{k,p}(\Omega)$ the closure of $C_0^\infty(\Omega)$ in $W^{k,p}(\Omega)$

$H^k(\Omega)$ $W^{k,2}(\Omega)$

$H_0^k(\Omega)$ $W_0^{k,2}(\Omega)$

$\|\cdot\|_{k,p,\Omega}$ the norm in $W^{k,p}(\Omega)$

$\|\cdot\|_{k,\Omega} \equiv \|\cdot\|_{k,2,\Omega}$ the norm in $H^k(\Omega)$

$C^k([0,T];X)$ the space of continuous X-valued functions in $[0,T]$ whose derivatives up to the order $k \in \mathbf{N}$ are continuous

$L^p(0,T;X)$ the space of measurable X-valued functions in $(0,T)$ such that $\int_0^T \|u\|_X^p\, dt < +\infty$, $p \in [1,\infty)$

$L^\infty(0,T;X)$ the space of measurable X-valued functions in $(0,T)$ such that $\displaystyle\inf_{\substack{\text{meas } M=0, \\ M\subset(0,T)}}\ \sup_{t\in(0,T)\backslash M}\|v(t)\|_X < +\infty$

$W^{1,p}(0,T;X)$ the subspace of $L^p(0,T;X)$ whose first generalized derivative belongs to $L^p(0,T;X)$

$W^{1,p}(0,T;V,H)$ the subspace of $L^p(0,T;V)$ whose first generalized derivative belong to $L^p(0,T;V^*)$ and $V \subset H \subset V^*$ forms the evolution triplet

$W(V)$ $W^{1,p}(0,T;V,H)$

$H^1(0,T;H)$ $W^{1,2}(0,T;H)$

$\mathcal{L}(X,Y)$ the space of all linear continuous mappings $A : X \to Y$

Functions

$F'(u;v)$ the directional derivative of $F : X \to Y$ at u and a direction v

$DF(u)$ Gâteaux or Fréchet derivative of $F : X \to Y$ at u

$f^0(u;v)$ the generalized derivative of a locally Lipschitz function $f : X \to \mathbf{R}$ at u and a direction v

$\bar\partial f(u)$ the generalized Clarke gradient of a locally Lipschitz function $f : X \to \mathbf{R}$

$f'(u;v)$ the directional derivative of a convex function $f : X \to \mathbf{R}$ at u and a direction v

I_K the indicator function of a closed convex set K

P_k the set of all polynomials of degree at most k

Other symbols

$\mathbf{A}, \mathbf{B}$	matrices
$\vec{u}, \vec{v}$	vectors
$\mathcal{T}_h, \mathcal{D}_h$	partitions of $\overline{\Omega}$ into finite elements
$\varepsilon(u) = \{\varepsilon_{ij}(u)\}$	the strain tensor corresponding to the displacement field u
σ, τ	stress tensors
ν	the unit outward normal vector to $\partial\Omega$
t	the tangential vector to $\partial\Omega$
T_ν, T_t	the normal, tangential component of a vector T
$\square$	end of proof

Let us consider a beam of the length l, subject to a vertical force f, clamped at the initial point $x = 0$ and supported by a spring at $x = l$.

Moreover let the beam be made of a homogenous material with the same shape of the cross section along the whole length. The deflection y of the beam satisfies the fourth order differential equation (after the normalization):

$$y^{(4)}(x) = f(x), \quad x \in (0, l). \tag{I.8}$$

The boundary conditions at $x = 0$ are standard:

$$y(0) = y'(0) = 0. \tag{I.9}$$

As far as conditions at $x = l$ is concerned, we have

$$y''(l) = 0 \quad \text{(no moment at } x = l), \tag{I.10}$$

while the remaining condition, expressing the relation between the deflection y and the reaction forces, given by $y^{(3)}(l)$ depends on physical characteristics of the spring itself.

Let us start with a simpler case, namely when the reaction-deflection relation is given by the function $b : \mathbf{R} \to \mathbf{R}$ defined as follows:

$$b(\xi) = \begin{cases} 0, & \xi \leq 0 \\ k\xi, & \xi > 0, \ k > 0 \text{ given} \end{cases}$$

and

$$y^{(3)}(l) = b(y(l)),$$

i.e.,

$$y^{(3)}(l) = ky_+(l), \tag{I.11}$$

where a_+ stands for the positive part of $a \in \mathbf{R}$. The condition (I.11) says that the reaction forces are equal to zero when there is no contact between the beam and the spring at $x = l$ and they are proportional to the deflection if $y(l) > 0$.

The variational formulation of the problem defined by (I.8)–(I.11) is classical: the total potential energy of the beam is given by the quadratic functional

$$J_0(\varphi) = \frac{1}{2} \int_0^l (\varphi'')^2 \, dx - \int_0^l f\varphi \, dx, \qquad (I.12)$$

while the energy of the spring is given by the expression

$$\frac{1}{2} k \left(\varphi_+(l)\right)^2. \qquad (I.13)$$

The total potential energy of the system "beam + spring" is

$$J(\varphi) = \frac{1}{2} \int_0^l (\varphi'')^2 \, dx + \frac{1}{2} k \left(\varphi_+(l)\right)^2 - \int_0^l f\varphi \, dx. \qquad (I.14)$$

According to the Lagrange principle of minimum of the total potential energy, the equilibrium state of the structure is characterized by a deflection, minimizing J over the space X of all *kinematically admissible deflections* with finite energy. From (I.9) and the form of J one deduces that the space X can be defined as follows:

$$X = \{\varphi : [0, l] \to \mathbf{R} \mid \varphi, \varphi' \text{ are absolutely continuous}$$
$$\text{in } [0, l], \quad \varphi(0) = \varphi'(0) = 0\}.$$

Thus we are looking for $y \in X$ such that

$$J(y) = \min_{\varphi \in X} J(\varphi). \qquad (I.15)$$

Using results which will be formulated later on, it is possible to prove that $y \in X$ satisfying (I.15) *exists and is unique*. One of main properties of J is the fact that J is *strictly convex* in X. In addition, J is *continuously differentiable* in X. Its gradient at $z \in X$ is given by

$$\langle \operatorname{grad} J(z), \varphi \rangle = \int_0^l z'' \varphi'' \, dx + kz_+(l)\varphi(l) - \int_0^l f\varphi \, dx \quad \forall \varphi \in X. \ (I.16)$$

A *necessary* and due to convexity also a *sufficient* condition for y to be a minimizer of J over X says that y has to be a *stationary point* of J in X, i.e., a point satisfying

$$\langle \operatorname{grad} J(y), \varphi \rangle = 0 \quad \forall \varphi \in X. \qquad (I.17)$$

From (I.16) it follows that (I.17) can be written in the form:

$$\int_0^l y'' \varphi'' \, dx + ky_+(l)\varphi(l) = \int_0^l f\varphi \, dx \quad \forall \varphi \in X, \qquad (I.18)$$

which is nothing else than the *principle of virtual work*.

Now, let us take a more complicated, nonlinear spring whose reaction-deflection relation is given by the "function" b defined as follows (see Fig.I.1):

$$b(\xi) = \begin{cases} 0, & \text{if } \xi \leq 0, \\ k\xi, & \text{if } \xi \in (0, a), \quad k, a > 0 \text{ given}, \\ [0, ka] & \text{if } \xi = a, \\ 0, & \text{if } \xi > a. \end{cases} \qquad (I.19)$$

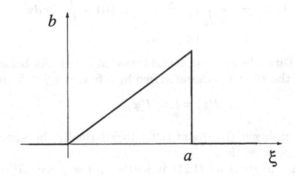

Figure I.1.

Contrary to the previous case, b is no longer a classical function but the multifunction, which is nonmonotone. At $x = a$ it admits values belonging to the whole closed interval $[0, ka]$. The physical interpretation of the spring with such a diagram can be stated as follows: the spring may sustain deflections not exceeding the value a. Then the spring is destroyed. Moreover, the reactions for the limit deflection are not specified. Only what we know is that they take values between 0 and ka.

It is straightforward to find the primitive function to (I.19). It is done by

$$j(\xi) = \begin{cases} 0, & \text{if } \xi \leq 0, \\ \frac{1}{2}k\xi^2, & \text{if } \xi \in (0, a), \\ \frac{1}{2}ka^2, & \text{if } \xi \geq a. \end{cases} \qquad (I.20)$$

One can easily check that j is Lipschitz in $\mathbf{R}$ since it holds

$$|j(\xi_1) - j(\xi_2)| \leq ka|\xi_1 - \xi_2| \quad \forall \xi_1, \xi_2 \in \mathbf{R}$$

and is continuously differentiable in $\mathbf{R}$, except the point $x = a$. Indeed:

$$j'_-(a) = ka \quad \text{(the left derivative)},$$
$$j'_+(a) = 0 \quad \text{(the right derivative)}.$$

As well we shall see later on, it is useful to define "a derivative" of j at the point a even if the classical one does not exist. We shall define a "generalized gradient

$\bar{\partial}j$" of j, which will coincide with the classical one (if this exists) and extends it in a reasonable way when there is lack of differentiability. In particular, if j is given by (I.20) then we define

$$\bar{\partial}j(\xi) = \begin{cases} j'(\xi) & \text{if } \xi \neq a \text{ (the classical derivative)}, \\ [0, ka], & \text{if } \xi = a, \end{cases}$$

i.e., $\bar{\partial}j(\xi) = b(\xi) \ \forall \xi \in \mathbf{R}$.

The total potential energy of the system "beam–spring" is now given by

$$\begin{aligned} J(\varphi) &= \frac{1}{2} \int_0^l (\varphi'')^2 \, dx + j(\varphi(l)) - \int_0^l f\varphi \, dx \qquad (I.21) \\ &\equiv J_0(\varphi) + j(\delta_l \varphi), \end{aligned}$$

where δ_l is the Dirac distribution concentrated at $x = l$. As before, the equilibrium state of the system is characterized by a function $y \in X$ such that

$$J(y) = \min_{\varphi \in X} J(\varphi). \qquad (I.22)$$

Although both problems (I.15) and (I.22) look formally the same, there is a great difference between them.

The functional J given by (I.21) is *neither convex, nor differentiable*, in general, due to the presence of the *locally Lipschitz function* j. Consequently, the whole problem is much more complicated from the theoretical as well as from the computational point of view.

Next we shall formulate the analogy of the principle of virtual work for (I.22). It is readily seen that J is continuously differentiable at any $z \in X$ satisfying $z(l) \neq a$.

Let us introduce the generalized gradient $\bar{\partial}J$ of J at $z \in X$ as follows:

$$\bar{\partial}J(z) = \operatorname{grad} J_0(z) + \mathcal{X}\delta_l, \qquad (I.23)$$

where $\mathcal{X} \in \bar{\partial}j(z(l))$, i.e., the generalized gradient is the sum of the singleton term $\operatorname{grad} J_0(z)$ and the multivalued term $\mathcal{X}\delta_l$.

In subsequent parts it will be shown that the *necessary* (but not *sufficient*) condition for $y \in X$ to be a minimizer of J over X is that y is the so-called *substationary point* of J in X, i.e., the point such that

$$0 \in \bar{\partial}J(y), \qquad (I.24)$$

where 0 is the zero element of the dual space X^*. From (I.23) it follows that (I.24) reads as follows:

$$0 \in \operatorname{grad} J_0(y) + \mathcal{X}\delta_l, \quad \mathcal{X} \in \bar{\partial}j(y(l)). \qquad (I.25)$$

Let $y \in X$ be a substationary point of J. Then (I.25) can be equivalently expressed by:

$$\begin{cases} \exists \mathcal{X} \in \bar{\partial}j(y(l)) \text{ such that} \\ \operatorname{grad} J_0(y) + \mathcal{X}\delta_l = 0 \quad \text{in } X^*. \end{cases} \qquad (I.26)$$

Taking into account the form of J_0 and multiplying $(I.26)_2$ by a test function $\varphi \in X$ we finally arrive at the problem:

$$\begin{cases} \text{Find } y \in X \text{ and } \mathcal{X} \in \bar{\partial}j(y(l)) \text{ such that} \\ \int_0^l y'' \varphi'' \, dx + \mathcal{X}\varphi(l) = \int_0^l f\varphi \, dx \quad \forall \varphi \in X \end{cases} \tag{I.27}$$

which presents the generalization of the principle of virtual work to our non-convex and nonsmooth problem. Using the integration by parts in $(I.27)_2$ we recover the differential equation (I.8), the boundary condition (I.10) and the reaction–displacement relation

$$y^{(3)}(l) = \mathcal{X} \in \bar{\partial}j(y(l)). \tag{I.28}$$

The boundary conditions (I.9) are satisfied a priori by the definition of X.

Problem (I.27) is a simple example of the so-called *hemivariational inequality*. To see difficulties arising when solving such problems, we construct a discretization of (I.27) enabling us to solve this problem numerically.

Let $\Delta_h : 0 \equiv x_0 < x_1 < ... < x_{n(h)} \equiv l$ be a partition of $[0,l]$. Define the finite-dimensional space X_h as follows:

$$X_h = \{\varphi_h \in C^1([0,l]) \mid \varphi|_{[x_{i-1},x_i]} \in P_3 \quad \forall i = 1, ..., n(h),$$
$$\varphi_h(0) = \varphi_h'(0) = 0\},$$

i.e., X_h is the space of once continuously differentiable functions in $[0,l]$ whose restrictions on any subinterval of Δ_h are cubic polynomials and satisfying the prescribed boundary condition at $x = 0$. It is well-known that any $\varphi_h \in X_h$ is uniquely determined by its values $\varphi_h(x_i)$, $\varphi_h'(x_i)$, $i = 1, ..., n$.

The approximation of (I.27) now reads as follows:

$$\begin{cases} \text{Find } y_h \in X_h \text{ and } \mathcal{X}_h \in \bar{\partial}j(y_h(l)) \text{ such that} \\ \int_0^l y_h'' \varphi_h'' \, dx + \mathcal{X}_h \varphi_h(l) = \int_0^l f\varphi_h \, dx \quad \forall \varphi_h \in X_h. \end{cases} \tag{I.29}$$

Let $\vec{y} = (y_1, y_2, ..., y_{2n-1}, y_{2n}) \in \mathbf{R}^{2n}$ be the vector whose components are related to the nodal values of y_h and y_h' as follows:

$$y_{2i-1} = y_h'(x_i), \quad i = 1, ..., n;$$
$$y_{2i} = y_h(x_i), \quad i = 1, ..., n.$$

Then (I.29) can be rewritten in the following equivalent form

$$\begin{cases} \text{Find } \vec{y} \in \mathbf{R}^{2n} \text{ and } \mathcal{X}_h \in \mathbf{R} \text{ such that} \\ \mathbf{A}\vec{y} + \begin{pmatrix} 0 \\ 0 \\ \vdots \\ 0 \\ \mathcal{X}_h \end{pmatrix} = \vec{F} \\ \text{and } \mathcal{X}_h \in \bar{\partial}j(y_{2n}), \end{cases} \tag{I.30}$$

where $\mathbf{A}$ is the stiffness matrix and $\vec{F}$ is the load vector. Taking into account the special form of (I.30), namely the fact that the nonlinearity concerns of the last component of $\vec{y}$, one can eliminate the first $(2n-1)$ components of $\vec{y}$. Thus we arrive at the following very simple problem in one variable, denoted by y:

$$\begin{cases} \text{Find } y, \, \mathcal{X} \in \mathbf{R} \text{ such that} \\ \mathcal{A}y + \mathcal{X} = \mathcal{F}, \quad \mathcal{A}, \mathcal{F} \in \mathbf{R} \text{ given,} \\ \mathcal{X} \in \bar{\partial}j(y). \end{cases} \qquad (I.31)$$

For the sake of simplicity let us suppose that $\mathcal{A} = 1$ and also let the parameters a and k, defining j be equal to 1. The solution of the multivalued problem

$$\begin{cases} \text{Find } y, \, \mathcal{X} \in \mathbf{R} \text{ such that} \\ y + \mathcal{X} = \mathcal{F}, \\ \mathcal{X} \in \bar{\partial}j(y), \end{cases} \qquad (I.32)$$

where

$$\bar{\partial}j(y) = \begin{cases} 0, & \text{if } y \in (-\infty, 0] \cup (1, \infty), \\ y, & \text{if } y \in (0, 1), \\ [0, 1], & \text{if } y = 1, \end{cases}$$

can be easily discovered. Indeed, one can find the following solutions:

$$y = \mathcal{F}, \, \mathcal{X} = 0 \qquad \qquad \text{if } \mathcal{F} \le 0$$

$$y = \frac{\mathcal{F}}{2}, \, \mathcal{X} = \frac{\mathcal{F}}{2} \qquad \qquad \text{if } \mathcal{F} \in (0, 1)$$

$$\left. \begin{array}{l} y = 1, \, \mathcal{X} = 0 \\ y = \frac{1}{2}, \, \mathcal{X} = \frac{1}{2} \end{array} \right\} \qquad \text{if } \mathcal{F} = 1$$

$$\left. \begin{array}{l} y = 1, \, \mathcal{X} = \mathcal{F} - 1 \\ y = \frac{\mathcal{F}}{2}, \, \mathcal{X} = \frac{\mathcal{F}}{2} \\ y = \mathcal{F}, \, \mathcal{X} = 0 \end{array} \right\} \qquad \text{if } \mathcal{F} \in (1, 2)$$

$$\left. \begin{array}{l} y = 1, \, \mathcal{X} = 1 \\ y = 2, \, \mathcal{X} = 0 \end{array} \right\} \qquad \text{if } \mathcal{F} = 2$$

$$y = \mathcal{F}, \, \mathcal{X} = 0 \qquad \qquad \text{if } \mathcal{F} > 2$$

Thus the number of solutions depends on the magnitude of $\mathcal{F}$.

Problem (I.27) is one of the simplest hemivariational inequalities. Such inequalities enable us to involve nonmonotone and multivalued constitutive relations into the mathematical model.

Remark I.1 *In contrast to the convex case, formulation (I.27) is more general than (I.22). The set of all solutions to (I.27) contains not only global mini-mizers but also other types of substationary points (see Fig.I.2). Moreover, the*

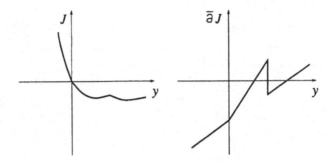

Figure I.2.   Graphs of J and $\bar{\partial}J$ when $\mathcal{F} = \frac{3}{2}$.

approximation of hemivariational inequalities, which will be based on the generalized principle of virtual work enables us to approximate simultaneously the second component $\mathcal{X}$, having a good mechanical meaning.

I Introductory Topics

1 MATHEMATICAL PRELIMINARIES

In order to keep the presentation as self-contained we decided to start with basic results, which will be needed in subsequent parts of the monograph. First, we give a survey of function spaces of Sobolev and Bochner type and their properties, then we mention elements of convex and nonconvex analysis. We recall also main results on the approximation of monotone problems of elliptic and parabolic type. Subsequent parts of this monograph will present an extension of these results to a larger class of problems, involving nonmonotone and nondifferentiable inclusions.

1.1 FUNCTIONAL SPACES AND THEIR PROPERTIES

Let $\Omega \subset \mathbf{R}^n$ be a domain, i.e., Ω is an *open* and *connected* subset in $\mathbf{R}^n$. Moreover throughout this book we deal with *bounded* domains only. Other restrictions will be imposed later.

We start with spaces of smooth functions. Let $n \in \mathbf{N}$. Any vector $\alpha = (\alpha_1, \alpha_2, ..., \alpha_n)$ whose components α_i belong to $\mathbf{N}_0$ for any $i = 1, ..., n$ is called a *multi-index* and the number $|\alpha| = \sum_{i=1}^n \alpha_i$ its *length*. If $u : \Omega \mapsto \mathbf{R}$ is a sufficiently smooth function then the symbol $D^\alpha u$ is defined as follows:

$$D^\alpha u = \frac{\partial^{|\alpha|} u}{\partial x_1^{\alpha_1} \cdots \partial x_n^{\alpha_n}},$$

i.e., we differentiate u α_i times with respect to x_i, $i = 1, ..., n$.

3

The space $C^k(\Omega)$

Let $k \in \mathbf{N}_0$. Then $C^k(\Omega)$ denotes the set of all continuous real functions defined in Ω, whose derivatives D^α up to the order k (i.e., all $\alpha \in \mathbf{N}_0^n$ such that $|\alpha| \leq k$) are continuous in Ω with the following convention of the notation: if $k = 0$ then $D^0 v(x) = v(x)$ for $x \in \Omega$ and $C^0(\Omega) \equiv C(\Omega)$. Further

$$C^\infty(\Omega) = \bigcap_{k=0}^{\infty} C^k(\Omega).$$

Let $k \in \mathbf{N}_0 \cup \{\infty\}$. Then $C_0^k(\Omega)$ stands for the subset of $C^k(\Omega)$ containing all functions vanishing in a neighbourhood of $\partial\Omega$. Again $C_0^0(\Omega) \equiv C_0(\Omega)$.

The space $C^k(\overline{\Omega})$

First let $k = 0$. By $C(\overline{\Omega}) \equiv C^0(\overline{\Omega})$ we denote the subset of $C(\Omega)$ containing all functions which can be continuously extended on the closure $\overline{\Omega}$.

Now let $k \in \mathbf{N}$. Then the symbol $C^k(\overline{\Omega})$ stands for the set of all real functions defined in Ω, the derivatives of which up to the order k can be continuously extended on $\overline{\Omega}$, i.e., a function $u \in C^k(\overline{\Omega})$ if and only if for any $\alpha \in \mathbf{N}_0^n$ such that $|\alpha| \leq k$ there exists a function $v_\alpha \in C(\overline{\Omega})$ such that

$$v_\alpha(x) = D^\alpha u(x) \qquad \text{for any } x \in \Omega.$$

For the sake of simplicity of notations we shall write $D^\alpha u$ instead of v_α even for $x \in \partial\Omega$.

We set

$$C^\infty(\overline{\Omega}) = \bigcap_{k=0}^{\infty} C^k(\overline{\Omega}).$$

It is well-known that for any $k \in \mathbf{N}_0$ the expression

$$\|u\|_{C^k(\overline{\Omega})} = \sum_{|\alpha| \leq k} \max_{x \in \overline{\Omega}} |D^\alpha u(x)| \tag{1.1}$$

defines the norm with respect to which $C^k(\overline{\Omega})$ is complete, i.e., $C^k(\overline{\Omega})$ is a Banach space. It is also readily seen that

$$\|u_j - u\|_{C^k(\overline{\Omega})} \to 0 \quad \text{iff} \quad D^\alpha u_j \rightrightarrows D^\alpha u \quad \text{(uniformly) in } \overline{\Omega}$$

for any $\alpha \in \mathbf{N}_0^n$, $|\alpha| \leq k$.

The space $C^{0,1}(\overline{\Omega})$

Let $u : \overline{\Omega} \to \mathbf{R}$. We say that u is *Lipschitz* in $\overline{\Omega}$ iff there exists a positive constant c such that

$$|u(x) - u(y)| \leq c\|x - y\| \qquad \text{for any } x, y \in \overline{\Omega}.$$

The set of all Lipschitz continuous functions in $\overline{\Omega}$ will be denoted by $C^{0,1}(\overline{\Omega})$.

Lebesgue spaces $L^p(\Omega)$, $p \in [1, \infty]$

First let $p \in [1, \infty)$. The set of all real measurable functions u in Ω satisfying

$$\int_\Omega |u|^p dx < \infty \qquad \text{(in the Lebesgue sense)}$$

will be denoted by $L^p(\Omega)$. The norm in $L^p(\Omega)$ is given by

$$\|u\|_{L^p(\Omega)} \equiv \left(\int_\Omega |u|^p dx \right)^{\frac{1}{p}}. \tag{1.2}$$

By $L^\infty(\Omega)$ we denote the set of all real, measurable and essentially bounded functions in Ω. More precisely: $u \in L^\infty(\Omega)$ if and only if u is measurable and the number

$$\|u\|_{L^\infty(\Omega)} \equiv \inf_{\substack{\text{meas } M=0, \\ M \subset \Omega}} \sup_{x \in \Omega \setminus M} |u(x)| \tag{1.3}$$

is finite. The corresponding infimum is considered over all subsets M of Ω whose Lebesgue measure is zero. Also (1.3) defines the norm in $L^\infty(\Omega)$. Now we recall some well-known results:

Theorem 1.1 *It holds that:*

(i) Let $p \in [1, \infty)$. Then $L^p(\Omega)$ equipped with the norm (1.2) is complete;

(ii) The space $L^\infty(\Omega)$ is complete with respect to the norm given by (1.3).

Theorem 1.2 *Let $p \in (1, \infty)$. Then $L^p(\Omega)$ is reflexive.*

This theorem has the following important consequence:

Theorem 1.3 *Let $p \in (1, \infty)$ and $\mathcal{N} \subset L^p(\Omega)$. Then the following statements are equivalent:*

– for any sequence $\{f_j\}$, $f_j \in \mathcal{N}$ there exists a weakly convergent subsequence in $L^p(\Omega)$, i.e., $\exists \{f_{j'}\} \subset \{f_j\}$ and $f \in L^p(\Omega)$ such that

$$\int_\Omega f_{j'} \varphi dx \longrightarrow \int_\Omega f \varphi dx \qquad \forall \varphi \in L^q(\Omega),$$

where $1/p + 1/q = 1$;

– $\exists c > 0 : \forall f \in \mathcal{N} : \|f\|_{L^p(\Omega)} \leq c$.

For $p = 1$ the situation is more involved. We have (see Ekeland and Temam, 1976):

Theorem 1.4 *(Dunford-Pettis) Let $\mathcal{N} \subset L^1(\Omega)$. The following statements are equivalent:*

– *for any sequence* $\{f_j\}$, $f_j \in \mathcal{N}$ *there exists a weakly convergent subsequence in* $L^1(\Omega)$, *i.e.*, $\exists \{f_{j'}\} \subset \{f_j\}$ *and* $f \in L^1(\Omega)$ *such that*

$$\int_\Omega f_{j'} \varphi dx \longrightarrow \int_\Omega f \varphi dx \qquad \forall \varphi \in L^\infty(\Omega);$$

– *for any* $\varepsilon > 0$ *there exists* $\lambda > 0$ *such that*

$$\forall f \in \mathcal{N} \qquad \int_{\{|f| \geq \lambda\}} |f| dx \leq \varepsilon;$$

– $\forall \varepsilon > 0 \ \exists \delta > 0$ *such that*

$$\int_\omega |f| dx \leq \varepsilon$$

for any $f \in \mathcal{N}$ *and any measurable subset* $\omega \subset \Omega$ *satisfying* meas $\omega \leq \delta$.

Now we mention basic results for Lebesgue spaces, which will be needed in the sequel (for details see Kufner et al., 1977 and Yosida, 1965).

Theorem 1.5 *(Egoroff's theorem) Let* $\{f_j\}$ *be a sequence of real measurable functions in* Ω, *which converges to a measurable function* f *a.e. in* Ω. *Then* $\{f_j\}$ *converges to* f *uniformly up to small sets, i.e., for any* $\varepsilon > 0$ *there exists a measurable set* ω, meas $\omega < \varepsilon$, *and* $\{f_j\}$ *converges to* f *uniformly in* $\Omega \setminus \omega$.

Theorem 1.6 *(a consequence of Fatou's lemma) Let* $\{f_j\}$ *be a sequence of measurable functions in* Ω *such that* $|f_j(x)| \leq \varphi(x)$ *for a.a.* $x \in \Omega$ *and* $j = 1, 2, ...$, *where* $\varphi \in L^1(\Omega)$. *Then*

$$-\infty < \int_\Omega \liminf_{j \to \infty} f_j dx \leq \liminf_{j \to \infty} \int_\Omega f_j dx$$

$$\leq \limsup_{j \to \infty} \int_\Omega f_j dx \leq \int_\Omega \limsup_{j \to \infty} f_j dx < +\infty.$$

Theorem 1.7 *(Mazur's lemma) Let* $\{f_j\}$ *be a sequence converging weakly to* f *in* $L^p(\Omega)$, $p \in [1, \infty]$. *Then there exists a sequence of convex combinations of* f_j, *which converges strongly to* f *in* $L^p(\Omega)$, *i.e., there exist sequences* $\{m_k\} \subset \mathbf{N}$ *and* $\{\mu_{j,k}\} \subset \mathbf{R}$ *such that*

$$\sum_{j=k}^{m_k} \mu_{j,k} = 1, \ \mu_{j,k} \geq 0, \ and \ \sum_{j=k}^{m_k} \mu_{j,k} f_j \to f, \ as \ k \to \infty.$$

Sobolev spaces

Before we introduce these spaces, we restrict the class of domains by imposing additional assumptions concerning their boundaries.

Definition 1.1 *We say that the domain* $\Omega \subset \mathbf{R}^n$ *has a Lipschitz continuous boundary* $\partial\Omega$ *if there exist real numbers* $\alpha, \beta > 0$ *such that for any point* $x^0 \in \partial\Omega$ *the Cartesian coordinate system can be rotated and shifted to* x^0 *in such a way that locally, around* x^0 *the boundary* $\partial\Omega$ *is represented by the graph of a Lipschitz function* $a : K_{n-1}(a) \to \mathbf{R}$, *where*

$$K_{n-1}(a) = \left\{ x' = (x_1, x_2, ..., x_{n-1}) \mid |x_i| < \alpha \quad \forall i = 1, 2, ..., n-1 \right\}$$

is an $(n-1)$*-dimensional cube and the points* $(x', a(x'))$ *belong to* $\partial\Omega$ *for any* $x' \in K_{n-1}(a)$. *All the points* (x', x_n) *such that* $x' \in K_{n-1}(a)$ *and* $a(x') < x_n < a(x') + \beta$ *lie inside of* Ω *while the points* (x', x_n), $x' \in K_{n-1}(a)$, $a(x') - \beta < x_n < a(x')$ *lie outside of* Ω. *Finally the number of such local Cartesian coordinate systems describing the whole boundary* $\partial\Omega$ *is finite, say* r.

Remark 1.1 *Sometimes the higher regularity of the boundary will be necessary. We say that* Ω *is of a class* C^m *iff the functions* a *describing* $\partial\Omega$ *belong to* $C^m(\overline{K}_{n-1}(a))$.

The fact that Ω has the Lipschitz boundary makes possible to define the outer unit normal vector ν at almost all points of the boundary. *Throughout this book we shall deal solely with domains having at least Lipschitz boundaries.*

Let $f \in L^p(\Omega)$, $p \in [1, \infty]$. An integrable function $g_\alpha \in L^1(\Omega)$ satisfying the integral identity:

$$\int_\Omega f D^\alpha \varphi \, dx = (-1)^{|\alpha|} \int_\Omega g_\alpha \varphi \, dx \qquad \forall \varphi \in C_0^\infty(\Omega) \qquad (1.4)$$

will be called the α-*th generalized derivative* of f. If such g_α exists, then it is unique and we use the notation $D^\alpha f$ instead of g_α.

The Sobolev space $W^{k,p}(\Omega)$, $k \in \mathbf{N}, p \in [1, \infty]$ is defined as a subspace of $L^p(\Omega)$ of functions whose all generalized derivatives up to the order k belong to $L^p(\Omega)$:

$$W^{k,p}(\Omega) = \left\{ v \in L^p(\Omega) \mid D^\alpha v \in L^p(\Omega) \quad \forall |\alpha| \leq k \right\}.$$

The spaces $W^{k,p}(\Omega)$ are endowed with the following norms:

$$\|u\|_{k,p,\Omega} \equiv \left(\sum_{|\alpha| \leq k} \int_\Omega |D^\alpha u|^p dx \right)^{\frac{1}{p}} \qquad \text{if } p \in [1, \infty) \qquad (1.5)$$

$$\|u\|_{k,\infty,\Omega} \equiv \sum_{|\alpha| \leq k} \|D^\alpha u\|_{L^\infty(\Omega)} \qquad \text{if } p = \infty. \qquad (1.6)$$

The expressions:

$$|u|_{k,p,\Omega} \equiv \left(\sum_{|\alpha| = k} \int_\Omega |D^\alpha u|^p dx \right)^{\frac{1}{p}}, \qquad p \in [1, \infty) \qquad (1.7)$$

$$|u|_{k,\infty,\Omega} \equiv \sum_{|\alpha| = k} \|D^\alpha u\|_{L^\infty(\Omega)}, \qquad p = \infty \qquad (1.8)$$

define the seminorms in the corresponding spaces* .

Below some basic properties of the Sobolev spaces which will be used in subsequent parts will be listed (for details see Nečas, 1967 and Adams, 1975).

Theorem 1.8 *The Sobolev space $W^{k,p}(\Omega)$, $k \in \mathbf{N}, p \in [1, \infty]$ is complete with respect to (1.5) when $p \in [1, \infty)$ and (1.6) if $p = \infty$.*

Theorem 1.9 *(the density result) Let $k \in \mathbf{N}$, $p \in [1, \infty)$. Then*

$$W^{k,p}(\Omega) = \overline{C^\infty(\overline{\Omega})},$$

where the closure is taken with respect to the norm (1.5).

One of very important properties of functions belonging to Sobolev spaces is the fact that one can speak of their boundary values. Before we recall this fact, we introduce the Lebesgue spaces $L^p(\partial\Omega)$, $p \in [1, \infty]$.

Let $\mathcal{I}$ be a finite collection of all Lipschitz functions a defined in $K_{n-1}(a)$ and fully describing $\partial\Omega$ (see Definition 1.1). Let $u : \partial\Omega \to \mathbf{R}$ be a function defined on $\partial\Omega$. We say that $u \in L^p(\partial\Omega)$ iff $u(x', a(x')) \in L^p(K_{n-1}(a))$ for any $a \in \mathcal{I}$. Let $\Gamma \subset \partial\Omega$ be a part of the boundary given by the graph of $a \in \mathcal{I}$. Then the surface integral of u along Γ is defined as follows:

$$\int_\Gamma u \, ds = \int_{K_{n-1}(a)} u(x', a(x')) \sqrt{1 + \sum_{i=1}^{n-1} \left(\frac{\partial a(x')}{\partial x_i}\right)^2} \, dx'.$$

Summing up integrals over all Γ's, covering $\partial\Omega$, we can define the value of $\int_{\partial\Omega} u \, ds$ (for more details see Nečas, 1967). Replacing u by $|u|^p$ we obtain the norm in $L^p(\partial\Omega)$ for $p \in [1, \infty)$ with the straightforward extension for $p = \infty$. Now we are ready to formulate:

Theorem 1.10 *(the trace theorem) Let $p \in [1, \infty]$. Then there exists a unique linear mapping $\gamma : W^{1,p}(\Omega) \to L^p(\partial\Omega)$ such that*

- $\gamma u = u$ *on $\partial\Omega$ for any $u \in C^\infty(\overline{\Omega})$;*

- $\exists c \equiv const. > 0$ *such that*

$$\|u\|_{L^p(\partial\Omega)} \le c\|u\|_{1,p} \qquad \forall u \in W^{1,p}(\Omega).$$

The linear continuous mapping γ introduced in the previous theorem is termed *a trace mapping.*

Let $k \in \mathbf{N}$, $p \in [1, \infty]$. We define the Sobolev space

$$W_0^{k,p}(\Omega) = \overline{C_0^\infty(\overline{\Omega})}$$

---

*When we do not have any risk of misunderstanding, the symbol Ω will be usually skipped.

the closure being understood in the sense of (1.5) if $p \in [1, \infty)$ or (1.6) when $p = \infty$. It is readily seen that $W_0^{k,p}(\Omega)$ is closed in $W^{k,p}(\Omega)$ and thus complete. When $k = 1$ the following simple characterization of $W_0^{1,p}(\Omega)$ holds:

Theorem 1.11 *Let $p \in [1, \infty)$. Then*

$$W_0^{1,p}(\Omega) = \{ v \in W^{1,p}(\Omega) \mid \gamma v = 0 \quad on \ \partial\Omega \}.$$

Similar characterization holds true also for $k > 1$ provided that the boundary $\partial\Omega$ is smooth enough (for details we refer to Nečas, 1967).

Special attention will be paid to the case when $p = 2$. Then the Sobolev spaces $W^{k,2}(\Omega)$, $W_0^{k,2}(\Omega)$ are the Hilbert spaces with the scalar product

$$(u, v)_{k,\Omega} \equiv \int_\Omega \sum_{|\alpha| \leq k} D^\alpha u D^\alpha v \, dx.$$

To keep notations simple, we shall use the symbols $H^k(\Omega)$, $H_0^k(\Omega)$ instead of $W^{k,2}(\Omega)$, $W_0^{k,2}(\Omega)$, respectively. The norm in $H^k(\Omega)$ will be denoted by $\| \cdot \|_{k,\Omega}$ instead of $\| \cdot \|_{k,2,\Omega}$, in what follows. This notation will be extended also to the case $k = 0$, i.e., $\| \cdot \|_{0,\Omega}$, $(\cdot, \cdot)_{0,\Omega}$ stands for the norm, the scalar product in $L^2(\Omega)$, respectively[†].

Imbedding theorems and compact imbeddings

Let X, Y be two linear normed spaces whose norms will be denoted by $\| \cdot \|_X$, $\| \cdot \|_Y$, respectively. We write $X \hookrightarrow Y$ iff $X \subset Y$ and the injection $i : X \to Y$ is continuous from X to Y, i.e.,

$$\exists c = \text{const} > 0 \text{ such that } \|iu\|_Y \leq c\|u\|_X \quad \forall u \in X.$$

If in addition, i is *compact*, we write $X \overset{\hookrightarrow}{\to} Y$. It holds:

Theorem 1.12 *Let $k \in \mathbf{N}$, $p \in [1, \infty)$. Then:*

— $\quad W^{k,p}(\Omega) \hookrightarrow L^{q*}(\Omega)$, *where* $\dfrac{1}{q^*} = \dfrac{1}{p} - \dfrac{k}{n}$ *if* $k < \dfrac{n}{p}$;

— $\quad W^{k,p}(\Omega) \overset{\hookrightarrow}{\to} L^q(\Omega)$, *for any* $q \in [1, q^*)$ *if* $k < \dfrac{n}{p}$;

— $\quad W^{k,p}(\Omega) \hookrightarrow L^q(\Omega)$, *for any* $q \in [1, \infty)$ *if* $k = \dfrac{n}{p}$;

— $\quad W^{k,p}(\Omega) \hookrightarrow C(\overline{\Omega})$, *if* $k > \dfrac{n}{p}$.

[†]When clear from the context, the symbol Ω will be usually omitted.

Similar theorem can be formulated for the trace mapping $\gamma : W^{1,p}(\Omega) \to L^p(\partial\Omega)$ introduced in Theorem 1.10. From this theorem it follows that $W^{1,p}(\Omega) \hookrightarrow L^p(\partial\Omega)$. This result however can be sharpened as follows from

Theorem 1.13 *Let $p \in [1,\infty)$ Then it holds:*

- $W^{1,p}(\Omega) \hookrightarrow L^{q*}(\partial\Omega)$, *where* $q^* = \dfrac{np - p}{n - p}$ *if* $1 \leq p < n$;

- $W^{1,p}(\Omega) \overset{\hookrightarrow}{\to} L^q(\partial\Omega)$, *for any* $q \in [1, q^*)$ *if* $1 \leq p < n$;

- $W^{1,p}(\Omega) \overset{\hookrightarrow}{\to} L^q(\partial\Omega)$, *for any* $q \in [1,\infty)$ *if* $p \geq n$.

The compact imbedding of the space $H^1(\Omega)$ into $L^2(\Omega)$ is known as the *Rellich's theorem.*

All the above results can be extended to vector-valued functions. If $X(\Omega)$ is a space of real functions $v : \Omega \to \mathbf{R}$ then the space of vector-valued functions $v = (v_1, ..., v_m)$ whose all the components belong to $X(\Omega)$ will be denoted by $X(\Omega; \mathbf{R}^m)$ in what follows. If τ is a $(n \times m)$ matrix function, whose elements belong to $X(\Omega)$ then we use the notation $\tau \in X(\Omega; \mathbf{R}^{n \times m})$. The norm of $v \in X(\Omega; \mathbf{R}^m)$, $\tau \in X(\Omega; \mathbf{R}^{n \times m})$, respectively, is defined in a usual way:

$$
\begin{cases}
\|v\|_{X(\Omega;\mathbf{R}^m)} = \displaystyle\sum_{i=1}^{m} \|v_i\|_{X(\Omega)}, & v = (v_i)_{i=1}^m \\[2mm]
\|\tau\|_{X(\Omega;\mathbf{R}^{n \times m})} = \displaystyle\sum_{\substack{i=1,...,n \\ j=1,...,m}} \|\tau_{ij}\|_{X(\Omega)}, & \tau = (\tau_{ij})_{\substack{i=1,...,n \\ j=1,...,m}} .
\end{cases}
\tag{1.9}
$$

Equivalent norms in Sobolev spaces

Let X be a linear, normed space and let $\| \cdot \|, [\| \cdot \|]$ be two norms in X. These norms are termed to be *equivalent* if there exist positive numbers c_1, c_2 such that

$$c_1 \|u\| \leq [\|u\|] \leq c_2 \|u\| \qquad \text{for any } u \in X.$$

In what follows we restrict ourselves to equivalent norms in $H^1(\Omega)$ only. Besides of the norm given by (1.5) with $k = 1$, $p = 2$ we may introduce other equivalent norms. We start with

Theorem 1.14 *(generalized Poincaré-Friedrichs inequality)* Let $\Gamma \subset \partial\Omega$ be such that its $(n\text{-}1)$-dimensional Lebesgue measure (in brief $meas_{n-1}\Gamma$) is positive. Then there exists a constant $c > 0$ such that

$$\|u\|_1 \leq c\Big(\int_\Omega \sum_{|\alpha|=1} |D^\alpha u|^2 dx + \int_\Gamma u^2 ds \Big)^{\frac{1}{2}} \tag{1.10}$$

holds for any $u \in H^1(\Omega)$.

Corollary 1.1 *Let* $\Gamma \subset \partial\Omega$, *meas*$_{n-1}\Gamma > 0$ *and*

$$V = \{v \in H^1(\Omega) \mid v = 0 \ on \ \Gamma\}.$$

Then from (1.10) it follows that

$$\|u\|_1 \leq c\Big(\int_\Omega \sum_{|\alpha|=1} |D^\alpha u|^2 dx\Big)^{\frac{1}{2}} \tag{1.11}$$

holds for any $u \in V$, *i.e., the right hand side of (1.11) is the equivalent norm to* $\|\cdot\|_1$ *in* V.

If $\Gamma \equiv \partial\Omega$ then (1.11) is known as the *Poincaré-Friedrichs inequality*.

Now let us pass to the vector-valued Sobolev spaces $H^1(\Omega; \mathbf{R}^n)$, $n = 2, 3$. By the symbol $\varepsilon(u) \in L^2(\Omega; \mathbf{R}^{n \times n})$ we denote the *linearized strain tensor*, corresponding to the displacement field $u \in H^1(\Omega; \mathbf{R}^n)$. The components of the matrix $\varepsilon(u)$ are given by

$$\varepsilon_{ij}(u) = \frac{1}{2}\Big(\frac{\partial u_i}{\partial x_j} + \frac{\partial u_j}{\partial x_i}\Big), \quad i, j = 1, ..., n.$$

It is readily seen that the mapping

$$u \mapsto \Big(\int_\Omega (\varepsilon_{ij}(u)\varepsilon_{ij}(u) + u^2)dx\Big)^{\frac{1}{2}} \tag{1.12}$$

defines the norm in $H^1(\Omega; \mathbf{R}^n)$. A nontrivial result says that (1.12) is an equivalent norm to the classical $H^1(\Omega; \mathbf{R}^n)$-norm. This follows from

Theorem 1.15 *(the second Korn's inequality) There exists a positive constant* c *such that*

$$\int_\Omega (\varepsilon_{ij}(u)\varepsilon_{ij}(u) + u^2)dx \geq c\|u\|_1^2$$

holds for any $u \in H^1(\Omega; \mathbf{R}^n)$.

Denote

$$[|u|]_1 \equiv \Big(\int_\Omega \varepsilon_{ij}(u)\varepsilon_{ij}(u)dx\Big)^{\frac{1}{2}}, \quad u \in H^1(\Omega; \mathbf{R}^n). \tag{1.13}$$

Then $[|u|]_1$ defines a seminorm in $H^1(\Omega; \mathbf{R}^n)$ only, since

$$[|u|]_1 = 0 \quad \text{if and only if } u \text{ is of the form}$$
$$u(x) = a \times x + b, \quad a, b \in \mathbf{R}^n.$$

Analogously to the Poincaré-Friedrichs inequality, (1.13) defines the norm on a properly defined subspace of $H^1(\Omega; \mathbf{R}^n)$ which is equivalent to $\|\cdot\|_1$. Indeed, this is a direct consequence of

Theorem 1.16 *(the first Korn's inequality) Let* $\Gamma \subset \partial\Omega$ *be such that* $meas_{n-1}$ $\Gamma > 0$ *and*

$$\mathbf{V} = \{v \in H^1(\Omega; \mathbf{R}^n) \mid v = 0 \quad on \ \Gamma\}.$$

Then there exists a positive constant c such that

$$\int_\Omega \varepsilon_{ij}(u)\varepsilon_{ij}(u)dx \geq c\|u\|_1^2$$

holds for any $u \in \mathbf{V}$.

Green's formula

Let $u, v \in C^1(\overline{\Omega})$. Then the following classical Green's formula holds:

$$\int_\Omega \frac{\partial u}{\partial x_i} v dx = -\int_\Omega u\frac{\partial v}{\partial x_i}dx + \int_{\partial\Omega} uv\nu_i ds, \tag{1.14}$$

where ν_i is the i-th component of the unit outward normal vector ν to $\partial\Omega$. Using the density result (Theorem 1.9) and Theorem 1.10 one can easily extend the validity of (1.14) to functions $u, v \in H^1(\Omega)$.

Now let $u \in C^2(\overline{\Omega})$, $v \in C^1(\overline{\Omega})$. On the basis of (1.14), we immediately obtain:

$$\int_\Omega \operatorname{grad} u. \operatorname{grad} v dx = -\int_\Omega \Delta u v dx + \int_{\partial\Omega} \frac{\partial u}{\partial \nu} v ds, \tag{1.15}$$

where the symbol $\operatorname{grad} = (\partial/\partial x_1, ..., \partial/\partial x_n)$ is the gradient, $\Delta = \partial^2/\partial x_1^2 + ... + \partial^2/\partial x_n^2$ is the laplacian and $\partial/\partial\nu$ stands for the normal derivative operator along $\partial\Omega$.

There is another consequence of the Green's formula which will be used in the case of the linear elasticity system. Let $\tau = (\tau_{ij})_{i,j=1}^n$ be a symmetric matrix, $\tau \in C^1(\overline{\Omega}; \mathbf{R}^{n\times n})$ and $v \in H^1(\Omega; \mathbf{R}^n)$. Then

$$\begin{aligned} \int_\Omega \tau_{ij}\varepsilon_{ij}(v)dx &= -\int_\Omega \tau_{ij,j}v_i dx + \int_{\partial\Omega} \tau_{ij}\nu_j v_i ds \\ &\equiv -\int_\Omega \operatorname{div}\tau.v dx + \int_{\partial\Omega} T_i v_i ds, \end{aligned} \tag{1.16}$$

where $\operatorname{div}\tau \in C(\overline{\Omega}; \mathbf{R}^n)$ is the vector, whose components are $\partial\tau_{ij}/\partial x_j$, $i = 1, ..., n$ and $T_i \equiv \tau_{ij}\nu_j$ is the i-th component of the stress vector T.

Next we introduce some spaces of functions defined in a time interval $(0, T)$, $T > 0$, taking values in a real Banach space X. These spaces are called *Bochner spaces*. Details can be found in Lions, 1969, Zeidler, 1990a or Yosida, 1965. We denote the norm in X by $\|\cdot\|_X$, the dual space of X by X^* and the duality pairing between X and X^* by $\langle\cdot,\cdot\rangle_X$.

The space $C^k([0,T]; X)$

Let $k \in \mathbf{N}_0$. By $C^k([0,T]; X)$ we denote the set of all continuous functions $u : [0,T] \to X$ whose (strong) derivatives up to the order k are continuous in $[0,T]$ and belong to X. The space $C^k([0,T]; X)$ is complete with respect to the norm

$$\|u\|_{C^k([0,T];X)} \equiv \sum_{i=0}^{k} \max_{t \in [0,T]} \|u^{(i)}(t)\|_X, \tag{1.17}$$

where $u^{(i)}$ is the i-th derivative of u. If $k = 0$ we use the following convention $C^0([0,T]; X) \equiv C([0,T]; X)$.

The space $L^p(0,T; X)$, $p \in [1,\infty]$

Let $p \in [1,\infty)$. The set of all measurable functions $u : (0,T) \to X$ for which

$$\int_0^T \|u(t)\|_X^p dt < \infty \qquad \text{(in the Lebesgue sense)}$$

is denoted by $L^p(0,T; X)$. This space is equipped with the norm

$$\|u\|_{L^p(0,T;X)} \equiv \left(\int_0^T \|u(t)\|_X^p dt \right)^{\frac{1}{p}}. \tag{1.18}$$

Recall that the function $u : (0,T) \to X$ is said to be measurable iff there exists a sequence $\{u_k\}$ of step functions (i.e., piecewise constant functions) $u_k : (0,T) \to X$ such that

$$\lim_{k \to \infty} u_k(t) = u(t) \quad \text{for a.a. } t \in (0,T).$$

The space $L^\infty(0,T; X)$ consists of all measurable functions $u : (0,T) \to X$, which are essentially bounded, that is, the number

$$\|u\|_{L^\infty(0,T;X)} \equiv \inf_{\substack{\text{meas } M=0 \\ M \subset (0,T)}} \sup_{t \in (0,T) \setminus M} \|u(t)\|_X \tag{1.19}$$

is finite (cf. the definition of the space $L^\infty(\Omega)$). We endow the space $L^\infty(0,T; X)$ with the norm $\| \cdot \|_{L^\infty(0,T;X)}$.

Let us collect some fundamental properties of the spaces $L^p(0,T; X)$.

Theorem 1.17 *It holds:*

(i) The space $L^p(0,T; X)$ is complete with respect to the norm (1.18) if $p \in [1,\infty)$ and (1.19) when $p = \infty$.

(ii) Let X be a Hilbert space with a scalar product $(\cdot, \cdot)_X$. Then $L^2(0,T; X)$ is also a Hilbert space equipped with the scalar product

$$(u,v)_{L^2(0,T;X)} \equiv \int_0^T (u(t), v(t))_X dt.$$

Theorem 1.18 *Let X be a reflexive, separable Banach space. Then it holds:*

(i) Let $p \in (1, \infty)$. Then the space $L^p(0, T; X)$ is reflexive and separable. Further, its dual space is

$$\left(L^p(0, T; X)\right)^* = L^q(0, T; X^*)$$

where $1/p + 1/q = 1$.

(ii) Let $p = 1$. Then the space $L^1(0, T; X)$ is separable and its dual space is

$$\left(L^1(0, T; X)\right)^* = L^\infty(0, T; X^*).$$

Theorem 1.18 has the following important consequence:

Theorem 1.19 *Let X be a reflexive, separable Banach space. Then the following statements hold:*

(i) Let $p \in (1, \infty)$ and $\{v_j\}$ be a bounded sequence in $L^p(0, T; X)$. Then there exists a weakly convergent subsequence $\{v_{j'}\} \subset \{v_j\}$ in $L^p(0, T; X)$, i.e., there exists $v \in L^p(0, T; X)$ and

$$\int_0^T \langle v_{j'}(t), w(t) \rangle_X \, dt \longrightarrow \int_0^T \langle v(t), w(t) \rangle_X \, dt \quad \forall w \in L^q(0, T; X^*)$$

where $1/p + 1/q = 1$.

(ii) Let $p = \infty$ and $\{v_j\}$ be a bounded sequence in $L^\infty(0, T; X)$. Then there exists a weak-$*$ convergent subsequence $\{v_{j'}\} \subset \{v_j\}$ in $L^\infty(0, T; X)$, i.e., there exists $v \in L^\infty(0, T; X)$ and

$$\int_0^T \langle v_{j'}(t), w(t) \rangle_X \, dt \longrightarrow \int_0^T \langle v(t), w(t) \rangle_X \, dt \quad \forall w \in L^1(0, T; X^*).$$

In the sequel we shall denote the weak convergence in $L^p(0, T; X)$, $p \in [1, \infty)$, by the symbol $\rightharpoonup$ and the weak-$*$ convergence in $L^\infty(0, T; X)$ by $\overset{*}{\rightharpoonup}$. Finally, we recall the following density result:

Proposition 1.1 *Let $p \in [1, \infty)$. The set of all polynomials $v : [0, T] \to X$ of the form*

$$v(t) = a_0 + a_1 t + \ldots + a_n t^n \tag{1.20}$$

with $a_i \in X$ for all i and $n = 0, 1, \ldots$ is dense in $L^p(0, T; X)$.

The spaces $W^{1,p}(0,T;X)$ *and* $W^{1,p}(0,T;V,H)$

In a similar way as in (1.4) we can define the generalized derivative in Bochner spaces. Let Y be another Banach space. Let $u \in L^p(0,T;X)$, $p \in [1,\infty]$. An integrable function $v \in L^1(0,T;Y)$ is called the *i-th generalized derivative* of u if it holds

$$\int_0^T u(t)\varphi^{(i)}(t)dt = (-1)^i \int_0^T v(t)\varphi(t)dt \qquad \forall \varphi \in C_0^\infty(0,T) \qquad (1.21)$$

where $\varphi^{(i)}$ is the i-th derivative of φ. The i-th generalized derivative of u will be denoted by $u^{(i)}$, again. The integrals in (1.21) are taken in the Bochner sense (see Yosida, 1965).

The next proposition indicates that the generalized gradients are compatible with weak limits.

Proposition 1.2 *Let* $p,q \in [1,\infty)$ *and* $i \in \mathbf{N}$. *Suppose that* X *is continuously imbedded in* Y. *Then from*

$$u_k^{(i)} = v_k \qquad \text{a.e. in } (0,T) \qquad \forall k = 1,2,\dots$$

and

$$\begin{aligned} u_k &\rightharpoonup u \qquad \text{in } L^p(0,T;X), \\ v_k &\rightharpoonup v \qquad \text{in } L^q(0,T;Y) \end{aligned}$$

it follows that

$$u^{(i)} = v \qquad \text{a.e. in } (0,T).$$

The previous result extends to the generalized derivatives and the weak-$*$ limits in $L^\infty(0,T;X)$.

Let $p \in [1,\infty]$. By $W^{1,p}(0,T;X)$ we denote the subspace of $L^p(0,T;X)$ containing functions whose first generalized derivatives also belong to $L^p(0,T;X)$, i.e.,

$$W^{1,p}(0,T;X) \equiv \{u \in L^p(0,T;X) \mid u' \in L^p(0,T;X)\}.$$

If $p = 2$ we use the notation $H^1(0,T;X)$ instead of $W^{1,2}(0,T;X)$. The space $W^{1,p}(0,T;X)$ is endowed with the norm

$$\|u\|_{W^{1,p}(0,T;X)} = \|u\|_{L^p(0,T;X)} + \|u'\|_{L^p(0,T;X)}.$$

Then $W^{1,p}(0,T;X)$ is the Banach space. Further, the following equivalent characterization of $W^{1,p}(0,T;X)$ holds:

Theorem 1.20 *Let* $u \in L^p(0,T;X)$, $p \in [1,\infty]$. *Then the following conditions are equivalent:*

(i) $u \in W^{1,p}(0,T;X)$;

(ii) There exists an absolutely continuous function u_0 from $[0,T]$ to X which is differentiable a.e. in $(0,T)$ and

$$\frac{d}{dt}u_0(t) = \lim_{\varepsilon \to 0}\frac{u_0(t+\varepsilon) - u_0(t)}{\varepsilon} \in L^p(0,T;X).$$

Moreover, $u(t) = u_0(t)$ and $u'(t) = \frac{d}{dt}u_0(t)$ for a.a. $t \in (0,T)$.

In order to define $W^{1,p}(0,T;V,H)$ we first introduce the so-called *evolution triplet*

$$"V \subseteq H \subseteq V^*",$$

in which V is a real separable, reflexive Banach space, H a real separable Hilbert space and V^* the dual space to V. By $\|\cdot\|_V$ and $|\cdot|_H$ we denote norms in V and H, respectively. The duality pairing between V and V^* is denoted by $\langle \cdot, \cdot \rangle_V$ and the scalar product in H by $(\cdot, \cdot)_H$. We assume that the space V is continuously and densely imbedded in H. Therefore, by identifying H and H^* also the imbedding $H \subseteq V^*$ makes sense and it is *continuous* and *dense*.

The following characterization of the generalized derivatives holds in the case of evolution triplets.

Proposition 1.3 *Let $p,q \in [1,\infty]$ and $u \in L^p(0,T;V)$. The following statements are equivalent:*

(i) there exists the generalized derivative $u^{(i)} \in L^q(0,T;V^*)$;

(ii) there is a function $w \in L^q(0,T;V^*)$ such that

$$\int_0^T (u(t),v)_H \varphi^{(i)}(t)dt = (-1)^i \int_0^T \langle w(t),v \rangle_V \varphi(t)dt$$

for all $v \in V$, $\varphi \in C_0^\infty(0,T)$ and $w = u^{(i)}$.

Let $p \in (1,\infty)$. The space $W^{1,p}(0,T;V,H)$ is defined as a subspace of $L^p(0,T;V)$ as follows:

$$W^{1,p}(0,T;V,H) = \{u \in L^p(0,T;V) \mid u' \in L^q(0,T;V^*)\}$$

where $1/p + 1/q = 1$. We equip $W^{1,p}(0,T;V,H)$ with the following norm

$$\|u\|_{W^{1,p}(0,T;V,H)} \equiv \|u\|_{L^p(0,T;V)} + \|u'\|_{L^q(0,T;V^*)}. \qquad (1.22)$$

Let us recall some important results of the space $W^{1,p}(0,T;V,H)$.

Theorem 1.21 *The space $W^{1,p}(0,T;V,H)$ is complete with respect to the norm (1.22).*

Remark 1.2 *In addition, let us suppose that V is a Hilbert space and $p = 2$. Then $W^{1,2}(0, T; V, H)$ equipped with the norm (1.22) (and with the corresponding scalar product) is a Hilbert space. In this case, we shall use the following simpler notation: $W^{1,2}(0, T; V, H) \equiv W(V)$ in what follows.*

Proposition 1.4 *The following continuous imbedding holds:*

$$W^{1,p}(0, T; V, H) \hookrightarrow C([0, T]; H).$$

In the sequel the values $u(t)$, $t \in [0, T]$, of the function $u \in W^{1,p}(0, T; V, H)$ will be understood in the sense of Proposition 1.4.

Proposition 1.5 *(integration by parts) Let $u, v \in W^{1,p}(0, T; V, H)$ and $t_1, t_2 \in [0, T]$. Then the following generalized integration by parts formula holds:*

$$\int_{t_1}^{t_2} \langle u'(t), v(t) \rangle_V \, dt$$

$$= (u(t_2), v(t_2))_H - (u(t_1), v(t_1))_H - \int_{t_1}^{t_2} \langle u(t), v'(t) \rangle_V \, dt.$$

Remark 1.3 *The counterpart of Proposition 1.1 holds in $W^{1,p}(0, T; V, H)$, $p \in (1, \infty)$, as well, i.e., the set of polynomials is dense in $W^{1,p}(0, T; V, H)$.*

Compact imbeddings

Next we state an important compact imbedding result in Bochner spaces (Lions, 1969):

Proposition 1.6 *Let $p, q \in (1, \infty)$. Let X, Z be real separable, reflexive Banach spaces and Y a real Banach space. Suppose that*

$$X \overset{\hookrightarrow}{\hookrightarrow} Y \hookrightarrow Z.$$

Then

$$\{u \in L^p(0, T; X) \mid u' \in L^q(0, T; Z)\} \quad \overset{\hookrightarrow}{\hookrightarrow} \quad L^p(0, T; Y).$$

Let $\Omega \subset \mathbf{R}^n$ be a bounded domain with the Lipschitz boundary $\partial\Omega$. Let V be a real Hilbert space such that $V \hookrightarrow H^1(\Omega)$. Then the following variant of the compactness result holds (Landes and Mustonen, 1987, Miettinen, 1996):

Proposition 1.7 *Let $\{u_k\}$ be a bounded sequence in $L^2(0, T; V) \cap L^\infty(0, T; L^1(\Omega))$. If $u_k(t)$ converges weakly to $u(t)$ in $L^1(\Omega)$ a.e. in $(0, T)$, then $u_{k'} \to u$ strongly in $L^2(0, T; L^2(\Omega))$ for some subsequence $\{u_{k'}\}$ of $\{u_k\}$.*

1.2   ELEMENTS OF NONSMOOTH ANALYSIS

Let X be a Banach space and X^* its dual. Further, let Y be another Banach space.

First, we recall some (classical) definitions of differential calculus (see Clarke, 1983). Let F be a mapping from X into Y. The *one-sided directional derivative* of F at a point u and a direction v is defined by

$$F'(u;v) = \lim_{t \to 0+} \frac{F(u+tv) - F(u)}{t} \tag{1.23}$$

assuming that the above limit (in Y) exists. The function F is said to be *Gâteaux differentiable* at a point u if there exists $DF(u) \in \mathcal{L}(X,Y)$ (the space of all linear and continuous mappings from X into Y) and $DF(u)v$ is equal to $F'(u,v)$ for all $v \in X$. If moreover

$$\lim_{\bar{u} \to u} \frac{\|F(\bar{u}) - F(u) - DF(u)(\bar{u} - u)\|_Y}{\|\bar{u} - u\|_X} = 0, \tag{1.24}$$

then F is said to be *Fréchet differentiable* at u. If the mapping $u \mapsto DF(u)$ is continuous at u, we say that F is *continuously differentiable* at u. Finally we introduce the concept of *strict differentiability*: F is declared to be strictly differentiable at u if there exists an element of $\mathcal{L}(X,Y)$, denoted by $D_sF(u)$ such that for each $v \in X$ one has

$$\lim_{\substack{\bar{u} \to u \\ t \to 0+}} \frac{F(\bar{u} + tv) - F(\bar{u})}{t} = D_sF(u)v \tag{1.25}$$

and the convergence is uniform with respect to v in compact sets. We summarize the relations among the above mentioned differentiability concepts in the following table:

continuously diff. $\implies$ Fréchet diff. $\implies$ Gâteaux diff.

continuously diff. $\implies$ strictly diff. $\implies$ Gâteaux diff.

Elements of convex analysis

The aim of this part is to introduce some basic results of convex analysis (Moreau, 1967, Rockafellar, 1969, Ekeland and Temam, 1976). A subset K of X is said to be *convex* iff

$$\lambda u + (1 - \lambda)v \in K \qquad \forall u, v \in K \text{ and } \forall \lambda \in (0,1).$$

Let $\bar{\mathbf{R}}$ be the extended real line, i.e., $\bar{\mathbf{R}} \equiv \mathbf{R} \cup \{+\infty\}$ and $f : X \to \bar{\mathbf{R}}$ be a *proper functional* in X, i.e., $f \not\equiv +\infty$ in X. A functional $f : X \to \bar{\mathbf{R}}$ is said to be *convex* in K iff:

$$f(\lambda u + (1 - \lambda)v) \le \lambda f(u) + (1 - \lambda)f(v) \qquad \forall u, v \in K \text{ and } \forall \lambda \in (0,1),$$

whenever the sum on the right hand side is defined.[†] If the strict inequality holds for any $u \neq v$ then f is said to be *strictly convex* in K. The set $D_{eff} = \{x \in X \mid f(x) < \infty\}$ is called the *effective domain* of f.

Definition 1.2 *An element* $w \in X^*$ *is said to be a subgradient of a convex functional* f *at a point* u *in which* $f(u)$ *is finite iff*

$$f(v) - f(u) \geq \langle w, v - u \rangle_X \qquad \forall v \in X.$$

The set of all subgradients of f *at* u *is called the subdifferential of* f *at* u *and will be denoted by* $\partial f(u)$, *in what follows. If* $f(u) = +\infty$ *we set* $\partial f(u) = \emptyset$.

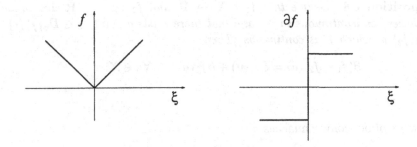

Figure 1.1. An example of a convex function $f : \mathbf{R} \to \mathbf{R}$ and its subdifferential.

Definition 1.3 *The normal cone* $N_K(u)$ *of a nonempty convex subset* K *at the point* $u \in K$ *is defined by*

$$N_K(u) = \{w \in X^* \mid \langle w, v - u \rangle_X \leq 0 \quad \forall v \in K\}.$$

An important example of the convex functional is the so-called *indicator function* I_K of the convex subset K defined by:

$$I_K(u) = \begin{cases} 0 & \text{iff } u \in K \\ +\infty & \text{elsewhere.} \end{cases}$$

It is easy to see that if $u \in K$ then the subdifferential of the indicator function I_K and the normal cone of the set K at u coincide, i.e

$$\partial I_K(u) = N_K(u). \tag{1.26}$$

Finally we recall the result concerning the additivity of subdifferentials:

[†]i.e., except the case $\infty + (-\infty)$

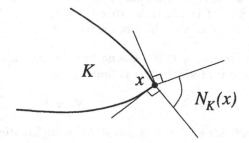

Figure 1.2.   An example of the normal cone $N_K(x)$ of a convex set K.

Proposition 1.8 *Assume that $f_1 : X \to \bar{\mathbf{R}}$ and $f_2 : X \to \bar{\mathbf{R}}$ are convex and lower semicontinuous in X, and that there exists a point $u_0 \in D_{eff}(f_1) \cap D_{eff}(f_2)$ at which f_1 is continuous. Then*

$$\partial(f_1 + f_2)(u) = \partial f_1(u) + \partial f_2(u) \qquad \forall u \in X.$$

Elements of nonconvex analysis

We restrict our presentation to *locally Lipschitz functionals*. A functional $f : X \to \mathbf{R}$ is said to be *Lipschitz near a point $u \in X$ of rank c* iff there exists a neighbourhood U of u such that

$$|f(v) - f(w)| \leq c\|v - w\|_X \qquad \forall v, w \in U, \tag{1.27}$$

where c is a positive constant. The functional f is *locally Lipschitz* iff it is Lipschitz near every point $u \in X$ and, moreover, *Lipschitz* if the constant c is independent of u.

Next we define the *generalized directional derivative* and the *generalized gradient* in the sense of F.H. Clarke (Clarke, 1983).

Definition 1.4 *Let f be Lipschitz near a point $u \in X$. The generalized directional derivative of f at u and the direction $v \in X$, denoted by $f^\circ(u; v)$, is given by*

$$f^\circ(u; v) = \limsup_{\substack{\bar{u} \to u \\ t \to 0+}} \frac{f(\bar{u} + tv) - f(\bar{u})}{t}.$$

Definition 1.5 *Let f be Lipschitz near a point $u \in X$. The generalized gradient of f at u, denoted by $\bar{\partial} f(u)$, is the subset of X^* defined by*

$$\bar{\partial} f(u) = \{ w \in X^* \mid f^\circ(u; v) \geq \langle w, v \rangle_X \ \forall v \in X \}.$$

Remark 1.4 *Suppose that X is finite dimensional and f is Lipschitz near a point $u \in X$. Then we have an equivalent characterization of the generalized* *gradient of f at u:*

$$\bar{\partial} f(u) = \overline{conv}\, \big\{ \lim_{j \to \infty} \nabla f(u_j) \mid u_j \to u,\ \nabla f(u_j)\ and\ lim\ exist \big\}$$

in which $\overline{conv}$ is the closed convex hull and $\nabla f(u_j)$ is the gradient of f at u_j.

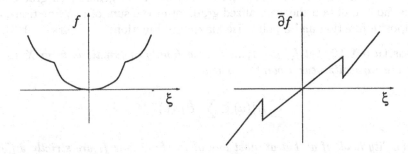

Figure 1.3. An example of a nonconvex, locally Lipschitz function $f : \mathbf{R} \to \mathbf{R}$ and its generalized gradient.

Below we list some basic properties of the generalized directional derivative and of the generalized gradient.

Proposition 1.9 *Let f be Lipschitz of rank c near a point $u \in X$. Then it holds:*

(i) For each $v \in X$ one has

$$f^{\circ}(u; v) = \max \big\{ \langle w, v \rangle_X \mid w \in \bar{\partial} f(u) \big\}; \tag{1.28}$$

(ii) The function $(u, v) \mapsto f^{\circ}(u; v)$ is upper semicontinuous, i.e.,

$$\underset{\substack{\bar{u} \to u \\ \bar{v} \to v}}{\text{limsup}}\, f^{\circ}(\bar{u}; \bar{v}) \leq f^{\circ}(u; v);$$

(iii) $\bar{\partial} f(u)$ is a nonempty, convex and weak-$*$ compact subset of X^* and $\|w\|_{X^*} \leq c$ for every $w \in \bar{\partial} f(u)$;

(iv) Let $\{u_j\}$ and $\{w_j\}$ be sequences in X and X^*, respectively, such that u_j converges to u, and that w is a cluster point of $w_j \in \bar{\partial} f(u_j)$ in the weak-$*$ topology. Then $w \in \bar{\partial} f(u)$.

(v) If X is finite-dimensional, then $\bar{\partial} f$ is upper semicontinuous at u in the sense of set-valued mappings.

For completeness, we recall the definition of the upper semicontinuity for a set-valued mapping (Aubin and Frankowska, 1990):

Definition 1.6 *Let T be a set-valued mapping from X to Y. Then T is said to be upper semicontinuous at $u \in X$ iff the following property holds: for any $\varepsilon > 0$ there exists $\delta > 0$ such that*

$$T(\bar{u}) \subset T(u) + \varepsilon B_Y \qquad \forall \bar{u} \in u + \delta B_X,$$

where B_X, B_Y are the unit balls in X, Y, respectively.

Now, we introduce three results from the calculus of generalized gradients, namely, how to obtain the generalized gradient of the sum of two functions, of a composite function and a pointwise maximum function (see Clarke, 1983).

Proposition 1.10 *Let f_i, $i = 1, ..., n$, be a family of functions from X to $\mathbf{R}$ which are Lipschitz near a point u. Then*

$$\bar{\partial}\left(\sum_{i=1}^{n} f_i\right)(u) \subset \sum_{i=1}^{n} \bar{\partial} f_i(u).$$

The equality holds if all but at most one of the functions f_i are strictly differentiable at u.

Theorem 1.22 *Let $F : X \to Y$ be a strictly differentiable mapping at a point u and let $g : Y \to \mathbf{R}$ be a Lipschitz function near $F(u)$. Then the composite function $f \equiv g \circ F : X \to \mathbf{R}$ is Lipschitz near u, and it holds that*

$$\bar{\partial} f(u) \subset \bar{\partial} g(F(u)) \circ D_s F(u). \tag{1.29}$$

The equality in (1.29) holds if, for example, $D_s F(u)$ is onto.

Proposition 1.11 *Let f_i, $i = 1, ..., n$, be a family of functions from X to $\mathbf{R}$ which are Lipschitz near a point u. The pointwise maximum function defined by*

$$f(v) = \max_{i=1,...,n} f_i(v)$$

is Lipschitz near u and

$$\bar{\partial} f(u) \subset \overline{conv}\{\bar{\partial} f_i(u) : i \in I(u)\}, \tag{1.30}$$

where $I(u)$ denotes the set of indices for which $f_i(u) = f(u)$.

We close this part by propositions, specifying the generalized gradient of a convex, bounded and of a strictly differentiable functional, respectively:

Proposition 1.12 *Let f be convex and bounded in some neighbourhood of $u \in X$. Then f is Lipschitz near u, the generalized gradient $\bar{\partial} f(u)$ coincides with the subdifferential $\partial f(u)$ and the generalized directional derivative $f^\circ(u; v)$ with the directional derivative $f'(u; v)$ for each v.*

Proposition 1.13 *The function $f : X \to \mathbf{R}$ is strictly differentiable at u if and only if f is Lipschitz near u and at the same time $\bar{\partial} f(u) = \{D_s f(u)\}$.*

Subdifferentials and generalized gradients of integral functionals

The aim of this part is to present two results determining subdifferentials and generalized gradients of functionals f expressed by means of integrals (Brézis, 1973, Aubin and Clarke, 1979). For the sake of simplicity we assume that the Banach space X is equal to $L^2(\Omega; \mathbf{R}^m)$ and Ω is an open bounded subset of $\mathbf{R}^n$. First, let $j : \mathbf{R}^m \to \bar{\mathbf{R}}$ be a convex, lower semicontinuous and proper function. Define an integral functional $f : L^2(\Omega; \mathbf{R}^m) \to \bar{\mathbf{R}}$ as follows:

$$f(u) = \begin{cases} \int_\Omega j(u(x))dx & \text{if } j(u) \in L^1(\mathbf{R}^m) \\ +\infty & \text{otherwise.} \end{cases}$$

Proposition 1.14 *Let j be as above. Then the functional f is convex, lower semicontinuous and proper, and the subdifferential $\partial f(u) \subset L^2(\Omega; \mathbf{R}^m)$ is characterized by the relation*

$$w \in \partial f(u) \Longleftrightarrow w(x) \in \partial j(u(x)) \quad \text{a.e. in } \Omega.$$

Now, let $j : \Omega \times \mathbf{R}^m$ be a function satisfying:

(i) The Carathéodory type conditions:

- *For all $\xi \in \mathbf{R}^m$ the function $x \mapsto j(x, \xi)$ is measurable in Ω;*
- *For almost all $x \in \Omega$ the function $\xi \mapsto j(x, \xi)$ is locally Lipschitz in $\mathbf{R}^m$;*

(ii) The function $j(\cdot, 0) \in L^1(\Omega)$;

(iii) There exists a positive constant c such that

$$\eta \in \bar{\partial} j(x, \xi) \implies |\eta| \leq c(1 + |\xi|)$$

for almost all $x \in \Omega$ and each $\xi \in \mathbf{R}^m$.

We consider the integral functional in $L^2(\Omega; \mathbf{R}^m)$ defined by

$$f(u) = \int_\Omega j(x, u(x))dx.$$

The following proposition holds:

Proposition 1.15 *Let j fulfill the conditions (i)-(iii). Then the functional f is locally Lipschitz in $L^2(\Omega; \mathbf{R}^m)$ and for each $u \in L^2(\Omega; \mathbf{R}^m)$ and $w \in \bar{\partial} f(u) \subset L^2(\Omega; \mathbf{R}^m)$, one has*

$$w(x) \in \bar{\partial} j(x, u(x)) \quad \text{a.e. in } \Omega.$$

Remark 1.5 *Propositions 1.14 and 1.15 remain valid if $L^2(\Omega; \mathbf{R}^m)$ is replaced by $L^2(\Gamma; \mathbf{R}^m)$, $\Gamma \subset \partial\Omega$.*

Optimality conditions

We state optimality conditions for general minimization problems in a Banach space X (Clarke, 1983, Aubin and Frankowska, 1990). We start with a convex case.

Theorem 1.23 *(the necessary and sufficient optimality condition) Let f : $X \to \bar{R}$ be a proper convex functional. Then $u^* \in X$ is a minimum point of f in X if and only if*

$$0 \in \partial f(u^*). \qquad (1.31)$$

In many situation the functional f has a special form, namely

$$f = \bar{f} + I_K, \qquad (1.32)$$

where $\bar{f} : X \to \bar{R}$ is a convex functional and I_K the indicator function of a nonempty, closed and convex subset K of X. Assuming that $\bar{f}$ is lower semicontinuous in X and there exists a point $u_0 \in D_{eff}(\bar{f}) \cap K$ at which $\bar{f}$ is continuous, the optimality condition (1.31) takes the form

$$0 \in \partial f(u^*) = \partial \bar{f}(u^*) + \partial I_K(u^*) = \partial \bar{f}(u^*) + N_K(u^*). \qquad (1.33)$$

Indeed, taking into account that I_K being the indicator function of the nonempty, closed and convex set K is convex, lower semicontinuos and proper in X, the above result follows from Proposition 1.8 and (1.26).

Next we consider the case when $f : X \to R$ is a locally Lipschitz, generally nonconvex functional.

Definition 1.7 *(substationary point) A point $u^* \in K$, where K is a nonempty, closed and convex subset of X, is called a substationary point of a locally Lipschitz functional $f : X \to R$ on K iff $0 \in \bar{\partial}f(u^*) + N_K(u^*)$. If $K = V$ the condition reduces to $0 \in \bar{\partial}f(u^*)$.*

Remark 1.6 *Note that due to the definition of the normal cone the substationary point condition $0 \in \bar{\partial}f(u^*) + N_K(u^*)$ can be written equivalently as follows: there exist $u^* \in K$ and $w^* \in \bar{\partial}f(u^*)$ such that*

$$\langle w^*, v - u^* \rangle_X \geq 0 \qquad \forall v \in K. \qquad (1.34)$$

Now, we are able to formulate the following necessary conditions for a minimizer of f:

Theorem 1.24 *(the necessary condition) Let $f : X \to R$ be a locally Lipschitz functional and K be a nonempty, closed and convex subset of X. Then*

  (i) Unconstrained case: If $u^* \in X$ is a minimizer of f on X, then u^* is a substationary point of f on X, i.e.

$$0 \in \bar{\partial}f(u^*). \qquad (1.35)$$

(ii) Constrained case: If $u^* \in K$ is a minimizer of f on K, then u^* is a substationary point of f on K, i.e

$$0 \in \bar{\partial} f(u^*) + N_K(u^*). \tag{1.36}$$

It is clear that conditions (1.35) and (1.36) can not be sufficient for a general nonconvex functional f, since f may have many local minima, maxima or other types of substationary points.

Remark 1.7 *Let us shortly comment the relation between (i) and (ii) in Theorem 1.24. We define a functional $\tilde{f} \equiv f + I_K : X \to \bar{\mathbf{R}}$, where f and K are as in Theorem 1.24. Using the definition of the generalized gradient for the extended real valued functionals (see Clarke, 1983), denoted by $\hat{\partial}\tilde{f}$, it is possible to show that the following analogy to the necessary optimality condition (1.35) holds for $\tilde{f}$ (see, e.g., Aubin and Frankowska, 1990):*

$$0 \in \hat{\partial}\tilde{f}(u^*),$$

provided that u^* is a minimizer of $\tilde{f}$ in X. Now it holds (see details in Clarke, 1983):

$$0 \in \hat{\partial}\tilde{f}(u^*) \subset \hat{\partial}f(u^*) + \hat{\partial}I_K(u^*) = \bar{\partial}f(u^*) + \partial I_K(u^*). \tag{1.37}$$

Then taking into account (1.26) we get (1.36). Thus, (ii) is a consequence of (i) in this more general setting.

Existence results

In this section we list classical results which will be needed later to show the solvability of some minimization problems or problems of finding substationary points.

Theorem 1.25 *(fundamental theorem of calculus of variations) Let X be a reflexive Banach space and K its nonempty weakly closed subset. Let $f : X \to \bar{\mathbf{R}}$ be a weakly lower semicontinuous functional. Moreover, suppose that f is coercive in K, i.e.*

$$\lim_{\substack{\|u\|_X \to +\infty \\ u \in K}} f(u) = +\infty. \tag{1.38}$$

Then the minimization problem

$$\text{Find } u^* \in K \text{ such that } \inf_{u \in K} f(u) = f(u^*) \tag{1.39}$$

admits at least one solution.

Remark 1.8 *Suppose that X and f are as in Theorem 1.25. Further, let K be a nonempty, closed and convex subset. Hence, K is weakly closed and the existence of at least one minimizer of f on K follows from Theorem 1.25.*

At the end of this section we assume that X is *finite dimensional*. In the sequel we shall need also the following results (see Aubin and Ekeland, 1984, Browder and Hess, 1972):

Theorem 1.26 *(a consequence of Brouwer fixed point theorem) Let $T : \mathbf{R}^n \to \mathbf{R}^n$ be a continuous (single-valued) mapping such that there exists $r > 0$ satisfying*

$$(Tu, u) > 0 \qquad \forall u, \ |u| = r. \tag{1.40}$$

Then there exists a point u^* such that $Tu^* = 0$ and $|u^*| \leq r$.

Theorem 1.27 *(a consequence of Kakutani fixed point theorem) Let T be an upper semicontinuous set-valued mapping from X to X^* such that Tu is a nonempty, bounded, closed and convex subset of X^* $\forall u \in X$. Moreover, let T be coercive in X, i.e., there exists a function $c : \mathbf{R}_+ \to \mathbf{R}$ with $\lim_{r \to +\infty} c(r) = +\infty$ such that for all $u \in X$, $w \in T(u)$ it holds that $\langle w, u \rangle_X \geq c(\|u\|_X)\|u\|_X$. Then the range $R(T) \equiv \{Tu \mid u \in X\} = X^*$.*

Corollary 1.2 *Let T be the same as in Theorem 1.27. Further, suppose that K is a nonempty, closed, convex subset of X and there exists $u_0 \in K$ such that T is coercive with respect to u_0 on K, i.e., a function $c : \mathbf{R}_+ \to \mathbf{R}$ with $\lim_{r \to +\infty} c(r) = +\infty$ exists and*

$$\langle w, v - u_0 \rangle_X \geq c(\|v\|_X)\|v\|_X \qquad \forall v \in K \text{ and } \forall w \in Tv.$$

Then there exist $u^* \in K$ and $w^* \in Tu^*$ satisfying

$$\langle w^*, v - u^* \rangle_X \geq 0 \qquad \forall v \in K.$$

1.3 EQUATIONS AND INEQUALITIES WITH MONOTONE OPERATORS

The aim of this part is to recall main results of the theory of differential equations and inequalities involving monotone operators. In the first part we focus on the static case, while the second part deals with time dependent problems. We restrict ourselves to very basic results especially in the theory of nonlinear equations. For more extensive treatment of nonlinear problems we refer to Lions, 1969 and Fučik and Kufner, 1980.

Elliptic Problems

We start with linear problems (see Nečas, 1967).

Let V be a real Hilbert space and V^* its dual. The norm, the scalar product in V will be denoted by $\| \cdot \|$, $(\cdot, \cdot)$, respectively. The value of $f \in V^*$ at $v \in V$ will be denoted by $\langle f, v \rangle$. The norm of $f \in V^*$ is defined in a standard way:

$$\|f\|_{V^*} = \sup_{\substack{v \in V \\ v \neq 0}} \frac{\langle f, v \rangle}{\|v\|}.$$

Let $a : V \times V \to \mathbf{R}$ be a bilinear form such that

$$\exists M > 0 \quad : \quad |a(u,v)| \le M\|u\|\|v\| \quad \forall u, v \in V \quad \text{(boundedness)}; \quad (1.41)$$

$$\exists \alpha > 0 \quad : \quad a(v,v) \ge \alpha\|v\|^2 \quad \forall v \in V \quad \text{(V-ellipticity)}. \quad (1.42)$$

By *an abstract linear elliptic equation* we call a triplet $\{V, a, f\}$, where $f \in V^*$. Any element $u \in V$ satisfying

$$a(u, v) = \langle f, v \rangle \quad \forall v \in V \quad (1.43)$$

will be called *a solution* of $\{V, a, f\}$.

As far as the existence and the uniqueness of the solution to $\{V, a, f\}$ is concerned, we have:

Theorem 1.28 *(Lax-Milgram) Let the bilinear form $a : V \times V \to \mathbf{R}$ be bounded and V-elliptic. Then there exists a unique solution u of (1.43) for any $f \in V^*$ and*

$$\|u\| \le \frac{1}{\alpha}\|f\|_{V^*}.$$

Throughout this section we shall assume that a satisfies (1.41) and (1.42).

The bilinear form a defines a linear mapping $A : V \to V^*$:

$$a(u, v) = \langle Au, v \rangle \quad \forall u, v \in V. \quad (1.44)$$

Thus the abstract linear elliptic equation $\{V, a, f\}$ is equivalent to the linear operator equation

$$Au = f \quad \text{in } V^*.$$

From (1.41) and (1.42) it easily follows that

$$\|A\|_{\mathcal{L}(V,V^*)} \le M, \qquad \|A^{-1}\|_{\mathcal{L}(V^*,V)} \le \frac{1}{\alpha},$$

i.e, $A \in \mathcal{L}(V, V^*)$, $A^{-1} \in \mathcal{L}(V^*, V)$.

Now, let K be a *nonempty, closed* and *convex* subset of V. By *an abstract elliptic inequality of the first kind* we call a triplet $\{K, a, f\}$, where $a : V \times V \to \mathbf{R}$ is a bilinear form and $f \in V^*$. Any element $u \in K$ satisfying

$$a(u, v - u) \ge \langle f, v - u \rangle \quad \forall v \in K \quad (1.45)$$

will be called a solution of $\{K, a, f\}$. The following analogy to Theorem 1.28 can be proven (Lions, 1969):

Theorem 1.29 *Let K has the property mentioned above and let $a : V \times V \to \mathbf{R}$ satisfy (1.41) and (1.42). Then $\{K, a, f\}$ has a unique solution u for any $f \in V^*$. If u_i are solutions to $\{K, a, f_i\}$, $i = 1, 2$, then*

$$\|u_1 - u_2\| \le \frac{1}{\alpha}\|f_1 - f_2\|_{V^*}.$$

Remark 1.9 *Let I_K be the indicator function of K. Then (1.45) is equivalent to*

$$\begin{cases} \text{Find } u \in V \text{ such that} \\ a(u, v - u) + I_K(v) - I_K(u) \geq \langle f, v - u \rangle \quad \forall v \in V. \end{cases} \qquad (1.46)$$

Using (1.44), the inequality (1.46) can be written in the form:

$$\begin{cases} \text{Find } u \in V \text{ such that} \\ I_K(v) - I_K(u) \geq \langle f - Au, v - u \rangle \quad \forall v \in V \end{cases}$$

or equivalently:

$$\begin{cases} \text{Find } u \in V \text{ such that} \\ f - Au \in \partial I_K(u), \end{cases}$$

where $\partial I_K(u)$ denotes the subdifferential of the indicator function I_K at the point u (see Definition 1.2). Notice that $u \in K$.

Formulation (1.46) is a special case of a more general setting of elliptic variational inequalities. Let $j : V \to \bar{\mathbf{R}}$ be a *convex, lower semicontinuous* and *proper* functional in V. By *an abstract elliptic inequality of the second kind* we call a quadruplet $\{K, a, f, j\}$. A function $u \in K$ satisfying

$$a(u, v - u) + j(v) - j(u) \geq \langle f, v - u \rangle \quad \forall v \in K \qquad (1.47)$$

will be called *a solution* of $\{K, a, f, j\}$. Recalling Proposition 1.8 and assuming that a point $u_0 \in D_{eff}(j) \cap K$ exists where either j or I_K is continuous (note that I_K is continuous at u_0 if $u_0 \in \text{int } K$), the inequality (1.47) can be written in the following equivalent form:

$$\begin{cases} \text{Find } u \in V \text{ such that} \\ f - Au \in \partial j(u) + \partial I_K(u). \end{cases} \qquad (1.48)$$

If a, K and j satisfy all the above mentioned assumptions, there exists a unique solution u of (1.47).

In many problems arising in practice, the bilinear form a is also *symmetric* in V, i.e.

$$a(u, v) = a(v, u) \quad \forall u, v \in V. \qquad (1.49)$$

In this case, problems (1.43),(1.45) and (1.47) can be equivalently characterized as an abstract minimization problem for the convex functional $J : V \to \mathbf{R}$ defined by

$$J(v) = \frac{1}{2} a(v, v) - \langle f, v \rangle \qquad (1.50)$$

or

$$J(v) = \frac{1}{2}a(v,v) + j(v) - \langle f, v \rangle \tag{1.51}$$

in the case of the inequality of the second kind.
Indeed, it holds:

Theorem 1.30 *In addition, let $a : V \times V \to \mathbf{R}$ be symmetric in V. Then*

(i) $u \in V$ solves (1.43) if and only if

$$J(u) = \min_{v \in V} J(v), \tag{1.52}$$

where J is given by (1.50);

(ii) $u \in K$ solves (1.45) or (1.47) if and only if

$$J(u) = \min_{v \in K} J(v), \tag{1.53}$$

where J is given by (1.50) or by (1.51) in the case of the variational inequality of the second kind.

Next we show how these abstract results can be used when formulating particular linear elliptic problems. Let[§]

$$Au \equiv -\frac{\partial}{\partial x_i}\left(a_{ij}\frac{\partial u}{\partial x_j}\right) + a_0 u \tag{1.54}$$

be the second order linear scalar differential operator. With any A the following bilinear form a defined on $H^1(\Omega) \times H^1(\Omega)$ will be associated:

$$a(u,v) = \int_\Omega \left(a_{ij}\frac{\partial u}{\partial x_j}\frac{\partial v}{\partial x_i} + a_0 uv\right) dx. \tag{1.55}$$

The coefficients a_{ij}, a_0 will satisfy the following assumptions:

$$\begin{cases} a_{ij}, a_0 \in L^\infty(\Omega) & \forall i,j = 1,...,n; \\ a_0 \geq 0 & \text{a.e. in } \Omega, \end{cases} \tag{1.56}$$

and

$$\exists \alpha_0 = \text{const.} > 0 \quad : \quad a_{ij}(x)\xi_i\xi_j \geq \alpha_0\xi_i\xi_i \tag{1.57}$$
$$\text{holds for any } \xi_i \in \mathbf{R} \text{ and a.a. } x \in \Omega.$$

From (1.56) and the Hölder's inequality it holds that

$$|a(u,v)| \leq M\|u\|_1\|v\|_1 \quad \forall u,v \in H^1(\Omega), \tag{1.58}$$

[§]in the sequel the summation convention will be used

where the positive constant M depends only on

$$\max_{i,j=1,\ldots,n} \left\{ \|a_{ij}\|_{L^\infty(\Omega)}, \|a_0\|_{L^\infty(\Omega)} \right\}.$$

Condition (1.57) yields:

$$a(v,v) \geq \alpha_0 |v|_1^2 + \int_\Omega a_0 v^2 dx \qquad \forall v \in H^1(\Omega). \qquad (1.59)$$

Suppose that there exists a positive number α_1 such that

$$a_0(x) \geq \alpha_1 > 0 \qquad \text{for a.a. } x \in \Omega. \qquad (1.60)$$

Then from this and (1.59) it follows that

$$a(v,v) \geq \alpha \|v\|_1^2 \qquad \forall v \in H^1(\Omega),$$

where $\alpha = \min\{\alpha_0, \alpha_1\}$. On the contrary if $a(x) \equiv 0$ then

$$a(v,v) \geq \alpha_0 |v|_1^2 \qquad \forall v \in H^1(\Omega), \qquad (1.61)$$

i.e., a is no longer elliptic on *the whole* space $H^1(\Omega)$. On the other hand, if we restrict ourselves to a properly chosen subspace V of $H^1(\Omega)$ in which the seminorm $|\cdot|_1$ becomes already a norm, then a is $H^1(\Omega)$-elliptic in V. We have already met such situation (see Corollary 1.1, Section 1.1).

Let

$$V = \{v \in H^1(\Omega) \mid v = 0 \quad \text{on } \Gamma_1\}, \qquad (1.62)$$

where $\Gamma_1 \subset \partial\Omega$ and $\text{meas}_{n-1}\Gamma_1 > 0$. Then due to the generalized Poincaré-Friedrichs inequality we have

$$a(v,v) \geq \alpha \|v\|_1^2 \qquad \forall v \in V.$$

Now we are able to formulate simple scalar boundary value problems. Let the boundary $\partial\Omega$ be split into two disjoint, measurable parts Γ_1, Γ_2 such that $\text{meas}_{n-1}\Gamma_1 > 0$. Let V be given by (1.62). Further let $h \in L^2(\Omega)$, $g \in L^2(\Gamma_2)$. We shall consider the linear elliptic problem $\{V, a, f\}$, with a given by (1.55) and $f \in V^*$ defined by

$$\langle f, v \rangle = \int_\Omega hv dx + \int_{\Gamma_2} gv ds. \qquad (1.63)$$

Using the trace theorem (see Theorem 1.10) we see that the expression on the right hand side of (1.63) really defines the linear, continuous functional over V, i.e., $f \in V^*$. A function $u \in V$ solves $\{V, a, f\}$ if and only if

$$\int_\Omega \left(a_{ij} \frac{\partial u}{\partial x_j} \frac{\partial \varphi}{\partial x_i} + a_0 u\varphi \right) dx = \int_\Omega h\varphi dx + \int_{\Gamma_2} g\varphi ds \qquad (1.64)$$

holds for any $\varphi \in V$. Since all the assumptions of the Lax–Milgram theorem are satisfied, there exists a unique solution u. Assuming that u is smooth enough, we can apply the Green's formula to the first integral in (1.64). We see that u is the solution of the following mixed Dirichlet–Neumann boundary value problem:

$$\begin{cases} Au = h & \text{in } \Omega \\ u = 0 & \text{on } \Gamma_1 \\ \partial u/\partial \nu_A = g & \text{on } \Gamma_2, \end{cases} \qquad (1.65)$$

where the symbol $\partial/\partial\nu_A$ stands for the conormal derivative operator defined as follows:

$$\partial u/\partial \nu_A = a_{ij}\frac{\partial u}{\partial x_j}\nu_i, \qquad (1.66)$$

with ν_i being the i-th component of the unit outer normal vector to $\partial\Omega$. If $\Gamma_1 \equiv \partial\Omega$, then (1.65) reduces to the homogenous Dirichlet boundary value problem. On the contrary, if $\Gamma_2 \equiv \partial\Omega$ then we obtain the Neumann problem. To guarantee the existence and the uniqueness of its solution for *any given data* $h \in L^2(\Omega)$, $g \in L^2(\partial\Omega)$ one has to suppose that a_0 satisfies (1.60).

Now we present a simple model example of a variational inequality. Let $\{K, a, f\}$ be given by the following data:

$$K = \{v \in H_0^1(\Omega) \mid v \geq 0 \text{ a.e. in } \Omega\},$$

$f \in L^2(\Omega)$ and a by (1.55). We solve the problem

$$\begin{cases} \text{Find } u \in K \text{ such that} \\ \displaystyle\int_\Omega \left(a_{ij}\frac{\partial u}{\partial x_j}\frac{\partial}{\partial x_i}(\varphi - u) + a_0 u(\varphi - u)\right)dx \\ \geq \displaystyle\int_\Omega f(\varphi - u)dx \quad \forall\varphi \in K. \end{cases} \qquad (1.67)$$

Since K is a nonempty, closed, convex subset of $H_0^1(\Omega)$ and the bilinear form is $H^1(\Omega)$–elliptic, there exists a unique solution of (1.67) as follows from Theorem 1.29. Assuming that u is smooth enough, we may apply the Green's formula to the first integral and we obtain:

$$\int_\Omega Au(\varphi - u)dx \geq \int_\Omega f(\varphi - u)dx \quad \forall\varphi \in K \qquad (1.68)$$

(The boundary integral vanishes because u and φ are equal to zero on $\partial\Omega$). Now we take the test function φ of the form $\varphi = u + \omega$, where $\omega \in C_0^\infty(\Omega)$, $\omega \geq 0$ in Ω. Then (1.68) yields

$$\int_\Omega (Au - f)\omega\, dx \geq 0 \qquad \forall\omega \in C_0^\infty(\Omega),\ \omega \geq 0. \qquad (1.69)$$

Hence

$$Au \geq f \qquad \text{a.e. in } \Omega. \tag{1.70}$$

Substituting $\varphi = 2u, 0$ into (1.68) we get

$$\int_\Omega (Au - f)u\, dx = 0.$$

Since at the same time the integrand is non–negative in Ω as follows from (1.70) and the definition of K, we finally deduce that

$$(Au - f)u = 0 \qquad \text{a.e. in } \Omega.$$

Summarizing: sufficiently smooth function u being the solution of $\{K, a, f\}$ satisfies the following relations a.e. in Ω:

$$u \geq 0, \quad Au \geq f, \quad (Au - f)u = 0. \tag{1.71}$$

The domain Ω can be split into two parts:

$$\begin{aligned}
\Omega_+ &= \{x \in \Omega \mid u(x) > 0\}, \\
\Omega_0 &= \{x \in \Omega \mid u(x) = 0\}.
\end{aligned}$$

From (1.71) it follows that $Au = f$ in Ω_+. The partition of Ω into Ω_+ and Ω_0 is one of the *unknowns* of our problem.

Remark 1.10 *If the coefficient matrix defining a is symmetric in Ω, i.e., $a_{ij}(x) = a_{ji}(x)$ a.e. in Ω, then the corresponding bilinear form a is symmetric in $H^1(\Omega)$. In this case, problem (1.65) is equivalent to the minimization of*

$$J(v) = \frac{1}{2} \int_\Omega \left(a_{ij} \frac{\partial v}{\partial x_j} \frac{\partial v}{\partial x_i} + a_0 v^2 \right) dx - \int_\Omega hv\, dx - \int_{\Gamma_2} gv\, ds$$

over the space V, while the solution of (1.67) is characterized as the minimum of

$$J(v) = \frac{1}{2} \int_\Omega \left(a_{ij} \frac{\partial v}{\partial x_j} \frac{\partial v}{\partial x_i} + a_0 v^2 \right) dx - \int_\Omega fv\, dx$$

over the convex set K.

We now pass to the case of elliptic systems. One of the most significant examples (at least in this book) is the so–called *linear elasticity system*, characterizing the equilibrium of a deformable body, made of an elastic material, obeying a linear Hooke's law.

Let the body be represented by a domain $\Omega \subset \mathbf{R}^n$, $n = 2, 3$. The body will be subject to a body force $F = (F_i)_{i=1}^n$ and to a surface load $P = (P_i)_{i=1}^n$ on a portion Γ_P of $\partial\Omega$. On the remaining part $\Gamma_u \equiv \partial\Omega \setminus \overline{\Gamma}_P$ the body will be fixed. Suppose that $\text{meas}_{n-1}\Gamma_u > 0$. The following three notions are important when formulating the problem: *the symmetric stress tensor* $\tau = (\tau_{ij})_{i,j=1}^n$, *the symmetric linearized strain tensor* $\varepsilon = (\varepsilon_{ij})_{i,j=1}^n$ *and the displacement field* $u = (u_i)_{i=1}^n$.

The equilibrium state of Ω is characterized by:

- *the equilibrium equation*:

$$\frac{\partial \tau_{ij}}{\partial x_j} + F_i = 0 \quad \text{in } \Omega, \ i = 1, ..., n; \tag{1.72}$$

- *the compatibility of τ with the surface load P*:

$$\tau_{ij}\nu_j = P_i \quad \text{on } \Gamma_P, \ i = 1, ..., n; \tag{1.73}$$

- *the stress-strain relation*:

$$\tau_{ij} = c_{ijkl}\varepsilon_{kl}, \quad i, j, k, l = 1, ..., n, \tag{1.74}$$

where the elasticity coefficients $c_{ijkl} \in L^\infty(\Omega)$ satisfy the symmetry conditions

$$c_{ijkl} = c_{jikl} = c_{klij} \quad \text{a.e. in } \Omega \tag{1.75}$$

and the ellipticity condition

$$\exists \alpha_0 : c_{ijkl}(x)\xi_{ij}\xi_{kl} \geq \alpha_0 \xi_{ij}\xi_{ij} \tag{1.76}$$

for any $\xi_{ij} = \xi_{ji} \in \mathbf{R}$ and a.a. $x \in \Omega$;

- *the strain-displacement relation*:

$$\text{there exists a deformation field } u \text{ such that} \tag{1.77}$$

$$\varepsilon \equiv \varepsilon(u) = (\varepsilon_{ij}(u))_{i,j=1}^n, \text{ where } \varepsilon_{ij}(u) = \frac{1}{2}\left(\frac{\partial u_i}{\partial x_j} + \frac{\partial u_j}{\partial x_i}\right) \forall i, j = 1, ..., n;$$

- *the kinematical boundary conditions*:

$$u_i = 0 \quad \text{on } \Gamma_u, \ i = 1, ..., n. \tag{1.78}$$

The equilibrium state of Ω is given by the deformation field u satisfying (1.78) and (1.72)–(1.74) with $\tau \equiv \tau(u) = (\tau_{ij}(u))_{i,j=1}^n$, where

$$\tau_{ij}(u) = c_{ijkl}\varepsilon_{kl}(u), \quad i, j = 1, ..., n$$

and $\varepsilon(u) = (\varepsilon_{ij}(u))_{i,j=1}^n$ is defined by (1.77).

Let

$$\mathbf{V} = \left\{ v \in H^1(\Omega; \mathbf{R}^n) \mid v = 0 \text{ on } \Gamma_u \right\}$$

and define

$$a(u, v) = \int_\Omega c_{ijkl}\varepsilon_{ij}(u)\varepsilon_{kl}(v)dx, \quad u, v \in \mathbf{V}; \tag{1.79}$$

$$\langle f, v \rangle = \int_{\Omega} F_i v_i dx + \int_{\Gamma_P} P_i v_i ds. \tag{1.80}$$

Since $c_{ijkl} \in L^{\infty}(\Omega)$ $\forall i, j, k, l, = 1, ..., n$, the bilinear form a is bounded and symmetric on $\mathbf{V} \times \mathbf{V}$, due to (1.75). From (1.76) it follows that

$$a(v, v) \geq \alpha_0 \int_{\Omega} \varepsilon_{ij}(v) \varepsilon_{ij}(v) dx \geq c\|v\|_1^2 \quad \forall v \in \mathbf{V},$$

making use of the first Korn's inequality (see Theorem 1.16). Let us suppose that $F \in L^2(\Omega; \mathbf{R}^n)$ and $P \in L^2(\Gamma_P; \mathbf{R}^n)$. Then (1.80) defines a linear and continuous functional on $\mathbf{V}$.

From the Lax-Milgram theorem it follows that the problem

$$\begin{cases} \text{Find } u \in \mathbf{V} \text{ such that} \\ a(u, v) = \langle f, v \rangle \quad \forall v \in \mathbf{V}, \end{cases} \tag{1.81}$$

with a, f given by (1.79),(1.80), respectively, has a unique solution. It is an easy exercise to show that a sufficiently regular solution u of (1.81) satisfies (1.72)–(1.74) and (1.78). Since a is symmetric on $\mathbf{V}$, (1.81) is equivalent to the minimization of

$$J(v) = \frac{1}{2} \int_{\Omega} c_{ijkl} \varepsilon_{ij}(v) \varepsilon_{kl}(v) dx - \int_{\Omega} F_i v_i dx - \int_{\Gamma_P} P_i v_i ds \tag{1.82}$$

on $\mathbf{V}$.

Now let us suppose that besides of Γ_u, Γ_P there is a part Γ_c of the boundary, along which Ω is *unilaterally* supported by a *rigid, frictionless* and *smooth* foundation. As before, the body is subject to body forces F and surface loads P on Γ_P (see Fig.1.4). One seeks for the equilibrium state of Ω.

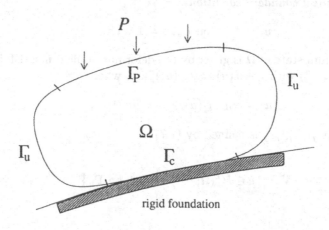

Figure 1.4.

The unilateral conditions along Γ_c can be written as follows (for their derivation and justification see Hlaváček et al., 1988 or Kikuchi and Oden, 1988):

$$u_\nu \equiv u.\nu \leq 0 \quad \text{(one–sided contact) on } \Gamma_c; \tag{1.83}$$

$$T_\nu \equiv \tau_{ij}(u)\nu_i\nu_j \leq 0 \quad \text{(only pressure may occur) on } \Gamma_c; \tag{1.84}$$

$$\text{if } u_\nu(x) < 0 \Rightarrow T_\nu(x) = 0 \quad \text{(no contact–no pressure) } x \in \Gamma_c; \tag{1.85}$$

$$T_t \equiv \tau_{ij}(u)\nu_i t_j = 0 \quad \text{on } \Gamma_c \text{ (no friction).} \tag{1.86}$$

The symbols u_ν, T_ν, T_t stand for the normal component of the displacement field u, the normal, the tangential component of the stress vector $T(u) = (T_i(u))_{i=1}^n$, respectively, where $T_i(u) \equiv \tau_{ij}(u)\nu_j$ and t is the unit tangential vector to $\partial\Omega$.

The set K of all kinematically admissible displacements is a subset of $H^1(\Omega; \mathbf{R}^n)$ defined by

$$K = \{v \in H^1(\Omega; \mathbf{R}^n) \mid v = 0 \text{ on } \Gamma_u, \ v_\nu \leq 0 \text{ on } \Gamma_c\}.$$

The equilibrium state of Ω is characterized by a displacement field $u \in K$, satisfying (1.72)–(1.74),(1.77),(1.78) and (1.83)–(1.86). This problem is known in the literature as the *Signorini problem*, and, the unilateral conditions (1.83)–(1.85) are called the *Signorini-Fichera type conditions*. This problem can be formulated in a weak form as the variational inequality of the first kind:

$$\begin{cases} \text{Find } u \in K \text{ such that} \\ a(u, v - u) \geq \langle f, v - u \rangle \quad \forall v \in K \end{cases} \tag{1.87}$$

or equivalently

$$\begin{cases} \text{Find } u \in K \text{ such that} \\ J(u) = \min_{v \in K} J(v), \end{cases} \tag{1.88}$$

where J is given by (1.82). Since K is a nonempty, closed and convex subset of $H^1(\Omega; \mathbf{R}^n)$ and a is V–elliptic as follows from the first Korn's inequality, (1.87) has a unique solution u. Applying the Green's formula (1.16) to the inner energy integral given by $a(v, v)$, one deduce that a sufficiently regular solution u satisfies (1.72)–(1.74),(1.77),(1.78) and (1.83)–(1.86).

Let I_ν be the indicator function of K:

$$I_\nu(v) = \begin{cases} 0 & \text{iff } v \in K \\ +\infty & \text{elsewhere.} \end{cases}$$

Then (see Remark 1.9), the variational inequality (1.87) is equivalent to

$$f - Au \in \partial I_\nu(u), \tag{1.89}$$

where $A \in \mathcal{L}(V, V^*)$ is the mapping defined by (1.44) with a given by (1.79). If u is smooth enough then the Green's formula (1.16) and the boundary conditions on Γ_u, Γ_P and the frictionless condition on Γ_c yield:

$$f - Au = -T_\nu(u), \tag{1.90}$$

i.e., $f - Au$ is a functional supported by the boundary Γ_c and equal to $-T_\nu(u)$. Using the local characterization (see Proposition 1.14), from (1.89) and (1.90) it follows that the unilateral conditions (1.83)–(1.85) can be equivalently expressed by

$$-T_\nu(u)(x) \in \partial j_\nu(u_\nu)(x) \quad \text{for a.a. } x \in \Gamma_c,$$

where j_ν is the indicator function of $\mathbf{R}_-$.

Now, let us take into account the influence of friction between Ω and the rigid support. We shall restrict ourselves to a simple case, namely to the so-called *given friction model*. Let g be a given positive number. Then (1.86) will be replaced by

$$\begin{aligned}
&|T_t(x)| \leq g \quad \text{on } \Gamma_c; \\
&\text{if } |T_t(x)| < g \implies u_t(x) \equiv (u.t)(x) = 0, \; x \in \Gamma_c; \\
&\text{if } |T_t(x)| = g \implies \exists \lambda(x) \geq 0 \text{ s.t. } u_t(x) = -\lambda(x)T_t(x),
\end{aligned} \quad (1.91)$$

where $T_t \equiv T_t(u) = \tau_{ij}(u)\nu_i t_j$.

The physical interpretation of (1.91) is the following: the tangential component of stresses T_t can not exceed the given slip stress g. If locally $|T_t|$ is strictly below g, then there is no slip, i.e., $u_t = 0$. On the contrary, if $|T_t|$ attains the value of g, then a slip may occur and its direction is opposite to T_t.

The mathematical model of the Signorini problem with a given friction leads to the elliptic inequality $\{K, a, f, j\}$ of the second kind, where K, a, f are the same as before and

$$j(v) \equiv g \int_{\Gamma_c} |v_t| ds$$

is the work of friction forces. The weak solution of this problem is defined as follows:

$$\begin{cases} \text{Find } u \in K \text{ such that} \\ a(u, v - u) + j(v) - j(u) \geq \langle f, v - u \rangle \quad \forall v \in K \end{cases} \quad (1.92)$$

or equivalently

$$\begin{cases} \text{Find } u \in K \text{ such that} \\ J(u) = \min_{v \in K} J(v) \end{cases} \quad (1.93)$$

where

$$J(v) = \frac{1}{2}a(v, v) + j(v) - \langle f, v \rangle. \quad (1.94)$$

Since j is proper, convex and continuous on $\mathbf{V}$, there exists a unique solution u of (1.92) (or (1.93)). Assuming that u is smooth enough one can use a similar approach as in the frictionless case and deduce that the friction conditions (1.91) can be written in the following local form

$$-T_t(u)(x) \in \partial j_t(u_t(x)),$$

where j_t is the convex function defined by $j_t(\xi) = g|\xi|$. Hence (see Fig.1.1)

$$\partial j_t(\xi) = \begin{cases} -g & \text{if } \xi \in (-\infty, 0) \\ [-g, g] & \text{if } \xi = 0 \\ g & \text{if } \xi \in (0, +\infty). \end{cases}$$

Until now elliptic equations and inequalities with linear mappings have been presented. Now let us pass to more general cases.

Let $T : W \to W^*$ be a mapping from a reflexive Banach space W into its dual W^*, which is nonlinear, in general. The duality pairing between W and W^* will be denoted by $\langle \cdot, \cdot \rangle$.

We shall look for solutions of the equation

$$T(u) = f \text{ in } W^* \iff \langle T(u), v \rangle = \langle f, v \rangle \quad \forall v \in W \tag{1.95}$$

or the inequality

$$u \in K : \langle T(u), v - u \rangle \geq \langle f, v - u \rangle \quad \forall v \in K, \tag{1.96}$$

where $f \in W^*$ is given and K is a nonempty, closed and convex subset of W.

In order to guarantee the existence and the uniqueness of the solution to (1.95) and (1.96), the mapping T is supposed to satisfy the following assumptions:

T is *strongly monotone* in W, i.e., there exists a strictly
increasing function $\alpha : [0, \infty) \to \mathbf{R}$ such that $\alpha(0) = 0$, $\qquad$ (1.97)
$\lim_{t \to \infty} \alpha(t) = \infty$ and for any $u, v \in W$ it holds:
$\langle T(u) - T(v), u - v \rangle \geq \alpha(\|u - v\|)\|u - v\|;$

T is *locally Lipschitz continuous* in W, i.e.,
there exists a positive constant $M(r)$ such that $\qquad$ (1.98)
$\|T(u) - T(v)\|_{W^*} \leq M(r)\|u - v\|$ holds for any $u, v \in B_r$,
where $B_r = \{v \in W \mid \|v\| \leq r\}$.

Then we have (Lions, 1969):

Theorem 1.31 *Let $T : W \to W^*$ be strongly monotone and locally Lipschitz in W. Then equation (1.95) and inequality (1.96) have a unique solution u for any right hand side $f \in W^*$.*

This theorem is a generalization of Theorems 1.28 and 1.29. Also Theorem 1.30 extends to the case of nonlinear mappings. We say that $T : W \to W^*$ is the *potential operator* if there exists a functional $\Phi : W \to \mathbf{R}$ which is Gâteaux differentiable at any point $v \in W$ and such that its Gâteaux derivative $D\Phi(v) = T(v)$ for any $v \in W$. Such a functional (if it exists) is called the *potential* of T. In this case, equation (1.95) and inequality (1.96) can be written in the following form

$$u \in W : \langle D\Phi(u), v \rangle = \langle f, v \rangle \quad \forall v \in W; \tag{1.99}$$

$$u \in K : \langle D\Phi(u), v - u \rangle \geq \langle f, v - u \rangle \quad \forall v \in K. \tag{1.100}$$

It is not surprising that both these problems are related to the minimization of Φ in W, K, respectively. Indeed, one has (Nečas and Hlaváček, 1981):

Theorem 1.32 *Let $\Phi : W \to \mathbf{R}$ be a potential of T such that:*

$$\text{for any } u, v, z \in W \text{ fixed, the function} \tag{1.101}$$
$$t \mapsto \langle D\Phi(u + tv), z \rangle \text{ is continuous in } \mathbf{R};$$

$$\langle D\Phi(u + v), v \rangle - \langle D\Phi(u), v \rangle \geq \alpha(\|v\|)\|v\| \quad \forall u, v \in W \tag{1.102}$$

with the function α, having the same properties as in (1.97). Then there exists unique solutions to the following minimization problems:

$$u \in W \ : \ \Phi(u) = \min_{v \in W} \Phi(v), \tag{1.103}$$

$$u \in K \ : \ \Phi(u) = \min_{v \in K} \Phi(v). \tag{1.104}$$

Moreover, (1.103) is equivalent to (1.99) and (1.104) to (1.100).

Time dependent problems

As in the elliptic case we start with linear problems.

Let V and H be real separable Hilbert spaces. By $\| \cdot \|$, $((\cdot, \cdot))$, V^* and $\langle \cdot, \cdot \rangle$ we denote the norm, the scalar product in V, the dual space to V and the duality pairing between V and V^*, respectively. The norm and the scalar product in H are denoted by $| \cdot |$ and $(\cdot, \cdot)$. We suppose that V is continuously and densely imbedded in H and, furthermore, we identify H with its dual H^*. Then

$$\text{``}V \subseteq H \subseteq V^*\text{''}$$

forms an evolution triplet and the Bochner space $W(V) \equiv W^{1,2}(0, T; V, H)$ is well-defined (see Section 1.1).

Let $a(t; \cdot, \cdot) : V \times V \to \mathbf{R}$ be an uniformly bounded and V-elliptic bilinear form with respect to $t \in [0, T]$, i.e.

$$\exists M > 0 : |a(t; u, v)| \leq M \|u\| \|v\| \quad \forall u, v \in V, \ \forall t \in [0, T]; \tag{1.105}$$

$$\exists \alpha > 0, \beta \geq 0 : a(t; v, v) \tag{1.106}$$
$$\geq \alpha \|v\|^2 - \beta |v|^2 \ \forall v \in V, \forall t \in [0, T].$$

In addition, we assume that

the function $t \mapsto a(t; u, v)$ is measurable in $(0, T)$ for every $u, v \in V$. (1.107)

Finally let

$$\text{the function } f \in L^2(0,T;V^*) \text{ and the initial value } u_0 \in H \qquad (1.108)$$

be given.

By an *abstract linear parabolic equation* we mean the problem:

$$\begin{cases} \text{Find } u \in W(V) \text{ satisfying} \\ \displaystyle\int_0^T \langle u'(t), v(t)\rangle dt + \int_0^T a(t; u(t), v(t)) dt \\ = \displaystyle\int_0^T \langle f(t), v(t)\rangle dt \qquad \forall v \in L^2(0,T;V) \end{cases} \qquad (1.109)$$

and the initial condition

$$u(0) = u_0. \qquad (1.110)$$

Remark 1.11 *Equation (1.109) is equivalent to*

$$\langle u'(t), v\rangle + a(t; u(t), v) = \langle f(t), v\rangle \quad \forall v \in V \text{ and a.a. } t \in (0,T).$$

In the sequel we shall use both formulations simultaneously.

The following basic existence and uniqueness result holds (Zeidler, 1990a):

Theorem 1.33 *Let the bilinear form $a(t; \cdot, \cdot) : V \times V \to \mathbf{R}$ satisfy (1.105)–(1.107). Then there exists a unique solution u of (1.109)–(1.110) for any $f \in L^2(0,T;V^*)$ and $u_0 \in H$.*

Remark 1.12 *Without loss of generality we may assume $\beta = 0$ in (1.106). Indeed, setting $u = \tilde{u}e^{\beta t}$ and substituting it into (1.109)–(1.110) we obtain the following transformed problem:*

$$\begin{cases} \text{Find } \tilde{u} \in W(V) \text{ such that} \\ \int_0^T \langle \tilde{u}'(t), v(t)\rangle dt + \int_0^T a(t; \tilde{u}(t), v(t)) dt \\ + \int_0^T \beta(\tilde{u}(t), v(t)) dt = \int_0^T \langle e^{-\beta t} f(t), v(t)\rangle dt \quad \forall v \in L^2(0,T;V) \end{cases} \qquad (1.111)$$

and

$$\tilde{u}(0) = u_0. \qquad (1.112)$$

By redefining $\tilde{a}(t; u, v) \equiv a(t; u, v) + \beta(u, v)$, $\tilde{u}_0 \equiv u_0$ and $\tilde{f} \equiv e^{-\beta t} f$ we see that the above problem is of the same form as the original one given by (1.109)–(1.110), and that $\tilde{a}(t; \cdot, \cdot)$ satisfies (1.106) with the constant $\beta = 0$.

In what follows, we shall suppose that the constant β in (1.106) is equal to 0.

Remark 1.13 *Let u_i, $i = 1, 2$, be solutions of (1.109)–(1.110) with the data $\{f_i, u_0^i\}$, $i = 1, 2$, respectively. Then after a simple calculation we obtain the* *following stability estimates:*

$$|u_1(t) - u_2(t)| \tag{1.113}$$

$$\leq \left(|u_0^1 - u_0^2|^2 + \frac{1}{\alpha} \|f_1 - f_2\|_{L^2(0,T;V^*)}^2 \right)^{\frac{1}{2}} \forall t \in (0,T];$$

$$\|u_1 - u_2\|_{L^2(0,T;V)} \tag{1.114}$$

$$\leq \frac{1}{\sqrt{\alpha}} \left(|u_0^1 - u_0^2|^2 + \frac{1}{\alpha} \|f_1 - f_2\|_{L^2(0,T;V^*)}^2 \right)^{\frac{1}{2}}.$$

For every $t \in [0,T]$ the bilinear form $a(t;\cdot,\cdot)$ defines a linear operator $A(t)$: $V \to V^*$ by

$$a(t;u,v) = \langle A(t)u, v \rangle \qquad \forall u, v \in V. \tag{1.115}$$

Clearly, from (1.105),(1.106) it follows that

$$\sup_{t \in [0,T]} \|A(t)\|_{\mathcal{L}(V,V^*)} \leq M, \qquad \sup_{t \in [0,T]} \|A(t)^{-1}\|_{\mathcal{L}(V^*,V)} \leq \frac{1}{\alpha}. \tag{1.116}$$

Hence, equation (1.109) can be rewritten in the following equivalent operator form:

$$u'(t) + A(t)u(t) = f(t) \qquad \text{in } V^*, \text{ for a.a. } t \in (0,T). \tag{1.117}$$

Analogously to the elliptic case we define parabolic variational inequalities:

Let K be a *nonempty, closed* and *convex* subset of V. By an *abstract parabolic variational inequality* of the *first kind* we mean the problem:

$$\begin{cases} \text{Find } u \in W(V) \cap L^2(0,T;K) \text{ satisfying} \\ \int_0^T \langle u'(t), v(t) - u(t) \rangle dt + \int_0^T a(t;u(t), v(t) - u(t)) dt \\ \geq \int_0^T \langle f(t), v(t) - u(t) \rangle dt \qquad \forall v \in L^2(0,T;K) \end{cases} \tag{1.118}$$

and the initial condition (1.110).

Now let $j : V \to \bar{R}$ be a *convex, lower semicontinuous* and *proper* functional in V. Then an *abstract parabolic variational inequality* of the *second kind* is defined by:

$$\begin{cases} \text{Find } u \in W(V) \cap L^2(0,T;K) \text{ satisfying} \\ \int_0^T \langle u'(t), v(t) - u(t) \rangle dt + \int_0^T a(t;u(t), v(t) - u(t)) dt \\ + \int_0^T j(v(t)) dt - \int_0^T j(u(t)) dt \\ \geq \int_0^T \langle f(t), v(t) - u(t) \rangle dt \qquad \forall v \in L^2(0,T;K) \end{cases} \tag{1.119}$$

and the initial condition (1.110).

Remark 1.14 *Similarly as in Remark 1.11 the parabolic inequality (1.118) has an equivalent form*

$$\begin{cases} \langle u'(t), v - u(t) \rangle + a(t;u(t), v - u(t)) \\ \geq \langle f(t), v - u(t) \rangle \quad \forall v \in K \text{ and a.a. } t \in (0,T). \end{cases}$$

The same holds for (1.119):

$$\begin{cases} \langle u'(t), v - u(t) \rangle + a(t; u(t), v - u(t)) + j(v) - j(u(t)) \\ \geq \langle f(t), v - u(t) \rangle \quad \forall v \in K \text{ and a.a. } t \in (0, T). \end{cases}$$

In order to ensure that the problems formulated above have solutions we need additional assumptions on a, f, u_0 and j. First, suppose that $a(t; \cdot, \cdot)$ is differentiable with respect to t in $(0, T)$ and its derivative $a'(t; \cdot, \cdot) : V \times V \to \mathbf{R}$ is an uniformly bounded bilinear form with respect to $t \in (0, T)$, i.e.,

$$\exists \bar{M} > 0 : |a'(t; u, v)| \leq \bar{M} \|u\| \|v\| \quad \forall u, v \in V, \forall t \in (0, T) \qquad (1.120)$$

and satisfies the measurability condition

$$\text{the function } t \mapsto a'(t; u, v) \text{ is measurable} \qquad (1.121)$$
$$\text{in } (0, T) \text{ for every } u, v \in V.$$

Further, let us suppose that

$$f, f' \in L^2(0, T; V^*) \quad \text{and} \quad u_0 \in K. \qquad (1.122)$$

For simplicity, the functional j is assumed to be finite in the following sense:

$$|j(v)| < +\infty \ \forall v \in K \text{ and } \left| \int_0^T j(v(t))dt \right| < +\infty \ \forall v \in L^2(0, T; K). \quad (1.123)$$

This condition guarantees that (1.119) is well-defined for all $v \in L^2(0, T; K)$. Otherwise, we restrict the test functions v to $L^2(0, T; K \cap D_{eff}(j))$. Finally, we suppose either that

$$A(0)u_0 - f(0) \in H \qquad (1.124)$$

for the inequality of the first kind or

$$\{A(0)u_0 - f(0) + \partial j(u_0)\} \cap H \neq \emptyset \qquad (1.125)$$

for the inequality of the second kind, where $A(0)$ is defined by (1.115) at $t = 0$. Then the following existence result holds (Barbu, 1993):

Theorem 1.34 *Let the bilinear forms $a(t; \cdot, \cdot)$, $a'(t; \cdot, \cdot)$ satisfy (1.105)–(1.107) and (1.120)–(1.121), respectively, and let (1.122) and (1.124) be valid. Then the abstract parabolic variational inequality of the first kind has a unique solution u. Moreover,*

$$u, u' \in L^2(0, T; V) \cap L^\infty(0, T; H).$$

Theorem 1.35 *Let the bilinear forms $a(t; \cdot, \cdot)$, $a'(t; \cdot, \cdot)$ satisfy the same conditions as in Theorem 1.34, and let (1.122), (1.123) and (1.125) be valid. Then* *the abstract parabolic variational inequality of the second kind has a unique solution u. Moreover,*

$$u, u' \in L^2(0, T; V) \cap L^\infty(0, T; H).$$

Remark 1.15 *It is straightforward to verify that the stability estimates (1.113) and (1.114) remain true for solutions of the inequality of the first kind. Also, if j is a Lipschitz function, one can derive the corresponding estimates in the case of the inequality of the second kind.*

Remark 1.16 *Denote by I_K the indicator function of the set K. Then (1.118) is equivalent to the following inclusion problem:*

$$\begin{cases} Find \ u \in W(V) \ such \ that \\ f(t) - u'(t) - A(t)u(t) \in \partial I_K(u(t)) & for \ a.a. \ t \in (0, T). \end{cases}$$

Similarly, (1.119) can be rewritten in the form:

$$\begin{cases} Find \ u \in W(V) \ such \ that \\ f(t) - u'(t) - A(t)u(t) \in \partial I_K(u(t)) + \partial j(u(t)) & for \ a.a. \ t \in (0, T). \end{cases}$$

If the bilinear form $a(t; \cdot, \cdot)$ is symmetric for every $t \in [0, T]$, i.e.,

$$a(t; u, v) = a(t; v, u) \quad \forall u, v \in V \text{ and } t \in [0, T], \tag{1.126}$$

some of the hypotheses of Theorems 1.34 and 1.35 can be weakened. Indeed, (1.124),(1.125) are unnecessary and instead of (1.122) we suppose that

$$f \in L^2(0, T; H) \quad \text{and} \quad u_0 \in K. \tag{1.127}$$

Then the following existence result holds (Barbu, 1993):

Theorem 1.36 *Let the bilinear forms $a(t; \cdot, \cdot)$, $a'(t; \cdot, \cdot)$ be as in Theorems 1.34 and 1.35, and let (1.123),(1.126) and (1.127) be satisfied. Then the parabolic variational inequalities of the first and of the second kind have a unique solution u and*

$$u \in H^1(0, T; H) \cap L^\infty(0, T; V).$$

Next we apply these abstract results to some concrete examples. As a model linear parabolic equation we consider:

$$\begin{cases} u'(t) + A(t)u(t) = h(t) & \text{in } \Omega_T = \Omega \times (0, T), \\ u = 0 & \text{on } \Gamma_1 \times (0, T), \\ \partial u / \partial \nu_{A(t)} = g(t) & \text{on } \Gamma_2 \times (0, T), \\ u(0) = u_0 \text{ in } \Omega, \end{cases} \tag{1.128}$$

where the operator $A(t)$, $t \in [0,T]$, is defined by

$$A(t)u(t) \equiv -\frac{\partial}{\partial x_i}\left(a_{ij}(t)\frac{\partial u(t)}{\partial x_j}\right) + a_i(t)\frac{\partial u(t)}{\partial x_i} + a_0(t)u(t) \quad (1.129)$$

and the conormal derivative $\partial/\partial \nu_{A(t)}$ by (1.66) for a.a. $t \in (0,T)$. The boundary $\partial\Omega$ consists of two disjoint, measurable sets Γ_1 and Γ_2 and the Hilbert space $V = \{v \in H^1(\Omega) : v = 0 \text{ on } \Gamma_1\}$. Setting $H \equiv L^2(\Omega)$ we see that $\{V, H, V^*\}$ forms an evolution triplet.

The bilinear form $a(t;\cdot,\cdot) : V \times V \to \mathbf{R}$, $t \in [0,T]$, associated with our problem has the form:

$$a(t;u,v) = \int_\Omega \left(a_{ij}(t)\frac{\partial u}{\partial x_j}\frac{\partial v}{\partial x_i} + a_i(t)\frac{\partial u}{\partial x_i}v + a_0(t)uv\right)dx. \quad (1.130)$$

We assume that

$$a_{ij}, a_i, a_0 \in L^\infty(\Omega_T), \quad (1.131)$$

and the coefficients a_{ij} satisfy the ellipticity condition (1.57) uniformly with respect to $t \in (0,T)$. Now, from (1.131) we have that

$$\max_{i=1,\ldots n}\{\|a_i\|_{L^\infty(\Omega_T)}\} \leq c_1$$

$$\|a_0\|_{L^\infty(\Omega_T)} \leq c_2,$$

and, therefore, by applying the Hölder's inequality we get:

$$\left|\int_\Omega a_i(t)\frac{\partial v}{\partial x_i}vdx\right| + \left|\int_\Omega a_0(t)v^2 dx\right|$$

$$\leq \frac{\alpha_0}{2}|v|_1^2 + \left(\frac{c_1^2}{2\alpha_0} + c_2\right)|v|_0^2 \quad \forall v \in V,$$

with α_0 from (1.57). From this

$$a(t;v,v) \geq \frac{\alpha_0}{2}|v|_1^2 - \left(\frac{c_1^2}{2\alpha_0} + c_2\right)|v|_0^2$$

$$\geq \frac{\alpha_0}{2}\|v\|_1^2 - \left(\frac{c_1^2}{2\alpha_0} + c_2 + \frac{\alpha_0}{2}\right)|v|_0^2 \quad \forall v \in V$$

and, thus, (1.106) is satisfied with $\alpha = \alpha_0/2$ and $\beta = c_1^2/2\alpha_0 + c_2 + \alpha_0/2$. Note that the assumption that $\text{meas}_{n-1}\Gamma_1 > 0$ is not necessary. The validity of (1.105) and (1.107) is obvious.

Further, let $h \in L^2(0,T;L^2(\Omega))$ and $g \in L^2(0,T;L^2(\Gamma_2))$. Then the right hand side f (in (1.109)) has the representation

$$\langle f(t),v\rangle = \int_\Omega h(t)vdx + \int_{\Gamma_2} g(t)vds \quad \forall v \in V. \quad (1.132)$$

Applying again the trace theorem (Theorem 1.10) we see that f belongs to $L^2(0, T; V^*)$. Finally, we suppose that $u_0 \in L^2(\Omega)$. Then from Theorem 1.33 we can conclude that our model problem is uniquely solvable.

As an example of the parabolic variational inequality of the first kind we consider the *parabolic obstacle problem*:

$$\begin{cases} u' - \Delta u = f & \text{in } \{(x,t) \in \Omega_T | u(x,t) > \psi(x)\}, \\ u' - \Delta u \geq f & \text{in } \Omega_T, \\ u(x,t) \geq \psi(x) & \text{in } \Omega_T, \\ u = 0 & \text{on } \Gamma_T = \partial\Omega \times (0,T), \\ u(x,0) = u_0(x) & \text{in } \Omega, \end{cases} \quad (1.133)$$

where $\Omega \subset \mathbf{R}^n$ is a domain with smooth boundary, $f \in L^2(0, T; L^2(\Omega))$, $\psi \in H^2(\Omega)$ is such that $\psi \leq 0$ a.e. on $\partial\Omega$ and $u_0 \in H_0^1(\Omega)$ is such that $u_0 \geq \psi$ a.e. in Ω. Setting $V = H_0^1(\Omega)$, $H = L^2(\Omega)$,

$$K = \{u \in V \mid u(x) \geq \psi(x) \text{ for a.a. } x \in \Omega\} \quad (1.134)$$

and defining the bilinear form $a : V \times V \to \mathbf{R}$ by

$$a(u,v) = \int_\Omega \nabla u \cdot \nabla v \, dx \quad \forall u, v \in V \quad (1.135)$$

and the linear form $f \in L^2(0, T; H)$ by

$$\langle f(t), v \rangle = \int_\Omega f(t) v \, dx \quad \forall v \in V, \quad (1.136)$$

we see that the weak formulation of (1.133) is of the form (1.118). It is easy to verify that all the assumptions of Theorem 1.36 are satisfied. Thus we can conclude that (1.133) has a unique solution $u \in H^1(0, T; H) \cap L^\infty(0, T; V)$. If moreover $f' \in L^2(0, T; V^*)$ and $A(0)u_0 - f(0) \in H$ then Theorem 1.34 says that $u' \in L^2(0, T; V) \cap L^\infty(0, T; H)$.

An important example of the parabolic variational inequality of the second kind is a temperature control through the boundary (or the interior) (see Duvaut and Lions, 1976). Let Ω be a domain in $\mathbf{R}^n$ whose temperature is regulated by the temperature on the boundary Γ. We have two reference temperatures h_1 and h_2, $h_1 < h_2$. Our aim is to regulate the temperature u on the boundary in such a way that it deviates as little as possible from the interval (h_1, h_2). This is realized by using an appropriate "thermostatic control devices" obeying the following rules:

(i) *if $u \in [h_1, h_2]$ then the temperature is in the desired interval and no corrections are necessary:*

$$\frac{\partial u}{\partial \nu} = 0;$$

(ii) if $u \notin [h_1, h_2]$ heat is added or reduced proportionally to the distance between u and the interval (h_1, h_2): if $u > h_2$ then

$$-\partial u/\partial \nu = k_2(u - h_2) \quad \text{if} \quad k_2(u - h_2) \leq g_2$$
$$-\partial u/\partial \nu = g_2 \quad \text{if} \quad k_2(u - h_2) > g_2$$

or if $u < h_1$ then

$$-\partial u/\partial \nu = k_1(u - h_1) \quad \text{if} \quad k_1(u - h_1) \geq g_1$$
$$-\partial u/\partial \nu = g_1 \quad \text{if} \quad k_1(u - h_1) < g_1.$$

The positive constants k_1, k_2 represent the efficiency of the control devices and the constants g_1, g_2 $(g_1 < 0 < g_2)$ the fact that their efficiency is limited. By introducing the convex function $\psi : \mathbf{R} \to \mathbf{R}$ (see Fig.1.5)

$$\psi(r) = \begin{cases} g_1 r & \text{if } r \leq h_1 + g_1/k_1 \\ k_1/2(r - h_1)^2 & \text{if } h_1 + g_1/k_1 < r < h_1 \\ 0 & \text{if } h_1 \leq r \leq h_2 \\ k_2/2(r - h_2)^2 & \text{if } h_2 < r < h_2 + g_2/k_2 \\ g_2 r & \text{if } r \geq h_2 + g_2/k_2 \end{cases} \qquad (1.137)$$

we can rewrite the conditions *(i)–(ii)* as follows:

$$-\frac{\partial u}{\partial \nu} \in \partial \psi(u). \qquad (1.138)$$

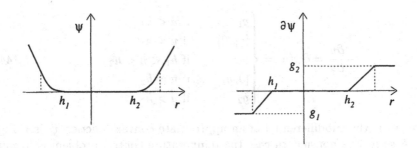

Figure 1.5.

Furthermore, we assume that the temperature u obeys the time-dependent heat equation

$$u'(t) - \Delta u(t) = g(t) \quad \text{in } \Omega \qquad (1.139)$$

and the initial condition

$$u(0) = u_0, \qquad (1.140)$$

with $g \in L^2(0, T; H)$ and $u_0 \in V$. As before $V = H^1(\Omega)$ and $H = L^2(\Omega)$. The functional $j : V \to \mathbf{R}$ is given by

$$j(v) = \int_\Gamma \psi(v) ds \quad \forall v \in V \tag{1.141}$$

which is proper, convex and lower semicontinuous in V.

Then the temperature control problem (1.138)–(1.140) is equal to the following parabolic variational inequality of the second kind ($K = V$):

$$\begin{cases} \text{Find } u \in W(V) \text{ such that} \\ \int_0^T \langle u'(t), v(t) - u(t) \rangle dt + \int_0^T \langle \nabla u(t), \nabla v(t) - \nabla u(t) \rangle dt \\ \quad + \int_0^T j(v(t)) dt - \int_0^T j(u(t)) dt \\ \geq \int_0^T \langle f(t), v(t) - u(t) \rangle dt \quad \forall v \in L^2(0, T; V) \end{cases} \tag{1.142}$$

and satisfying the initial condition (1.140). It is easy to see that the functional j is Fréchet differentiable (note that ψ is continuously differentiable). Thus (1.142) transforms to the equation

$$\int_0^T \langle u'(t), v(t) \rangle dt + \int_0^T \langle \nabla u(t), \nabla v(t) \rangle dt \tag{1.143}$$

$$+ \int_0^T \langle Dj(u(t)), v(t) \rangle dt = \int_0^T \langle f(t), v(t) \rangle dt \quad \forall v \in L^2(0, T; V),$$

where Dj is the Fréchet derivative of j.

Let us consider the limit case in which $k_1 = k_2 = +\infty$. Then the temperature control conditions *(i)–(ii)* are described by the multi-valued function

$$-\frac{\partial u}{\partial \nu} = \partial \psi(u) = \begin{cases} g_1 & \text{if } u < h_1 \\ [g_1, 0] & \text{if } u = h_1 \\ 0 & \text{if } h_1 < u < h_2 \\ [0, g_2] & \text{if } u = h_2 \\ g_2 & \text{if } u > h_2, \end{cases} \tag{1.144}$$

where $\partial \psi$ is the subdifferential of an appropriate convex function ψ (see Fig. 1.6). Also in this nonsmooth case the temperature control problem is formulated by (1.142). In both cases we can use Theorem 1.36.

So far we have assumed that the operators $A(t)$, $t \in [0, T]$, are linear. Now we pass to nonlinear operators. As before we introduce an evolution triplet $\{V, H, V^*\}$, where V, H are real separable Hilbert spaces. Let $\{A(t) : t \in [0, T]\}$ be a family of nonlinear operators from V to V^* satisfying the following assumptions:

(i) $A(t)$ is *monotone* in V for every $t \in [0, T]$, i.e.,

$$\langle A(t)u - A(t)v, u - v \rangle \geq 0 \quad \forall u, v \in V, t \in [0, T];$$

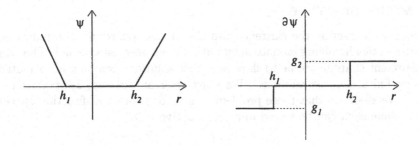

Figure 1.6.

(ii) $A(t)$ is *demicontinuous* in V for every $t \in [0,T]$, i.e.,

$$u_k \to u \text{ strongly in } V \implies A(t)u_k \rightharpoonup A(t)u \text{ weakly in } V^*, \forall t \in [0,T];$$

(iii) The function $t \mapsto A(t)$ is weakly measurable, i.e., the function

$$t \mapsto \langle A(t)u, v \rangle$$

is measurable on $(0,T)$ for all $u, v \in V$;

(iv) $A(t)$ is *uniformly coercive* in V with respect to $t \in [0,T]$:

$$\exists \alpha > 0, \beta \in \mathbf{R} : \langle A(t)u, u \rangle \geq \alpha \|u\|^2 + \beta \quad \forall u \in V, \forall t \in [0,T];$$

(v) $A(t)$ is *uniformly bounded* in V with respect to $t \in [0,T]$:

$$\exists M > 0 : \|A(t)u\|_{V^*} \leq M(1 + \|u\|) \quad \forall u \in V, \forall t \in [0,T].$$

By *an abstract nonlinear parabolic equation* (in an operator form) we mean the problem:

$$\begin{cases} \text{Find } u \in W(V) \text{ such that} \\ u'(t) + A(t)u(t) = f(t) \quad \text{in } V^* \quad \text{for a.a. } t \in (0,T) \\ \text{and } u(0) = u_0. \end{cases} \quad (1.145)$$

The assumptions stated above guarantee the following existence result (an extension of Theorem 1.33) (Zeidler, 1990b):

Theorem 1.37 *Let $A(t) : V \to V^*$, $t \in [0,T]$, be operators satisfying (i)-(v). Then for every $f \in L^2(0,T;H)$ and $u_0 \in H$ the equation (1.145) has a unique solution.*

Similarly, it is also possible to extend Theorems 1.34–1.36 to the case of nonlinear operators, but this is out of the scope of this presentation.

1.4 APPROXIMATION OF EQUATIONS AND INEQUALITIES OF MONOTONE TYPE

In the previous section the existence and the uniqueness results for equations and inequalities involving monotone operators have been established. The aim of the present chapter will be to show how such solutions can be approximated by using finite element methods. We start with an abstract approximation theory. We shall see that these problems when formulated in function spaces of finite dimension can be solved in a constructive way.

Approximation of elliptic equations

Let us consider an abstract *linear* elliptic equation $\{V, a, f\}$, where V is a Hilbert space, $a : V \times V \to \mathbf{R}$ is a bounded, V-elliptic bilinear form and $f \in V^*$. Let $S \subset V$ be a *finite dimensional* subspace of V, $\dim S = n$ and let $\{\varphi_i\}_{i=1}^n$ be its basis. Now, instead of $\{V, a, f\}$ we shall consider the elliptic problem $\{S, a, f\}$, whose solution is denoted by u_S:

$$u_S \in S : a(u_S, v) = \langle f, v \rangle \quad \forall v \in S. \tag{1.146}$$

Due to the Lax-Milgram theorem there exists a unique solution u_S of $\{S, a, f\}$. Since S is finite dimensional, u_S can be written in the form

$$u_S = \sum_{j=1}^n \alpha_j \varphi_j, \tag{1.147}$$

for an appropriate choice of the coefficients $\alpha_1, ..., \alpha_n$. Inserting this expression into (1.146) and using the fact that the necessary and sufficient condition for (1.146) to be satisfied for all $v \in S$ is to be satisfied for any basis function φ_i, we obtain:

$$a(\sum_{j=1}^n \alpha_j \varphi_j, \varphi_i) = \langle f, \varphi_i \rangle \quad \forall i = 1, ..., n. \tag{1.148}$$

This leads to the following *linear system of algebraic equations*:

$$\sum_{j=1}^n a(\varphi_j, \varphi_i) \alpha_j = \langle f, \varphi_i \rangle \quad \forall i = 1, ..., n \tag{1.149}$$

or in the matrix form:

$$\mathbf{A}\vec{\alpha} = \vec{F}, \quad \vec{\alpha} = (\alpha_1, ..., \alpha_n) \in \mathbf{R}^n, \tag{1.150}$$

where $\mathbf{A} = (a_{ij})_{i,j=1}^n$ with $a_{ij} = a(\varphi_j, \varphi_i)$ is the so-called *stiffness matrix* and $\vec{F} = (F_i)_{i=1}^n$ with $F_i = \langle f, \varphi_i \rangle$ is the *load vector*. Thus the case of linear elliptic equations in a finite dimensional subspace S of V reduces to the classical problem of *linear algebra*, namely to solve (1.150).

A natural question arises, namely to estimate the error between the solutions u, u_S of $\{V, a, f\}$, $\{S, a, f\}$, respectively. We have the following fundamental result (see Ciarlet, 1978):

Lemma 1.1 *(Cea's lemma) Let $a : V \times V \to \mathbf{R}$ be a bounded and V-elliptic bilinear form. Then*

$$\|u - u_S\| \leq \frac{M}{\alpha} \inf_{v \in S} \|u - v\|, \qquad (1.151)$$

where M, α are the constants from (1.41) and (1.42).

Proof: From the definitions of $\{V, a, f\}$, $\{S, a, f\}$ and the fact that $S \subset V$ it follows that

$$a(u, v) = \langle f, v \rangle \quad \forall v \in S$$
$$a(u_S, v) = \langle f, v \rangle \quad \forall v \in S.$$

From this we see that

$$a(u - u_S, v) = 0 \quad \forall v \in S. \qquad (1.152)$$

Let $v \in S$ be an arbitrary element. From (1.41) and (1.42) we obtain that

$$\alpha \|u - u_S\|^2 \leq a(u - u_S, u - u_S)$$
$$= a(u - u_S, u - v) + a(u - u_S, v - u_S) = a(u - u_S, u - v)$$
$$\leq M \|u - u_S\| \|u - v\|$$

making use of (1.152) and the fact that S is a linear set. From this we easily arrive at (1.151). □

Remark 1.17 *If a is symmetric in V then $\{S, a, f\}$ is equivalent to the minimization of J given by (1.50) over S:*

$$u_S \in S : J(u_S) = \min_{v \in S} J(v) \qquad (1.153)$$

or expressed in the algebraic form:

$$\vec{\alpha}^* \in \mathbf{R}^n : \mathcal{J}(\vec{\alpha}^*) = \min_{\vec{\alpha} \in \mathbf{R}^n} \mathcal{J}(\vec{\alpha}), \qquad (1.154)$$

where

$$\mathcal{J}(\vec{\alpha}) = \frac{1}{2}(\vec{\alpha}, \mathbf{A}\vec{\alpha}) - (\vec{F}, \vec{\alpha}),$$

with $\mathbf{A}$, $\vec{F}$ being the stiffness matrix, the load vector, respectively.

The approach when the infinite dimensional space V is replaced by its finite dimensional subspace S is known as the *Galerkin method* or as the *Ritz method* (see (1.153)) when a is symmetric.

From (1.151) we see that if $u \notin S$ then the error $\|u - u_S\|$ is positive. One of the basic problems of the approximation theory can be formulated as follows: are we able to reduce this error below a given tolerance? Clearly, only one space S is not sufficient.

Let $h > 0$ be a discretization parameter (the meaning of h will be seen later). With any $h > 0$ we associate a finite dimensional subspace V_h of V, $\dim V_h = n(h)$ where $n(h) \to \infty$ for $h \to 0+$. Let $u_h \in V_h$ be a unique solution of $\{V_h, a, f\}$. Then on the basis of Lemma 1.1 we know that the error between u and u_h can be estimated as follows:

$$\|u - u_h\| \leq \frac{M}{\alpha} \inf_{v_h \in V_h} \|u - v_h\|. \tag{1.155}$$

From this the following convergence result easily follows:

Theorem 1.38 *Let the family $\{V_h\}$, $h \to 0+$ possesses the following approximation property:*

$$\forall v \in V \; \exists \{v_h\}, \; v_h \in V_h \text{ such that } \|v - v_h\| \to 0, \; h \to 0+. \tag{1.156}$$

Then

$$\|u - u_h\| \to 0, \quad h \to 0+,$$

i.e., the Ritz-Galerkin method is convergent.

Now let us pass to nonlinear elliptic equations. Let $T : W \to W^*$ be a mapping from a reflexive Banach space W into its dual W^*, which is *strongly monotone* and locally Lipschitz continuous in W (see (1.97),(1.98)).

Let $S \subset W$ be a finite dimensional subspace of W. Analogously to the linear case we solve the problem:

$$u_S \in S : \langle T(u_S), v \rangle = \langle f, v \rangle \quad \forall v \in S \tag{1.157}$$

for $f \in W^*$ given or equivalently

$$\langle T(\sum_{j=1}^{n} \alpha_j \varphi_j), \varphi_i \rangle = \langle f, \varphi_i \rangle \quad \forall i = 1, ..., n, \tag{1.158}$$

where again $\{\varphi_j\}_{j=1}^{n}$ are the basis functions of S. In contrast to the linear problem, (1.158) represents *a nonlinear system of algebraic equations* for unknown coefficients $\alpha_1, ... \alpha_n \in \mathbf{R}$. Also the analogy of the Cea's lemma holds true. Indeed:

Lemma 1.2 *Let T be strongly monotone and locally Lipschitz continuous in W. Then*

$$\alpha(\|u - u_S\|) \leq C \inf_{v \in S} \|u - v\|, \tag{1.159}$$

where α is the function from (1.97) and C is a constant depending on $\|f\|_{W^*}$ and $\|T(0)\|_{W^*}$.

Proof is parallel to this one of Lemma 1.1.

Remark 1.18 *If T is the potential operator with the potential Φ, satisfying (1.101) and (1.102), problem (1.157) is equivalent to*

$$u_S \in S : \Phi(u_S) = \min_{v \in S} \Phi(v), \tag{1.160}$$

i.e., to the minimization problem for a convex but non-quadratic functional.

For the same reasons as before we introduce a family $\{W_h\}$, $h \to 0+$ of finite dimensional subspaces of W and define the following nonlinear problems on W_h:

$$u_h \in W_h : \langle T(u_h), v_h \rangle = \langle f, v_h \rangle \quad \forall v_h \in W_h. \tag{1.161}$$

From (1.159) and the properties of α we easily deduce the following convergence result:

Theorem 1.39 *Let T be the same as above and suppose that*

$$\forall v \in W \; \exists \{v_h\}, \; v_h \in W_h \text{ such that } \|v - v_h\| \to 0, \; h \to 0+. \tag{1.162}$$

Then

$$\|u - u_h\| \to 0, \quad h \to 0+. \tag{1.163}$$

Approximation of elliptic variational inequalities

In this section we briefly recall main results of the abstract approximation theory of elliptic inequalities. For more details see Hlaváček et al., 1988.

Let $\{K, a, f\}$ be an elliptic inequality of the first kind, where K is a nonempty, closed and convex subset of a real Hilbert space V, $f \in V^*$ and $a : V \times V \to \mathbf{R}$ be a bounded and elliptic bilinear form in V. Our goal is to approximate its solutions u. The idea is the same as in the case of elliptic equations. Suppose that Q is another nonempty, closed and convex subset of a finite dimensional subspace $S \subset V$. Let us consider the inequality $\{Q, a, f\}$ with the solution u_Q, i.e.,

$$u_Q \in Q : a(u_Q, v - u_Q) \geq \langle f, v - u_Q \rangle \quad \forall v \in Q. \tag{1.164}$$

We start with the algebraic representation of (1.164). Let $\{\varphi_i\}_{i=1}^n$ be a basis of S. Then S is isometrically isomorphic with $\mathbf{R}^n$. The corresponding isomorphism which associates with any $v \in S$ its coordinates with respect to $\{\varphi_i\}_{i=1}^n$ will be denoted by $\mathcal{T}_S$ while $\mathcal{T}_S^{-1}$ stands for its inverse. It is readily seen that $\mathcal{U} \equiv \mathcal{T}_S^{-1}(Q)$ is a nonempty, closed and convex subset of $\mathbf{R}^n$ and that (1.164) can be equivalently expressed as follows:

$$\vec{\alpha}^* \in \mathcal{U} : (\mathbf{A}\vec{\alpha}^*, \vec{\alpha} - \vec{\alpha}^*) \geq (\vec{F}, \vec{\alpha} - \vec{\alpha}^*) \quad \forall \vec{\alpha} \in \mathcal{U}, \qquad (1.165)$$

where $\mathbf{A}, \vec{F}$ is the stiffness matrix, the load vector, respectively.

If the bilinear form a is symmetric in V, (1.164) is equivalent to

$$u_Q \in Q : J(u_Q) = \min_{v \in Q} J(v), \qquad (1.166)$$

where J is given by (1.50). The algebraic representation of (1.166) leads to the following *nonlinear mathematical programming problem*:

$$\vec{\alpha}^* \in \mathcal{U} : \mathcal{J}(\vec{\alpha}^*) = \min_{\vec{\alpha} \in \mathcal{U}} \mathcal{J}(\vec{\alpha}), \qquad (1.167)$$

where

$$\mathcal{J}(\vec{\alpha}) = \frac{1}{2}(\vec{\alpha}, \mathbf{A}\vec{\alpha}) - (\vec{F}, \vec{\alpha})$$

is the quadratic function. This problem can be solved by using methods of numerical minimization, whose particular choice strongly depends on the character of the set $\mathcal{U}$, defining the constraints.

Now we shall derive the error estimate $\|u - u_Q\|$. Let us note that the set Q used for the definition of the approximate problem $\{Q, a, f\}$ *is not* necessarily a part of the original set K. We have the following generalization of the Cea's lemma:

Lemma 1.3 *Let $a : V \times V \to \mathbf{R}$ be a bounded and V-elliptic bilinear form and u, u_Q be unique solutions to $\{K, a, f\}$, $\{Q, a, f\}$, respectively. Then*

$$\begin{aligned}
\alpha\|u - u_Q\|^2 &\leq a(u - u_Q, u - u_Q) \\
&\leq \langle f, u - v_Q \rangle + \langle f, u_Q - v \rangle + a(u - u_Q, u - v_Q) \qquad (1.168) \\
&\quad + a(u, v - u_Q) + a(u, v_Q - u)
\end{aligned}$$

holds for any $v_Q \in Q$ and any $v \in K$.

Proof: From the definition of $\{K, a, f\}$, $\{Q, a, f\}$ it follows that

$$a(u, u) \leq \langle f, u - v \rangle + a(u, v) \quad \forall v \in K,$$
$$a(u_Q, u_Q) \leq \langle f, u_Q - v_Q \rangle + a(u_Q, v_Q) \quad \forall v_Q \in Q.$$

Therefore

$$\begin{aligned}
\alpha\|u - u_Q\|^2 &\leq a(u - u_Q, u - u_Q) = a(u, u) + a(u_Q, u_Q) \\
&\quad - a(u_Q, u) - a(u, u_Q) \leq \langle f, u - v \rangle + a(u, v) \\
&\quad + \langle f, u_Q - v_Q \rangle + a(u_Q, v_Q) - a(u_Q, u) - a(u, u_Q) \\
&= \langle f, u - v_Q \rangle + \langle f, u_Q - v \rangle + a(u - u_Q, u - v_Q) \\
&\quad + a(u, v - u_Q) + a(u, v_Q - u) \quad \forall v_Q \in Q, \forall v \in K.
\end{aligned}$$

□

Remark 1.19 *If $Q \subset K$ then (1.168) can be simplified. Indeed: inserting $v = u_Q \in K$ into the right hand side of (1.168) we obtain*

$$\alpha \|u - u_Q\|^2 \leq a(u - u_Q, u - u_Q) \leq \langle f, u - v_Q \rangle \qquad (1.169)$$
$$+ a(u - u_Q, u - v_Q) + a(u, v_Q - u) \quad \forall v_Q \in Q.$$

In order to reduce the error between u and u_Q below an a priori given tolerance, again only one convex set Q is not sufficient. As in the case of equations, with any discretization parameter $h > 0$ we associate a nonempty, closed and convex subset K_h of a finite dimensional space V_h. Denote by u, u_h the solution to $\{K, a, f\}$, $\{K_h, a, f\}$, respectively. On the basis of Lemma 1.3 the error between $\|u - u_h\|$ can be estimated as follows:

$$\alpha \|u - u_h\|^2 \leq \langle f, u - v_h \rangle + \langle f, u_h - v \rangle + a(u - u_h, u - v_h) \quad (1.170)$$
$$+ a(u, v - u_h) + a(u, v_h - u) \quad \forall v_h \in K_h, \ \forall v \in K$$

with the following modification when $K_h \subset K \ \forall h > 0$:

$$\alpha \|u - u_h\|^2 \leq \langle f, u - v_h \rangle + a(u - u_h, u - v_h) \qquad (1.171)$$
$$+ a(u, v_h - u) \quad \forall v_h \in K_h.$$

Remark 1.20 *If $K_h \subset K$ for any $h > 0$ then K_h is termed an internal approximation of K, otherwise we say that K_h is an external one.*

Another possible application of (1.170), (1.171) is the following convergence result:

Theorem 1.40 *Let $a : V \times V \to \mathbf{R}$ be a bounded and V-elliptic bilinear form. Let the family $\{K_h\}$ possess the following properties:*

$$\forall v \in K \ \exists \{v_h\}, v_h \in K_h \text{ such that } \|v_h - v\| \to 0, \ h \to 0+; \ (1.172)$$

$$\text{from } v_h \rightharpoonup v \text{ (weakly) in } V, v_h \in K_h \text{ it follows that } v \in K. \ (1.173)$$

Then

$$\|u - u_h\| \to 0, \quad h \to 0+.$$

Proof: It is readily seen that $\{u_h\}$ is bounded in V. Hence there exist: a subsequence $\{u_{h'}\} \subset \{u_h\}$ and an element $u^* \in V$ such that

$$u_{h'} \rightharpoonup u^* \quad \text{in } V, \ h' \to 0+.$$

From (1.173) it follows that $u^* \in K$. On the other hand the solution u of $\{K, a, f\}$ can be approximated by elements of K_h:

$$\exists \{\bar{v}_h\}, \bar{v}_h \in K_h, \text{ such that } \|u - \bar{v}_h\| \to 0, \ h \to 0+.$$

Inserting $v := u^*$, $v_h := \bar{v}_h$ into (1.170) we see that all terms on the right hand side tend to zero and consequently

$$\|u - u_{h'}\| \to 0, \ h \to 0+.$$

We proved that *any* weakly convergent subsequence of $\{u_h\}$ strongly converges to u, which is unique. From this we may conclude that the *whole* sequence $\{u_h\}$ tends to u in the norm of V. $\qquad\qquad\qquad\qquad\qquad\qquad\square$

Remark 1.21 *If $K_h \subset K$ for any $h > 0$ then (1.173) is automatically satisfied. This follows from the fact that any convex set is closed if and only if is weakly closed.*

These results can be easily extended to variational inequalities of the second kind. Let $\{K, a, f, j\}$ be an inequality of the second kind and $\{K_h, a, f, j\}$ its approximation. Then it holds:

$$\alpha\|u - u_h\|^2 \leq \langle f, u - v_h \rangle + \langle f, u_h - v \rangle \qquad (1.174)$$
$$+ a(u - u_h, u - v_h) + a(u, v - u_h) + a(u, v_h - u)$$
$$+ j(v) - j(u_h) + j(v_h) - j(u) \quad \forall v_h \in K_h, \ \forall v \in K$$

or when $K_h \subset K \ \forall h > 0$:

$$\alpha\|u - u_h\|^2 \leq \langle f, u - v_h \rangle + a(u - u_h, u - v_h) \qquad (1.175)$$
$$+ a(u, v_h - u) + j(v_h) - j(u) \quad \forall v_h \in K_h.$$

Also Theorem 1.40 remains valid. Finally these results can be easily extended to elliptic inequalities involving strongly monotone and locally Lipschitz mappings T in a reflexive Banach space W. If u, u_h are solutions of

$$u \in K : \langle T(u), v - u \rangle \geq \langle f, v - u \rangle \quad \forall v \in K; \qquad (1.176)$$

$$u_h \in K_h : \langle T(u_h), v_h - u_h \rangle \geq \langle f, v_h - u_h \rangle \quad \forall v_h \in K_h, \qquad (1.177)$$

then

$$\alpha(\|u - u_h\|)\|u - u_h\| \leq \langle f, u - v_h \rangle + \langle f, u_h - v \rangle \qquad (1.178)$$
$$+ \langle T(u_h) - T(v), v_h - u \rangle + \langle T(u), v - u_h \rangle + \langle T(u), v_h - u \rangle$$

holds for any $v \in K$ and any $v_h \in K_h$, where α is a function with the properties formulated in (1.97).

Approximation of parabolic equations and variational inequalities

In this subsection we introduce the basic ideas and results concerning the abstract approximation theory for parabolic problems. We shall restrict ourselves to the so-called *standard Galerkin method* (see Glowinski et al., 1981, Thomée, 1984). We start with the *semidiscrete* approximation theory (called also the continuous-time Galerkin method), in which only the space variables are discretized. After that we turn our attention to the *fully discrete* approximation theory.

Semidiscrete approximation. Let V and H be real separable Hilbert spaces such that they form an evolution triplet

$$"V \subseteq H \subseteq V^*".$$

For the sake of simplicity, we assume that the bilinear form $a : V \times V \to \mathbf{R}$ *does not depend on time t.* Suppose that a satisfies (1.105) and (1.106) (because of Remark 1.12 we may assume that the constant β in (1.106) is equal to 0), $f \in L^2(0, T; V^*)$ and $u_0 \in H$. We describe the approximation of

$$\begin{cases} \text{Find } u \in W(V) \text{ such that} \\ \langle u'(t) + Au(t) - f(t), v \rangle = 0 \quad \forall v \in V, \text{ for a.a. } t \in (0, T) \\ \text{and} \quad u(0) = u_0 \end{cases} \qquad (1.179)$$

with A defined by (1.44).

Let $S \subset V$ be a finite dimensional subspace of V, $\dim S = n$ and $\{\varphi_i\}_{i=1}^n$ be its basis. Let us denote by u_{0S} an approximation of u_0 from S. Then the *semidiscrete approximation* (the space variables are discretized, but the time variable t is continuous) of (1.179) reads as follows:

$$\begin{cases} \text{Find } u_S \in H^1(0, T; S) \text{ such that} \\ \langle u_S'(t) + Au_S(t) - f(t), v \rangle = 0 \quad \forall v \in S, \text{ for a.a. } t \in (0, T) \\ \text{and} \quad u_S(0) = u_{0S}. \end{cases} \qquad (1.180)$$

Inserting the expansion

$$u_S(t) = \sum_{j=1}^n \alpha_j(t)\varphi_j$$

into (1.180) we arrive at the system of ordinary differential equations:

$$\begin{cases} \text{Find } \alpha(t) = (\alpha_1(t), ..., \alpha_n(t)) \in H^1(0, T; \mathbf{R}^n) \text{ such that} \\ \sum_{j=1}^n \alpha_j'(t)\langle \varphi_j, \varphi_i \rangle + \sum_{j=1}^n \alpha_j(t)\langle A\varphi_j, \varphi_i \rangle = \langle f(t), \varphi_i \rangle \\ \forall i = 1, ..., n, \text{ for a.a. } t \in (0, T) \text{ and} \\ \alpha_j(0) = \alpha_{0j} \quad \forall j = 1, .., n \quad (u_{0S} = \sum_{j=1}^n \alpha_{0j}\varphi_j), \end{cases} \qquad (1.181)$$

or in the matrix formulation:

$$\begin{cases} \mathbf{M}\vec{\alpha}'(t) + \mathbf{A}\vec{\alpha}(t) = \vec{F}(t) \\ \text{and} \quad \vec{\alpha}(0) = \vec{\alpha}_0, \end{cases} \tag{1.182}$$

where $\mathbf{M} = (m_{ij})_{i,j=1}^n$, $m_{ij} = \langle \varphi_j, \varphi_i \rangle$, is the so-called *mass matrix* and $\mathbf{A}$, $\vec{F}$ the stiffness matrix, the load vector, respectively. Let us mention that the duality pairing $\langle \varphi_i, \varphi_j \rangle$ is realized by the scalar product in H, i.e., $m_{ij} = (\varphi_i, \varphi_j)$.

Next we shall analyse how the semidiscrete problems approximate the original one. For that purpose we introduce a family $\{V_h\}$ of finite dimensional subspaces of V ($h > 0$ a discretization parameter, $\dim V_h = n(h)$ and $n(h) \to \infty$ as $h \to 0+$). Denote by $u_h \in L^2(0,T;V_h)$ a unique solution of (1.180) with the initial condition $u_h(0) = u_{0h} \in V_h$. The following well-known convergence result holds (Zeidler, 1990b, Glowinski et al., 1981):

Theorem 1.41 *Let the family* $\{V_h\}$, $h \to 0+$ *satisfy (1.156) and*

$$|u_{0h} - u_0| \to 0, \quad h \to 0+ . \tag{1.183}$$

Then

$$\|u - u_h\|_{L^2(0,T;V)} \to 0,$$
$$\|u - u_h\|_{C([0,T];H)} \to 0, \quad h \to 0+ .$$

Proof: For clarity we divide the proof into several steps.
Step I: A priori estimates:
By choosing $v = u_h(s)$ in (1.180) we have

$$\langle u'_h(s), u_h(s) \rangle + \langle A u_h(s), u_h(s) \rangle = \langle f(s), u_h(s) \rangle \tag{1.184}$$

for any $s \in (0,T)$. Then we estimate (1.184) term by term. The V-ellipticity of a and the inequality

$$ab \leq \varepsilon a^2 + \frac{1}{4\varepsilon} b^2, \quad \forall a, b \in \mathbf{R}, \ \forall \varepsilon > 0 \tag{1.185}$$

yield

$$\langle u'_h(s), u_h(s) \rangle = \frac{1}{2} \frac{d}{ds} |u_h(s)|^2;$$
$$\langle A u_h(s), u_h(s) \rangle \geq \alpha \|u_h(s)\|^2;$$
$$\langle f(s), u_h(s) \rangle \leq \|f(s)\|_* \|u_h(s)\| \leq \frac{1}{2\alpha} \|f(s)\|_*^2 + \frac{\alpha}{2} \|u_h(s)\|^2.$$

Substituting these inequalities into (1.184) and integrating the resulting inequality over the interval $(0,t)$, $t \leq T$, we obtain the stability estimate

$$\frac{1}{2} |u_h(t)|^2 + \frac{\alpha}{2} \|u_h\|^2_{L^2(0,t;V)} \leq \frac{1}{2} |u_h(0)|^2 + \frac{1}{2\alpha} \|f\|^2_{L^2(0,t;V^*)} \tag{1.186}$$

for any $t \in (0, T]$. Clearly (1.186) implies that

$$\max_{t \in [0,T]} |u_h(t)| \leq \text{const.},$$

$$\|u_h\|_{L^2(0,T;V)} \leq \text{const.}$$

From Theorem 1.19 we conclude that there exist: a subsequence $\{u_h\}$ (in order to simplify our notations we denote the subsequences by the same symbols as the original sequences in what follows) and a function $\tilde{u} \in L^\infty(0,T;H) \cap L^2(0,T;V)$ such that

$$u_h \overset{*}{\rightharpoonup} \tilde{u} \quad \text{in } L^\infty(0,T;H), \tag{1.187}$$

$$u_h \rightharpoonup \tilde{u} \quad \text{in } L^2(0,T;V), \ h \to 0+. \tag{1.188}$$

Step II: $\tilde{u}$ solves (1.179):

Let $\phi \in C_0^\infty((0,T))$ and $v \in V$ be given. From (1.156) the existence of a sequence $\{v_h\}$, $v_h \in V_h$, such that $v_h \to v$ in V follows. Denote $\Phi(x,t) = v(x)\phi(t)$ and $\Phi_h(x,t) = v_h(x)\phi(t)$. Then we have

$$\Phi_h \to \Phi \quad \text{in } L^2(0,T;V), \tag{1.189}$$

$$\Phi_h' \to \Phi' \quad \text{in } L^2(0,T;V^*), \ h \to 0+. \tag{1.190}$$

Further, integration by parts implies (see Proposition 1.5):

$$\int_0^T \langle u_h'(t), \Phi_h(t) \rangle dt = -\int_0^T \langle u_h(t), \Phi_h'(t) \rangle dt. \tag{1.191}$$

Inserting $v_h = \Phi_h(t)$ in (1.180), then integrating over the interval $(0,T)$ and using (1.191) we obtain:

$$-\int_0^T \langle u_h(t), \Phi_h'(t) \rangle dt + \int_0^T \langle Au_h(t), \Phi_h(t) \rangle dt = \int_0^T \langle f(t), \Phi_h(t) \rangle dt.$$

Letting $h \to 0+$ we see that

$$-\int_0^T \langle \tilde{u}(t), \Phi'(t) \rangle dt + \int_0^T \langle A\tilde{u}(t), \Phi(t) \rangle dt = \int_0^T \langle f(t), \Phi(t) \rangle dt$$

taking into account (1.187)–(1.190), or equivalently

$$\int_0^T \langle \tilde{u}(t), v \rangle \phi'(t) dt \tag{1.192}$$

$$= -\int_0^T \langle -A\tilde{u}(t) + f(t), v \rangle \phi(t) dt \quad \forall v \in V \text{ and } \forall \phi \in C_0^\infty((0,T)).$$

Hence, from Proposition 1.3 we deduce that

$$\tilde{u}' \in L^2(0,T;V^*) \quad \text{and} \quad \tilde{u}' = -A\tilde{u} + f. \tag{1.193}$$

It remains to show that

$$\tilde{u}(0) = u_0. \qquad (1.194)$$

For this purpose we choose $\phi \in C^\infty([0,T])$ such that $\phi(0) = 1$ and $\phi(T) = 0$ and repeat the same considerations as above. Then together with the assumption (1.183) we obtain

$$-\int_0^T \langle \tilde{u}(t), \Phi'(t) \rangle dt - (u_0, v) + \int_0^T \langle A\tilde{u}(t), \Phi(t) \rangle dt \qquad (1.195)$$

$$= \int_0^T \langle f(t), \Phi(t) \rangle dt$$

Using that $\tilde{u}' = -A\tilde{u} + f$ and integrating by parts

$$\int_0^T \langle \tilde{u}(t), \Phi'(t) \rangle dt = -\int_0^T \langle \tilde{u}'(t), \Phi(t) \rangle dt - (\tilde{u}(0), v) \qquad (1.196)$$

we get:

$$(\tilde{u}(0) - u_0, v) = 0 \quad \forall v \in V.$$

From the density of V in H we infer $\tilde{u}(0) = u_0$.

In addition, since the solution of (1.179) is unique, we conclude that $\tilde{u} \equiv u$ and the whole sequence $\{u_h\}$ tends to u in the sense (1.187), (1.188).

Step III: Strong convergence in $C([0,T]; H)$.

We have to show that

$$\|u_h - u\|_{C([0,T];H)} \equiv \max_{t \in [0,T]} |u_h(t) - u(t)| \to 0, \quad h \to 0 +. \qquad (1.197)$$

First note that the polynomials $p_h : [0,T] \to V_h$

$$p_h(t) = a_{h0} + a_{h1}t + \dots + a_{hn}t^n$$

with $a_{hi} \in V_h$ for all $i = 0, \dots, n$, $n \in \mathbf{N}$ and $h > 0$ are dense in $W(V)$ as follows from Remark 1.3 and the fact that $\cup_{h>0} V_h$ is dense in V. Thus, there exists a sequence $\{\bar{u}_h\}$ of polynomials, $\bar{u}_h : [0,T] \to V_h$ converging strongly to u in $W(V)$ as $h \to 0+$. Moreover, since the imbedding $W(V) \subset C([0,T], H)$ is continuous we have that

$$\|u - \bar{u}_h\|_{C([0,T];H)} \to 0, \quad h \to 0 +. \qquad (1.198)$$

Therefore, to prove (1.197) it is enough to show that

$$\|u_h - \bar{u}_h\|_{C([0,T];H)} \to 0, \quad h \to 0 +. \qquad (1.199)$$

If $t = 0$ then (1.183) and (1.198) imply

$$|u_h(0) - \bar{u}_h(0)| \to 0, \quad h \to 0 +. \qquad (1.200)$$

Let $t \in (0, T]$ be given. Since u_h, u is a solution of (1.180), (1.179), respectively, we get:

$$\langle u'_h(t), u_h(t) - \bar{u}_h(t) \rangle = \langle f(t) - Au_h(t), u_h(t) - \bar{u}_h(t) \rangle$$
$$= \langle u'(t) + Au(t) - Au_h(t), u_h(t) - \bar{u}_h(t) \rangle.$$

From this, (1.106) (with $\beta = 0$) and (1.200) we have that

$$\frac{1}{2}|u_h(t) - \bar{u}_h(t)|^2 - \frac{1}{2}|u_h(0) - \bar{u}_h(0)|^2 \qquad (1.201)$$

$$= \int_0^t \langle u'_h - \bar{u}'_h, u_h - \bar{u}_h \rangle ds$$

$$= \int_0^t \langle u' + A(u - u_h) - \bar{u}'_h, (u_h - u) + (u - \bar{u}_h) \rangle ds$$

$$\leq \int_0^t \left(\langle u' - \bar{u}'_h, u_h - \bar{u}_h \rangle + \langle A(u - u_h), u - \bar{u}_h \rangle \right) ds$$

$$\leq \|u' - \bar{u}'_h\|_{L^2(0,T;V^*)} \|u_h - \bar{u}_h\|_{L^2(0,T;V)}$$
$$+ \|Au - Au_h\|_{L^2(0,T;V^*)} \|u - \bar{u}_h\|_{L^2(0,T;V)}$$

$$\leq C\|u - \bar{u}_h\|_{W(V)} \to 0$$

uniformly with respect $t \in (0, T]$ as $h \to 0+$ (note that $\{u_h\}$, $\{\bar{u}_h\}$ are bounded in $L^2(0, T; V)$ and $\{Au_h\}$ is bounded in $L^2(0, T; V^*)$). Then (1.199) follows from (1.200) and (1.201).

Step IV: Strong convergence in $L^2(0, T; V)$.

Let $\{\bar{u}_h\}$ be as in step III, so that

$$\|u - \bar{u}_h\|_{L^2(0,T;V)} \to 0, \quad h \to 0+.$$

Therefore, it only remains to prove that

$$\|u_h - \bar{u}_h\|_{L^2(0,T;V)} \to 0, \quad h \to 0+.$$

Indeed, from (1.106) (with $\beta = 0$), step III and (1.188) it follows that

$$\alpha\|u_h - \bar{u}_h\|_{L^2(0,T;V)}^2$$

$$\leq \int_0^T \langle A(u_h - \bar{u}_h), u_h - \bar{u}_h \rangle dt$$

$$= \int_0^T \langle Au_h, u_h \rangle dt - \int_0^T \langle Au_h, \bar{u}_h \rangle dt - \int_0^T \langle A\bar{u}_h, u_h - \bar{u}_h \rangle dt$$

$$= \int_0^T \langle f - u'_h, u_h \rangle dt - \int_0^T \langle Au_h, \bar{u}_h \rangle dt - \int_0^T \langle A\bar{u}_h, u_h - \bar{u}_h \rangle dt$$

$$= \frac{1}{2}\left(|u_h(0)|^2 - |u_h(T)|^2\right) + \int_0^T \langle f, u_h \rangle dt$$

$$- \int_0^T \langle Au_h, \bar{u}_h \rangle dt - \int_0^T \langle A\bar{u}_h, u_h - \bar{u}_h \rangle dt$$

$$\to \frac{1}{2}\left(|u(0)|^2 - |u(T)|^2\right) + \int_0^T \langle f - Au, u \rangle dt = 0$$

as $h \to 0+$. This completes the proof of Theorem 1.41.   □

In order to obtain the error estimates between the solutions of the semidiscrete and continuous problem it is necessary to impose some additional regularity assumptions on u_h, u and u_0. Next we shall suppose that u belongs to $H^1(0,T;V)$ (note that u_h satisfies automatically this assumption) and $u_0 \in V$. In addition, we define the so-called *Ritz projection* p_h from V onto V_h, which is the "orthogonal projection" with respect the "inner product" $\langle A\cdot,\cdot\rangle$:

$$\langle Ap_h v, w_h\rangle = \langle Av, w_h\rangle \quad \forall w_h \in V_h. \tag{1.202}$$

Here and in what follows we assume that A is *symmetric*. Then the following error estimate holds (Thomée, 1984):

Theorem 1.42 *Let u_h, u be solutions of (1.180), (1.179), respectively. Then there exists a constant $C > 0$ such that*

$$|u_h(t) - u(t)| \le C|u_{0h} - u_0| \tag{1.203}$$
$$+C\left(|p_h u_0 - u_0| + \int_0^t |p_h u'(s) - u'(s)|ds\right) \quad \forall t \in [0,T]$$

and

$$\|u_h(t) - u(t)\| \le C\|u_{0h} - u_0\| + C\left(\|p_h u_0 - u_0\|\right) \tag{1.204}$$
$$+\|p_h u(t) - u(t)\| + \left(\int_0^t |p_h u'(s) - u'(s)|^2 ds\right)^{\frac{1}{2}}\right) \quad \forall t \in [0,T].$$

Proof: Denote

$$u_h - u = (u_h - p_h u) + (p_h u - u) \equiv \theta_h + \rho_h. \tag{1.205}$$

First we estimate θ_h. Using the fact that u_h and u are solutions of (1.180) and (1.179), respectively, we obtain the key relation

$$\langle \theta_h'(s), v\rangle + \langle A\theta_h(s), v\rangle \tag{1.206}$$
$$= \left(\langle u_h'(s), v\rangle + \langle Au_h(s), v\rangle\right) - \langle Ap_h u(s), v\rangle - \langle (p_h u)'(s), v\rangle$$
$$= \langle f(s), v\rangle - \langle Au(s), v\rangle - \langle (p_h u)'(s), v\rangle$$
$$= -\langle (p_h u)'(s) - u'(s), v\rangle = -\langle \rho_h'(s), v\rangle \quad \forall v \in V_h.$$

Inserting $v = \theta_h(s)$ in (1.206) and taking into account (1.106) we obtain

$$\frac{1}{2}\frac{d}{ds}|\theta_h(s)|^2 + \alpha\|\theta_h(s)\|^2 \le |\rho_h'(s)||\theta_h(s)|. \tag{1.207}$$

Thus, we have

$$\frac{d}{ds}|\theta_h(s)| \le |\rho_h'(s)| \tag{1.208}$$

implying

$$|\theta_h(t)| \leq |\theta_h(0)| + \int_0^t |\rho'_h(s)| ds \qquad (1.209)$$

$$\leq |u_{0h} - u_0| + |p_h u_0 - u_0| + \int_0^t |p_h u'(s) - u'(s)| ds$$

Here we have also used the fact that the time derivative and the operator p_h commute, i.e., $(p_h u)' = p_h u'$ and that $u \in H^1(0, T; V)$. For ρ_h we get easily the estimate

$$|\rho_h(t)| = |\rho_h(0) + \int_0^t \rho'_h(s) ds|$$

$$\leq |p_h u_0 - u_0| + \int_0^t |p_h u'(s) - u'(s)| ds.$$

This together with (1.209) proves (1.203).

To prove (1.204) we choose $v = \theta'_h(s)$ in (1.206). This yields

$$|\theta'_h(s)|^2 + \frac{1}{2}\frac{d}{ds}\langle A\theta_h(s), \theta_h(s)\rangle \leq |\rho'_h(s)||\theta'_h(s)| \qquad (1.210)$$

$$\leq \frac{1}{2}|\rho'_h(s)|^2 + \frac{1}{2}|\theta'_h(s)|^2.$$

We integrate (1.210) over the interval $(0, t)$ and use (1.105), (1.106). Hence, we have

$$\|\theta_h(t)\|^2 \leq \frac{M}{\alpha}\|\theta_h(0)\|^2 + \frac{1}{\alpha}\int_0^t |\rho'_h(s)|^2 ds \qquad (1.211)$$

$$\leq \frac{M}{\alpha}(\|u_{0h} - u_0\| + \|p_h u_0 - u_0\|)^2 + \frac{1}{\alpha}\int_0^t |p_h u'(s) - u'(s)|^2 ds.$$

From this we easily arrive at the second error estimate (1.204). $\qquad \square$

Next, we introduce *semidiscrete approximations* of parabolic variational inequalities.

Let K be a nonempty, closed and convex subset of V and let Q be another nonempty, closed and convex subset of a finite dimensional subspace $S \subset V$. Let a and f be as before. Then the semidiscrete approximation of the parabolic variational inequality of the first kind (1.118) is defined as follows:

$$\begin{cases} \text{Find } u_Q \in H^1(0, T; S) \cap L^2(0, T; Q) \text{ such that} \\ \langle u'_Q(t) + Au_Q(t), v - u_Q(t)\rangle \geq \langle f(t), v - u_Q(t)\rangle \\ \forall v \in Q, \text{ for a.a. } t \in (0, T) \\ \text{and} \quad u_Q(0) = u_{0Q} \end{cases} \qquad (1.212)$$

where u_{0Q} is an approximation of u_0 from Q. The corresponding algebraic representation of (1.212) reads as follows:

$$\begin{cases} \text{Find } \vec{\alpha}^* \in H^1(0,T;\mathbf{R}^n) \cap L^2(0,T;\mathcal{U}) \text{ such that} \\ (\mathbf{M}\vec{\alpha}^{*\prime}(t) + \mathbf{A}\vec{\alpha}^*(t), \vec{\alpha} - \vec{\alpha}^*(t)) \geq (\vec{F}(t), \vec{\alpha} - \vec{\alpha}^*(t)) \\ \forall \vec{\alpha} \in \mathcal{U}, \text{ for a.a. } t \in (0,T) \\ \text{and} \quad \vec{\alpha}^*(0) = \vec{\alpha}_0. \end{cases} \tag{1.213}$$

Here $\mathcal{U} \equiv T_S^{-1}(Q)$ and $\vec{\alpha}_0 \equiv T_S^{-1}(u_{0Q})$ (recall that T_S is the isomorphism between $\mathbf{R}^n$ and S). The meaning of other symbols is the same as in (1.182).

Now we establish an analogue of Theorem 1.41 for the approximation (1.212). We introduce a family $\{V_h\}$ of finite dimensional subspaces $V_h \subset V$ and a family $\{K_h\}$ of nonempty, closed and convex subsets $K_h \subset V_h$. For simplicity we assume that $0 \in K$ and $0 \in K_h$ for all $h > 0$. We denote by u, u_h the solution of (1.118), (1.212), respectively (with $S = V_h$ and $Q = K_h$). Then it holds (Glowinski et al., 1981):

Theorem 1.43 Let $a : V \times V \to \mathbf{R}$ be a bounded, V-elliptic and symmetric bilinear form and $f \in L^2(0,T;H)$. Let the family $\{K_h\}$ satisfy (1.172), (1.173), the sequence $\{u_{0h}\}$, $u_{0h} \in K_h$, satisfy (1.183) and be bounded in V. Moreover, assume that V is also compactly imbedded in H. Then

$$\|u - u_h\|_{L^2(0,T;V)} \to 0,$$
$$\|u - u_h\|_{C([0,T];H)} \to 0, \quad h \to 0+ .$$

Proof: Step I: A priori estimates.

Setting $v = 0 \in K_h$ in (1.212), we have:

$$\langle u_h'(s), u_h(s) \rangle + \langle A u_h(s), u_h(s) \rangle \leq \langle f(s), u_h(s) \rangle. \tag{1.214}$$

Then by repeating step I in the proof of Theorem 1.41 we obtain

$$\max_{t \in [0,T]} |u_h(t)| \leq \text{const.}, \tag{1.215}$$

$$\|u_h\|_{L^2(0,T;V)} \leq \text{const.} \tag{1.216}$$

Next we derive the second *a priori* estimates. Let $s \in (0,T]$ be given and $k > 0$ be such that $s - k \in (0,T)$. Now we substitute $v = u_h(s - k) \in K_h$ into the inequality (1.212) and divide it by k. This implies

$$-\langle u_h'(s), \frac{u_h(s-k) - u_h(s)}{-k} \rangle - \langle A u_h(s), \frac{u_h(s-k) - u_h(s)}{-k} \rangle \tag{1.217}$$
$$\geq -\langle f(s), \frac{u_h(s-k) - u_h(s)}{-k} \rangle.$$

Letting $k \to 0+$ in (1.217) and making use of Theorem 1.20 we have

$$\langle u_h'(s), u_h'(s) \rangle + \langle A u_h(s), u_h'(s) \rangle \leq \langle f(s), u_h'(s) \rangle. \tag{1.218}$$

Then using the relations (do not forget that $f \in L^2(0,T;H)$):

$$\langle Au_h(s), u_h'(s) \rangle = \frac{1}{2}\frac{d}{dt}\langle Au_h(s), u_h(s) \rangle$$

$$\langle f(s), u_h'(s) \rangle \le \frac{1}{2}|f(s)|^2 + \frac{1}{2}|u_h'(s)|^2$$

and integrating (1.218) over the time interval $(0,t)$ we get

$$\|u_h'\|_{L^2(0,t;H)}^2 + \langle Au_h(t), u_h(t) \rangle \le \|f\|_{L^2(0,t;H)}^2 + \langle Au_h(0), u_h(0) \rangle.$$

Now from (1.105), (1.106) and the fact that $\{\|u_{0h}\|\}$ is bounded we have that

$$\|u_h'\|_{L^2(0,T;H)} \le \text{const.}, \tag{1.219}$$

$$\max_{t \in [0,T]} \|u_h(t)\| \le \text{const.} \tag{1.220}$$

Taking into account (1.215), (1.216), (1.219), (1.220) and Theorem 1.19 we see that one can find a subsequence of $\{u_h\}$ (still denoted by the same sequence) such that

$$u_h \overset{*}{\rightharpoonup} \tilde{u} \quad \text{in } L^\infty(0,T;V) \text{ and } L^\infty(0,T;H), \tag{1.221}$$

$$u_h \rightharpoonup \tilde{u} \quad \text{in } L^2(0,T;V), \tag{1.222}$$

$$u_h' \rightharpoonup \tilde{w} \quad \text{in } L^2(0,T;H), \quad h \to 0+. \tag{1.223}$$

Finally, Proposition 1.2 implies $\tilde{w} = \tilde{u}'$ and from Proposition 1.6 (with $X = V$ and $Y = Z = H$) it follows that

$$u_h \to \tilde{u} \quad \text{in } L^2(0,T;H), \quad h \to 0+. \tag{1.224}$$

Step II: $\tilde{u}$ solves (1.118).

First we show that $\tilde{u}(t) \in K$ for a.a. $t \in (0,T)$. Indeed, from (1.224) we infer that (passing again to an appropriate subsequence)

$$u_h(t) \to \tilde{u}(t) \quad \text{in } H \text{ for a.a. } t \in (0,T).$$

This together with (1.220) implies

$$u_h(t) \rightharpoonup \tilde{u}(t) \quad \text{in } V \text{ for a.a. } t \in (0,T). \tag{1.225}$$

From (1.173) it follows that $\tilde{u}(t) \in K$ for a.a. $t \in (0,T)$.

Let $\Phi \in L^2(0,T;K)$ be given. Then there exists a sequence $\{\Phi_h\}$, $\Phi_h \in L^2(0,T;K_h)$ such that

$$\Phi_h \to \Phi \quad \text{in } L^2(0,T;V), \quad h \to 0+ \tag{1.226}$$

as follows from (1.172). Therefore

$$\int_0^T \langle u_h', \Phi_h - u_h \rangle dt + \int_0^T \langle Au_h, \Phi_h - u_h \rangle dt \tag{1.227}$$

$$\ge \int_0^T \langle f, \Phi_h - u_h \rangle dt.$$

Letting $h \to 0+$ in (1.227), taking into account (1.221)–(1.226) and the fact that

$$\liminf_{h \to 0+} \int_0^T \langle Au_h, u_h \rangle dt \geq \int_0^T \langle A\tilde{u}, \tilde{u} \rangle dt$$

we obtain that

$$\int_0^T \langle \tilde{u}', \Phi - \tilde{u} \rangle dt + \int_0^T \langle A\tilde{u}, \Phi - \tilde{u} \rangle dt$$

$$\geq \int_0^T \langle f, \Phi - \tilde{u} \rangle dt.$$

The initial condition $\tilde{u}(0) = u_0$ follows from step III below and the fact that $u_h(0) \to u_0$ in H. Hence, $\tilde{u}$ solves (1.118), and because the solution is unique one has $\tilde{u} \equiv u$ and all the convergences (1.221)–(1.224) hold for the whole sequence $\{u_h\}$.

Step III: Strong convergence in $C([0, T]; H)$.

This is a consequence of (1.183), (1.224) and the fact that $\{\|u_h'\|_{L^2(0,T;H)}\}$ is bounded. Indeed, we have

$$\frac{1}{2} |u_h(t) - u(t)|^2 - \frac{1}{2} |u_h(0) - u(0)|^2 \tag{1.228}$$

$$= \int_0^t (u_h' - u', u_h - u) ds$$

$$\leq \|u_h' - u'\|_{L^2(0,T;H)} \|u_h - u\|_{L^2(0,T;H)} \to 0$$

uniformly with respect to $t \in (0, T]$ as $h \to 0+$.

Step IV: Strong convergence in $L^2(0, T; V)$.

Recalling step IV in the proof of Theorem 1.41 it is sufficient to show that

$$\|u_h - \bar{u}_h\|_{L^2(0,T;V)} \to 0, \quad h \to 0+, \tag{1.229}$$

where $\{\bar{u}_h\}$ is a sequence of polynomials with values in K_h and converging strongly to u in $L^2(0, T; V)$. Such a sequence exists because of (1.172). In view of (1.106),(1.222),(1.224) and the assumption $f \in L^2(0, T; H)$ we can estimate

$$\alpha \|u_h - \bar{u}_h\|_{L^2(0,T;V)}^2$$

$$\leq \int_0^T \langle A(u_h - \bar{u}_h), u_h - \bar{u}_h \rangle dt$$

$$\leq \int_0^T \langle u_h' - f, \bar{u}_h - u_h \rangle dt - \int_0^T \langle A\bar{u}_h, u_h - \bar{u}_h \rangle dt$$

$$\to 0, \quad h \to 0+.$$

This completes the proof of Theorem 1.43. $\qquad\square$

Next we turn our attention to parabolic variational inequalities of the second kind with a symmetric bilinear form a.

Let $j : V \to \bar{R}$ be a convex, lower semicontinuous and proper functional. Recalling the formulation (1.119) we can extend the semidiscrete approximation (1.212), (1.213) to variational inequalities of the second kind in an obvious way. Let us sketch that a counterpart of Theorem 1.43 holds again. We assume that j is also Gâteaux differentiable in V (in a nondifferentiable case we have to approximate j by a sequence of convex Gâteaux differentiable functionals $\{j_h\}$ having appropriate properties, see, e.g., Duvaut and Lions, 1976). Then (1.119) is equal to

$$\begin{cases} \langle u'(t) + Au(t) + Dj(u(t)), v - u(t) \rangle \geq \langle f(t), v - u(t) \rangle \\ \forall v \in K \text{ and for a.a. } t \in (0, T), \end{cases} \tag{1.230}$$

where $Dj(u(t))$ is the Gâteaux derivative of j at $u(t)$. Below we show the necessary modifications of the proof of Theorem 1.43 to get the result.

As in step I we arrive at the inequalities

$$\langle u_h'(s) + Au_h(s) + Dj(u_h(s)), u_h(s) \rangle \leq \langle f(s), u_h(s) \rangle \tag{1.231}$$

and

$$\langle u_h'(s) + Au_h(s) + Dj(u_h(s)), u_h'(s) \rangle \leq \langle f(s), u_h'(s) \rangle \tag{1.232}$$

instead of (1.214) and (1.218). For the sake of simplicity we assume that $Dj(0) = 0$. First recall that the Gâteaux derivative of a convex functional is monotone. Thus, we have

$$\langle Dj(u_h(t)), u_h(t) \rangle = \langle Dj(u_h(t)) - Dj(0), u_h(t) - 0 \rangle \geq 0. \tag{1.233}$$

Secondly, the differentiability and the convexity of j imply

$$\int_0^t \langle Dj(u_h(s)), u_h'(s) \rangle ds = \int_0^t \frac{d}{ds} j(u_h(s)) ds \tag{1.234}$$
$$= j(u_h(t)) - j(u_h(0)) \geq -C(\|u_h(t)\| + 1).$$

Using (1.233) and (1.234) in (1.231) and (1.232), respectively, and repeating the analysis of step I we obtain the a priori estimates (1.215), (1.216), (1.219), (1.220) and the convergence results (1.221)–(1.224).

In step II–IV it is enough to note that

$$\liminf_{h \to 0+} \int_0^T j(u_h(t)) dt \geq \int_0^T \liminf_{h \to 0+} j(u_h(t)) dt \geq \int_0^T j(u(t)) dt$$

which is due to Fatou's lemma, the weak lower semicontinuity of j and (1.225). The rest of the proof is the same as before.

Fully discrete approximation. Let S be a finite dimensional subspace of V. Let Δ_k be an equidistant partition of $[0, T]$ into $m = m(k)$ subintervals of length $k = T/m$. We denote by λ^{i+1} the characteristic functions of the intervals

$]ik, (i + 1)k]$, $i = 0, ..., m - 1$. The symbol $L^2(\Delta_k; X)$ stands for all piecewise constant functions v on the partition Δ_k with values in X, i.e., $v \in L^2(\Delta_k; X)$ iff $v = \sum_{i=1}^{m} v^i \lambda^i$, $v^i \in X$. The time derivative $u'(t)$ is approximated by the following difference quotients:

$$u'(t) \approx \frac{u(t + k) - u(t)}{k} \quad \text{or} \quad u'(t) \approx \frac{u(t) - u(t - k)}{k}.$$

On the other hand, the right hand side f is approximated by a family of functions $\{f_k^i\}_{i=0}^{m}$, $f_k^i \in V^*$. Using these notations we introduce several *fully discrete approximations* of parabolic equations.

We start with the *implicit scheme*:

$$\begin{cases} \text{Find } u_{S,k} \equiv \sum_{i=1}^{m} u_{S,k}^i \lambda^i \in L^2(\Delta_k; S) \text{ such that} \\ \left\langle \dfrac{u_{S,k}^{i+1} - u_{S,k}^i}{k}, v \right\rangle + \left\langle A u_{S,k}^{i+1}, v \right\rangle = \left\langle f_k^{i+1}, v \right\rangle \\ \forall v \in S, \ i = 0, ..., m - 1 \\ \text{and } \ u_{S,k}^0 = u_{0S} \in S. \end{cases} \quad (1.235)$$

Other convenient approximations are: the *Crank-Nicolson scheme* defined by (1.235) with the following modification in $(1.235)_2$:

$$\left\langle \frac{u_{S,k}^{i+1} - u_{S,k}^i}{k}, v \right\rangle + \left\langle A \frac{u_{S,k}^{i+1} + u_{S,k}^i}{2}, v \right\rangle = \left\langle \frac{f_k^{i+1} + f_k^i}{2}, v \right\rangle \quad (1.236)$$

and the *explicit scheme*:

$$\left\langle \frac{u_{S,k}^{i+1} - u_{S,k}^i}{k}, v \right\rangle + \left\langle A u_{S,k}^i, v \right\rangle = \left\langle f_k^i, v \right\rangle. \quad (1.237)$$

Using the general θ-scheme we can rewrite $(1.235)_2$, (1.236), (1.237) in one expression:

$$\left\langle \frac{u_{S,k}^{i+1} - u_{S,k}^i}{k}, v \right\rangle + \left\langle A u_{S,k}^{i+\theta}, v \right\rangle = \left\langle f_k^{i+\theta}, v \right\rangle, \quad (1.238)$$

where $u_{S,k}^{i+\theta} = \theta u_{S,k}^{i+1} + (1 - \theta) u_{S,k}^i$ and $f_k^{i+\theta} = \theta f_k^{i+1} + (1 - \theta) f_k^i$, $\theta \in [0, 1]$. If $\theta = 1$, $\frac{1}{2}$ or 0, (1.238) defines the implicit, the Crank-Nicolson and the explicit scheme, respectively.

Recalling the definition of the stiffness matrix $\mathbf{A}$, the mass matrix $\mathbf{M}$ and the load vector $\vec{F}$ we can express the θ-scheme as follows:

$$\begin{cases} \text{Find } \vec{\alpha}^{(i+1)} \in \mathbf{R}^n \text{ such that} \\ \mathbf{M} \dfrac{\vec{\alpha}^{(i+1)} - \vec{\alpha}^{(i)}}{k} + \theta \mathbf{A} \vec{\alpha}^{(i+1)} + (1 - \theta) \mathbf{A} \vec{\alpha}^{(i)} = \theta \vec{F}^{(i+1)} + (1 - \theta) \vec{F}^{(i)} \end{cases} \quad (1.239)$$

for all $i = 0, ..., m - 1$ and $\vec{\alpha}^{(0)} = \vec{\alpha}_0$.

From (1.239) we see that at each time level $(i+1)k$, $i = 0, ..., m-1$, we have to solve the linear system of algebraic equations:

$$(\mathbf{M} + k\theta\mathbf{A})\vec{a}^{(i+1)} = k\vec{F}^{(i+\theta)} + (\mathbf{M} - k(1-\theta)\mathbf{A})\vec{a}^{(i)}, \qquad (1.240)$$

where $\vec{F}^{(i+\theta)} = \theta\vec{F}^{(i+1)} + (1-\theta)\vec{F}^{(i)}$.

Next we briefly mention the convergence results for the above fully discrete schemes.

Let $\{V_h\}$ be a family of finite dimensional subspaces of V. We denote by $u_{h,k}^{\theta} \equiv \sum_{i=0}^{m-1} u_{h,k}^{i+\theta}\lambda^{i+1} \in L^2(\Delta_k; V_h)$ with $u_{h,k}^{i+\theta}$ being a unique solution of the θ-scheme (with $S = V_h$ and $u_{0S} = u_{0h}$) and by u the unique solution of the continuous problem (1.109). Assume that $\{V_h\}, \{u_{0h}\}$ satisfy (1.156),(1.183), respectively. As a family $\{\{f_k^i\}_{i=0}^{m(k)}\}_k$ is concerned, we impose the following assumptions:

the sequence $\{f_k^{\theta}\}$, where $f_k^{\theta} \equiv \sum_{i=0}^{m-1} f_k^{i+\theta}\lambda^{i+1}$, $\theta \in [0,1]$, is uniformly bounded in $L^2(\Delta_k; V^*)$ with respect to k and θ, i.e.,   (1.241)
$$\exists C > 0 : \|f_k^{\theta}\|_{L^2(\Delta_k, V^*)} \leq C \quad \forall k, \theta;$$

the sequence $\{f_k^{\theta}\}$ possesses the convergence property:
$$v_h \to v \text{ in } L^2(0,T;V), \; v_h \in L^2(0,T;V_h), \text{ as } h \to 0+ \qquad (1.242)$$
$$\implies \int_0^T \langle f_k^{\theta}, v_h\rangle dt \to \int_0^T \langle f, v\rangle dt \text{ as } h, k \to 0+.$$

Further, we assume that the following *inverse inequality* holds between the V- and H-norms:

$$\|v\| \leq s(h)|v| \quad \forall v \in V_h, \qquad (1.243)$$

where $s(h)$ is a positive constant depending on h.

Then it is possible to show the following (Glowinski et al., 1981):

Theorem 1.44 *Let the assumptions stated above be satisfied. Then it holds:*

(i) If $\theta \in [\frac{1}{2}, 1]$ the θ-scheme is stable and convergent in the sense that the sequence $\{u_{h,k}^{\theta}\}$ tends to u weakly in $L^2(0,T;V)$ as $h, k \to 0+$;

(ii) If $\theta \in [0, \frac{1}{2})$ and (1.243) is satisfied then the θ-scheme is stable and convergent (in the above sense) under the condition $ks(h)^2 < C$, where $C > 0$ is large enough.

We omit the proof, since it is rather long and technical and, moreover, we get it as a byproduct when proving the corresponding result for fully discrete approximations of parabolic hemivariational inequalities (cf. Chapter 4).

Remark 1.22 *We shall see later that the constant C in Theorem 1.44 (ii) can be chosen as $\frac{\alpha}{2(1-\theta)M^2}$ (cf. (4.24)), where α, M are the constants from (1.105) and (1.106).*

Due to this theorem we say that the implicit and Crank-Nicolson schemes are *unconditionally stable* while the explicit scheme is *conditionally stable*.

Remark 1.23 *In order to get better convergence result for $\{u^\theta_{h,k}\}$ we need some additional assumptions. For example, if V is compactly imbedded in H, A is symmetric, $\{u_{0h}\}$ is a bounded sequence in V, and $f \in L^2(0, T; H)$ (in (1.241) and (1.242) V^* is replaced by H) we get that the sequence $\{u^\theta_{h,k}\}$ tends to u strongly in $L^2(0, T; V)$ and in $C([0, T]; H)$. It is possible to establish error estimates for $(u^\theta_{h,k} - u)$ if $u^\theta_{h,k}$ and u possess some further regularity. We refer to Thomée, 1984 and Neittaanmäki and Tiba, 1994 and references therein to see that type of results.*

These fully discrete schemes have obvious generalizations for parabolic variational inequalities (Glowinski et al., 1981, Trémoliéres, 1972). Let $Q \subset V$ be a nonempty, closed and convex approximation of the constraint set K. Then the θ-scheme for the parabolic variational inequality of the first kind reads as follows:

$$\begin{cases} \text{Find } u^\theta_{Q,k} \in L^2(\Delta_k; Q) \text{ such that} \\ \left\langle \dfrac{u^{i+1}_{Q,k} - u^i_{Q,k}}{k}, v - u^{i+\theta}_{Q,k} \right\rangle + \left\langle Au^{i+\theta}_{Q,k}, v - u^{i+\theta}_{Q,k} \right\rangle \geq \left\langle f^{i+\theta}_k, v - u^{i+\theta}_{Q,k} \right\rangle \quad (1.244) \\ \forall v \in Q, \ i = 0, ..., m-1 \quad \text{and} \quad u^0_{Q,k} = u_{Q0} \in Q. \end{cases}$$

Using the same notations as before problem (1.244) reduces to the following system of algebraic inequalities at each time level $(i+1)k$, $i = 0, ..., m-1$:

$$\begin{cases} \text{Find } \vec{\alpha}^{(i+1)} \in \mathbf{R}^n \text{ such that} \\ ((\mathbf{M} + k\theta\mathbf{A})\vec{\alpha}^{(i+1)}, \vec{\alpha} - \vec{\alpha}^{(i+1)}) \geq \\ (k\vec{F}^{(i+\theta)} + (\mathbf{M} - k(1-\theta)\mathbf{A})\vec{\alpha}^{(i)}, \vec{\alpha} - \vec{\alpha}^{(i+1)}) \quad \forall \vec{\alpha} \in \mathcal{U}. \end{cases} \quad (1.245)$$

Again, $\vec{F}^{(i+\theta)} = \theta\vec{F}^{(i+1)} + (1-\theta)\vec{F}^{(i)}$. We refer to Glowinski et al., 1981 and Trémoliéres, 1972 to see basic convergence results for the above approximations.

Finally, we mention that also for the parabolic variational inequality of the second kind we can define the θ-scheme with obvious modifications (see again Glowinski et al., 1981 and Trémoliéres, 1972).

The finite element approximation of equations and inequalities of monotone type

The aim of this section is to specify the construction of finite dimensional function spaces used in the abstract approximation theory. We shall describe very briefly one of the most popular methods, widely used in practice, namely the *finite element method*. We restrict our presentation to the simplest case, namely to linear elements in two and three dimensions which will be used in subsequent chapters. For more details we refer to Ciarlet, 1978. We start with 2-D case.

Let $\Omega \subset \mathbf{R}^2$ be a *polygonal* domain. We divide Ω into a finite number of non-degenerate, closed triangles T_i, $i \in \mathcal{I}$ in such a way that

$$\overline{\Omega} = \cup_{i \in \mathcal{I}} T_i \tag{1.246}$$

$$\overset{\circ}{T}_i \cap \overset{\circ}{T}_j = \emptyset \text{ for } i \neq j \text{ (i.e., the interiors of different} \tag{1.247}$$
triangles are disjoint)

$$\text{if } T_i \cap T_j \equiv M_{ij} \neq \emptyset \ (i \neq j) \text{ then } M_{ij} \text{ is} \tag{1.248}$$
either a vertex or the whole side of T_i, T_j.

Any partition of Ω, satisfying (1.246)–(1.248) will be called the *triangulation* of $\overline{\Omega}$ and will be denoted by $\mathcal{T}_h$ in what follows. The subscript h stands for the *norm* of the triangulation defined as follows:

$$h = \max_T h_T,$$

where $h_T \equiv \text{diam} T$ and the maximum is taken over all T realizing the triangulation of $\overline{\Omega}$. There is another parameter, also characterizing $\mathcal{T}_h$, namely

$$\rho = \min_{T \in \mathcal{T}_h} \rho_T,$$

where $\rho_T \equiv$ maximal radius of a ball inscribed in T.

Definition 1.8 *Let* $\{\mathcal{T}_h\}$, $h \to 0+$ *be a family of triangulations of* $\overline{\Omega}$. *We say that* $\{\mathcal{T}_h\}$ *is regular if there exists a constant* $\alpha > 0$ *such that*

$$\frac{h_T}{\rho_T} \leq \alpha$$

holds for any triangle $T \in \mathcal{T}_h$ *and any* $h > 0$. *If* $h/\rho \leq \alpha$ *for any* $h > 0$ *then* $\{\mathcal{T}_h\}$ *is said to be strongly regular.*

In what follows we shall deal only with regular, eventually strongly regular families of triangulations. Yet another condition concerning of $\{\mathcal{T}_h\}$ will be usually required. In real life problems different boundary conditions are prescribed along the boundary $\partial\Omega$. Thus $\partial\Omega$ can be decomposed into several parts Γ_j, such that crossing from one part to another, boundary conditions change. If such situation occurs then the family $\{\mathcal{T}_h\}$ has to be *consistent* with such a decomposition of $\partial\Omega$, i.e., any Γ_j is the union of the *whole* sides of triangles $T \in \mathcal{T}_h$ having a nonempty intersection with Γ_j whose one-dimensional Lebesgue measure is positive,

A similar partition can be also realized in 3-D case. Assume that $\Omega \subset \mathbf{R}^3$ is a *polyhedron*. Then the corresponding decomposition of $\overline{\Omega}$ (which will be also called the triangulation) is made of a finite number of polyhedra. Besides of (1.246), (1.247) the following modification of (1.248) has to be satisfied:

$$\text{if } T_i \cap T_j \equiv M_{ij} \neq \emptyset \ (i \neq j) \text{ then } M_{ij} \text{ is either} \tag{1.249}$$
a vertex or the whole side or the whole face of T_i, T_j.

Also the symbols h_T, ρ_T, h and ρ have the same meaning as in 2-D case. The notion of a regular, strongly regular and consistent (with respect to the partition of $\partial\Omega$ generated by different boundary conditions) family of triangulations can be introduced as before with appropriate modifications.

Any $T \in \mathcal{T}_h$ is called an *element*. Let $T \in \mathcal{T}_h$ be given and denote by $P_1(T)$ the set of all linear functions whose domain of definition is T, i.e.

$$p \in P_1(T) \quad \text{iff} \quad p(x,y) = \alpha_0 + \alpha_1 x + \alpha_2 y, \ (x,y) \in T, \text{ in 2D}$$

or

$$p(x,y,z) = \alpha_0 + \alpha_1 x + \alpha_2 y + \alpha_3 z, \ (x,y,z) \in T, \text{ in 3D},$$

$\alpha_i \in \mathbf{R}$, $i = 1,2,(3)$. Any $p \in P_1(T)$ is uniquely determined by its values at the vertices of T. If v is a continuous function in T, i.e., $v \in C(T)$ then *its linear lagrange interpolate on* T, denoted by $\Pi_T v$ in what follows, is a unique function from $P_1(T)$ satisfying:

$$\Pi_T v(a) = v(a) \quad \forall a \in \mathcal{N}_T,$$

where $\mathcal{N}_T$ is the set of all vertices of T. The values of $p \in P_1(T)$ at the points of $\mathcal{N}_T$ are called *the degrees of freedom*. A natural question arises, namely how to estimate the error between v and $\Pi_T v$. The following important approximation result holds:

Lemma 1.4 *Let $T \in \mathbf{R}^n$, $n = 2,3$ be a triangle or a polyhedra. Then there exists a constant $\hat{c} > 0$ such that*

$$|v - \Pi_T v|_{m,p,T} \le \hat{c}\frac{h_T}{\rho_T^m}|v|_{1,p,T} \tag{1.250}$$

holds for any $v \in W^{1,p}(T)$ with $p \in (n,\infty]$ and $m = 0,1$ or

$$|v - \Pi_T v|_{m,p,T} \le \hat{c}\frac{h_T^2}{\rho_T^m}|v|_{2,p,T} \tag{1.251}$$

holds for any $v \in W^{2,p}(T)$ with $p \in (\frac{n}{2},\infty]$ and $m = 0,1,2$.

Let $\Omega \subset \mathbf{R}^n$, $n = 2,3$ be a polygonal (polyhedral) domain and let $\{\mathcal{T}_h\}$, $h \to 0+$ be a regular family of triangulations of $\overline{\Omega}$. With any $\mathcal{T}_h$ the following finite dimensional space X_h will be associated:

$$X_h = \{v_h \in C(\overline{\Omega}) \mid v_h|_T \in P_1(T) \ \forall T \in \mathcal{T}_h\}. \tag{1.252}$$

The continuity of v_h between adjacent elements $T,T' \in \mathcal{T}_h$ is ensured by prescribing the same function values at the vertices belonging to $T \cap T'$. It is readily seen that X_h is a subspace of $W^{1,p}(\Omega)$ for any $p \in [1,\infty]$.

Let $v \in C(\overline{\Omega})$. Then one can define its *piecewise linear lagrange interpolate* $r_h v$ as a function from X_h given by

$$(r_h v)|_T = \Pi_T(v|_T) \quad \forall T \in \mathcal{T}_h.$$

On the basis of Lemma 1.4 and using the assumption on the regularity of $\{\mathcal{T}_h\}$ we arrive at the following fundamental result:

Theorem 1.45 *Let $\Omega \subset \mathbf{R}^n$, $n = 2, 3$, be a polygonal (polyhedral) domain. Then there exists a constant $\hat{c} > 0$ such that for any regular family $\{\mathcal{T}_h\}$, $h \to 0+$ of triangulations of $\overline{\Omega}$ one has:*

$$|v - r_h v|_{m,p,\Omega} \leq \hat{c} h^{1-m} |v|_{1,p,\Omega} \qquad (1.253)$$

holds for any $v \in W^{1,p}(\Omega)$ with $p \in (n, \infty]$ and $m = 0, 1$, or

$$|v - r_h v|_{m,p,\Omega} \leq \hat{c} h^{2-m} |v|_{2,p,\Omega} \qquad (1.254)$$

holds for any $v \in W^{2,p}(\Omega)$ with $p \in (\frac{n}{2}, \infty]$ and $m = 0, 1, 2$.

Remark 1.24 *In most applications appearing in the subsequent parts we use (1.254) with $p = 2$:*

$$|v - r_h v|_{m,\Omega} \leq \hat{c} h^{2-m} |v|_{2,\Omega} \qquad (1.255)$$

holds for any $v \in H^2(\Omega)$, $m = 0, 1, 2$.

Let $\Gamma \subset \partial\Omega$ be a nonempty, open part in $\partial\Omega$ and define the space

$$V = \{v \in H^1(\Omega) \mid v = 0 \text{ on } \Gamma\}. \qquad (1.256)$$

Let $\{\mathcal{T}_h\}$, $h \to 0+$ be a regular family of triangulations of polygonal (polyhedral) domain $\overline{\Omega}$ which is *consistent* with the decomposition of $\partial\Omega$ into Γ and $\partial\Omega \setminus \overline{\Gamma}$. With any such $\mathcal{T}_h$ we associate a finite dimensional space $V_h \subset V$ defined as follows:

$$V_h = \{v_h \in C(\overline{\Omega}) \mid v_h|_T \in P_1(T) \quad \forall T \in \mathcal{T}_h, \, v_h = 0 \text{ on } \Gamma\}. \qquad (1.257)$$

The condition $v_h = 0$ on Γ will be realized by setting $v_h(A)$ equal to zero at any vertex A of $\mathcal{T}_h$ lying on $\overline{\Gamma}$. From the definition of r_h it is readily seen that if $v \in V \cap C(\overline{\Omega})$ then $r_h v \in V_h$.

Another result which will be used in subsequent chapters is the so-called *inverse inequality* valid for elements of X_h:

Lemma 1.5 *(Inverse inequality) Let $\{\mathcal{T}_h\}$, $h \to 0+$ be a strongly regular family of triangulations of a polygonal (polyhedral) domain $\overline{\Omega}$. Then there exists a constant $\hat{c} > 0$ such that:*

$$\|v_h\|_{1,\Omega} \leq \hat{c} h^{-1} \|v_h\|_{0,\Omega} \qquad (1.258)$$

holds for any $v_h \in X_h$ and any $h > 0$.

Now we shall apply the abstract approximation results presented in the previous sections to the approximation of equations and inequalities by using finite element spaces introduced above.

Let $\{V, a, f\}$ be a *linear elliptic equation* of the second order, where V is given by (1.256), $a : V \times V \to \mathbf{R}$ be a bounded, V-elliptic bilinear form defined by (1.55) and $f \in V^*$. Let $\{\mathcal{T}_h\}$, $h \to 0+$, be a regular family of triangulations of a polygonal (polyhedral) domain $\overline{\Omega}$, consistent with the decomposition of $\partial\Omega$ into Γ and $\partial\Omega \setminus \overline{\Gamma}$ and define the space V_h by (1.257). Finally, let u, u_h denote the solutions of $\{V, a, f\}$, $\{V_h, a, f\}$, respectively. Then one has

Theorem 1.46 *It holds:*

$$- \text{if } u \in V \cap H^2(\Omega) \text{ then} \tag{1.259}$$

$$\|u - u_h\|_{1,\Omega} \le ch|u|_{2,\Omega},$$

where c is a positive constant independent of h;

$$- \text{if } V \cap H^2(\Omega) \text{ is dense in } V \text{ then} \tag{1.260}$$

$$\|u - u_h\|_{1,\Omega} \to 0 \quad \text{as } h \to 0+,$$

when no regularity of u is available.

Proof: In order to prove (1.259) we use (1.155) and (1.255):

$$\|u - u_h\|_{1,\Omega} \le \frac{M}{\alpha}\|u - r_h u\|_{1,\Omega} \le ch|u|_{2,\Omega}, \tag{1.261}$$

provided that $u \in H^2(\Omega)$. If there are no information on the smoothness of u, we use Theorem 1.38. We have to show that any function $v \in V$ can be approximated by elements of V_h in the sense of (1.156).

Let $v \in V$ be given. Owing to the density assumption (1.260), for any $\varepsilon > 0$ there exists a function $\bar{v} \in H^2(\Omega) \cap V$ such that

$$\|v - \bar{v}\|_{1,\Omega} \le \frac{\varepsilon}{2}. \tag{1.262}$$

Since $\bar{v}$ is already continuous, as follows from the imbedding of $H^2(\Omega)$ in $C(\overline{\Omega})$ one can construct its piecewise linear lagrange interpolate $r_h\bar{v} \in V_h$ and to use (1.255) again:

$$\|\bar{v} - r_h\bar{v}\|_{1,\Omega} \le ch|\bar{v}|_{2,\Omega} \le \frac{\varepsilon}{2}$$

for $h > 0$ sufficiently small. From this, (1.262) and the triangle inequality we see that

$$\|v - r_h\bar{v}\|_{1,\Omega} \le \|v - \bar{v}\|_{1,\Omega} + \|\bar{v} - r_h\bar{v}\|_{1,\Omega} \le \varepsilon$$

provided $h > 0$ is sufficiently small. $\square$

Remark 1.25 *In real life problems the density assumptions (1.260) is always satisfied. One has even more, namely $C^\infty(\overline{\Omega}) \cap V$ is dense in V (for the proof see Doktor, 1973).*

One of typical features of the finite element method is that the stiffness matrix $\mathbf{A}$ is sparse since supports of basis functions of V_h are small compared with the size of Ω. For more computational aspects we refer to (Axelsson and Barker, 1984, Křižek and Neittaanmäki, 1990).

Now let us pass to the approximation of *elliptic inequalities*. We start with the free boundary value problem $\{K, a, f\}$ defined as follows:

$$K = \{v \in H_0^1(\Omega) \mid v \geq \varphi \text{ a.e. in } \Omega\}, \tag{1.263}$$

where $\varphi \in C(\overline{\Omega})$ is given and such that $\varphi \leq 0$ on $\partial\Omega$, $a : H_0^1(\Omega) \times H_0^1(\Omega) \to \mathbf{R}$ is given by (1.55)–(1.57) and $f \in L^2(\Omega)$. Let $\{T_h\}$, $h \to 0+$ be a regular family of triangulations of a polygonal domain $\Omega \subset \mathbf{R}^2$ and define the space V_h by (1.257) with $\Gamma \equiv \partial\Omega$. Finally, let

$$K_h = \{v_h \in V_h \mid v_h(A) \geq \varphi(A) \quad \forall A \in \mathcal{N}_h\}, \tag{1.264}$$

where $\mathcal{N}_h$ stands for the set of all *interior nodes* of T_h. Let us notice that K_h is the external approximation of K, in general, since the inequality constraints are prescribed only at the nodes of $\mathcal{N}_h$.

Let u_h be a unique solution of $\{K_h, a, f\}$:

$$u_h \in K_h : a(u_h, v_h - u_h) \geq (f, v_h - u_h)_{0,\Omega} \quad \forall v_h \in K_h. \tag{1.265}$$

Moreover let the bilinear form a be symmetric. Then (1.265) is equivalent to

$$u_h \in K_h : J(u_h) = \min_{v_h \in K_h} J(v_h) \tag{1.266}$$

with J given by (1.50).

Denote by T_S the isomorphism between V_h and $\mathbf{R}^{n(h)}$ ($\dim V_h = n(h)$) and by T_S^{-1} its inverse. Then the set $\mathcal{U} \equiv T_S^{-1}(K_h)$ is a closed convex subset of $\mathbf{R}^{n(h)}$ given by

$$\mathcal{U} = \{\vec{\alpha} \in \mathbf{R}^{n(h)} \mid \alpha_i \geq \varphi(A_i) \quad \forall A_i \in \mathcal{N}_h, \ i = 1, ..., n(h)\}$$

and (1.266) leads to the *quadratic programming problem*:

$$\vec{\alpha}^* \in \mathcal{U} : J(\vec{\alpha}^*) = \min_{\vec{\alpha} \in \mathcal{U}} J(\vec{\alpha}), \tag{1.267}$$

where

$$J(\vec{\alpha}) = \frac{1}{2}(\vec{\alpha}, \mathbf{A}\vec{\alpha}) - (\vec{F}, \vec{\alpha}).$$

Our goal will be to estimate $\|u - u_h\|_{1,\Omega}$. We shall see that the analysis is considerably more involved than in the case of equations.

Theorem 1.47 *Suppose that the solution $u \in K \cap H^2(\Omega)$, the obstacle $\varphi \in H^2(\Omega)$ and the coefficients a_{ij} of the bilinear form a belong to $W^{1,\infty}(\Omega)$. Then*

$$\|u - u_h\|_{1,\Omega} \leq ch, \quad h \to 0+, \tag{1.268}$$

where c is a positive constant depending only on u, f and φ.

Proof: Since K_h is the external approximation of K, the error $\|u - u_h\|_{1,\Omega}$ will be estimated by means of (1.170). The Green's formula yields:

$$a(u, v - u_h) = (Au, v - u_h)_{0,\Omega} \quad \forall v \in K,$$
$$a(u, v_h - u) = (Au, v_h - u)_{0,\Omega} \quad \forall v_h \in K_h,$$

where $Au \equiv -\sum_{i,j} \frac{\partial}{\partial x_i}(a_{ij} \frac{\partial u}{\partial x_j}) + a_0 u$, making use of the regularity assumptions on u, a_{ij} and the fact that K, K_h are subsets of $H_0^1(\Omega)$. From this we see that (1.170) takes the following form:

$$\alpha \|u - u_h\|_{1,\Omega}^2 \leq (Au - f, v - u_h)_{0,\Omega} + (Au - f, v_h - u)_{0,\Omega} \quad (1.269)$$
$$+ a(u - u_h, u - v_h) \quad \forall v \in K, \forall v_h \in K_h.$$

As v_h we take a function $v_h = r_h u$, i.e., v_h is the piecewise lagrange linear interpolate of u. Then the second and the third term on the right hand side of (1.269) can be estimated as follows:

$$
\begin{aligned}
|(Au - f, r_h u - u)_{0,\Omega}| &\leq \|Au - f\|_{0,\Omega} \|r_h u - u\|_{0,\Omega} \quad (1.270) \\
&\leq ch^2 |u|_{2,\Omega}
\end{aligned}
$$

$$
\begin{aligned}
|a(u - u_h, u - r_h u)| &\leq M\varepsilon \|u - u_h\|_{1,\Omega}^2 + \frac{M}{4\varepsilon} \|u - r_h u\|_{1,\Omega}^2 \quad (1.271) \\
&\leq M\varepsilon \|u - u_h\|_{1,\Omega}^2 + \frac{cM}{4\varepsilon} h^2 |u|_{2,\Omega}^2,
\end{aligned}
$$

where $\varepsilon > 0$ is given (here we used the inequality (1.185)).

The most difficult is to estimate the first term $(Au - f, v - u_h)_{0,\Omega}$. We define

$$v = \sup\{\varphi, u_h\}.$$

Then $v \in H^1(\Omega)$ and since $\varphi \leq 0$ on $\partial\Omega$, v is equal to zero on $\partial\Omega$. Moreover $v \geq \varphi$ a.e. in Ω, so that $v \in K$. Denote by

$$\Omega_- = \{x \in \Omega \mid u_h(x) \leq \varphi(x)\}$$
$$\Omega_+ = \{x \in \Omega \mid u_h(x) > \varphi(x)\}.$$

We have

$$(Au - f, v - u_h)_{0,\Omega} = (Au - f, \varphi - u_h)_{0,\Omega_-}. \quad (1.272)$$

From the definition of K_h it follows that

$$u_h(A_i) \geq \varphi(A_i) \quad \forall A_i \in \mathcal{N}_h, \ i = 1, ..., n(h)$$

and consequently

$$u_h \geq r_h \varphi \quad \text{in } \Omega, \quad (1.273)$$

since both u_h and $r_h\varphi$ are piecewise linear on $\mathcal{T}_h$. From $Au \geq f$ in Ω (see (1.71)) and (1.272),(1.273) it follows that

$$(Au - f, \varphi - u_h)_{0,\Omega_-} \leq (Au - f, \varphi - r_h\varphi)_{0,\Omega_-}$$
$$\leq \|Au - f\|_{0,\Omega_-}\|\varphi - r_h\varphi\|_{0,\Omega_-} \leq c\|\varphi - r_h\varphi\|_{0,\Omega}$$
$$\leq ch^2|\varphi|_{2,\Omega}. \tag{1.274}$$

From this, (1.269),(1.270),(1.271) we arrived at the assertion of the theorem choosing ε in (1.271) sufficiently small.   □

Now we prove the convergence of approximate solutions u_h to u without any additional regularity assumptions concerning of u. For the sake of simplicity of our presentation we suppose that $\varphi \in H^2(\Omega) \cap H_0^1(\Omega)$. If it is so, the sets K, K_h can be written in the form:

$$K = \varphi + K^+,$$
$$K_h = r_h\varphi + K_h^+,$$

where

$$K^+ = \{v \in H_0^1(\Omega) \mid v \geq 0 \text{ a.e. in } \Omega\},$$
$$K_h^+ = \{v_h \in V_h \mid v_h(A_i) \geq 0, \ A_i \in \mathcal{N}_h, \ i = 1, ..., n(h)\}.$$

To prove that u_h tends to u we have to verify (1.172) and (1.173). We start with the former.

Let $v \in K$ be given. Then v can be split:

$$v = \varphi + w, \quad w \in K^+.$$

Now define

$$v_h \equiv r_h\varphi + w_h,$$

where the sequence $\{w_h\}$, $w_h \in K_h^+$ is such that

$$w_h \to w \quad \text{in } H^1(\Omega). \tag{1.275}$$

Such a sequence can be easily constructed on the basis of the following density result:

$$\overline{\{w \in C_0^\infty(\Omega) \mid w \geq 0 \text{ in } \Omega\}} = K^+$$

(see Glowinski, 1984). Indeed: first the function w is approximated by $\bar{w} \in C_0^\infty(\Omega)$, $\bar{w} \geq 0$ in Ω and then $\bar{w}$ by $r_h\bar{w}$.

Having (1.275) at our disposal, the triangle inequality yields

$$\|v - v_h\|_{1,\Omega} \leq \|\varphi - r_h\varphi\|_{1,\Omega} + \|w - w_h\|_{1,\Omega} \to 0, \quad h \to 0+$$

as follows from (1.255) and (1.275).

It remains to verify (1.173). Let $\{v_h\}$, $v_h \in K_h$ be a sequence such that

$$v_h \rightharpoonup v \quad \text{in } H^1(\Omega). \tag{1.276}$$

To prove that $v \in K$ it is sufficient to show that $v \geq \varphi$ a.e. in Ω. Let us write:

$$v_h = r_h \varphi + w_h, \quad w_h \in K_h^+. \tag{1.277}$$

Since at the same time

$$r_h \varphi \to \varphi, \quad h \to 0+ \text{ in } H^1(\Omega),$$

also the sequence $\{w_h\}$ is weakly convergent in $H^1(\Omega)$ to an element w:

$$w_h \rightharpoonup w, \quad h \to 0+ \text{ in } H^1(\Omega).$$

From (1.277) we see that $w = v - \varphi$ in Ω. On the other hand, K_h^+ is the internal approximation of K which is weakly closed. Hence $w \geq 0$ a.e. in Ω.

We can formulate the following result:

Theorem 1.48 *Let the obstacle $\varphi \in H^2(\Omega) \cap H_0^1(\Omega)$. Then for any regular system $\{T_h\}$ of triangulations of the polygonal domain $\Omega \subset \mathbf{R}^2$ it holds that*

$$\|u - u_h\|_{1,\Omega} \to 0, \quad h \to 0+,$$

where u, u_h are solutions of $\{K, a, f\}$, $\{K_h, a, f\}$, respectively.

Now we sketch the approximation of the frictionless Signorini problem in 2D, the variational formulation of which has been presented in Section 1.3. Let us recall that the convex set $K \subset H^1(\Omega; \mathbf{R}^2)$ is defined as follows:

$$K = \{v \in H^1(\Omega; \mathbf{R}^2) \mid v = 0 \text{ on } \Gamma_u, \ v.\nu \leq 0 \text{ on } \Gamma_c\},$$

where Γ_c is a part of $\partial\Omega$ along which the body Ω is unilaterally supported by a rigid foundation.

Let Ω be a polygonal domain and let $\{T_h\}$, $h \to 0+$ be a regular family of triangulations of $\overline{\Omega}$, which is consistent with the decomposition of $\partial\Omega$ into Γ_u, Γ_P and Γ_c. We define:

$$\mathbf{V}_h = \{v_h \in C(\overline{\Omega}; \mathbf{R}^2) \mid v_h|_T \in (P_1(T))^2 \quad \forall T \in T_h, \ v_h = 0 \text{ on } \Gamma_u\}$$

and its closed convex subset

$$\mathbf{K}_h = \{v_h \in \mathbf{V}_h \mid v_h.\nu \leq 0 \text{ on } \Gamma_c\}.$$

Now, we describe how to realize the unilateral condition in $\mathbf{K}_h$.

Let A be a contact node, i.e., a node of T_h such that $A \in \overline{\Gamma}_c \setminus \overline{\Gamma}_u$. If A is such that the outward unit normal vector is well defined, we set

$$(v_h.\nu)(A) \leq 0.$$

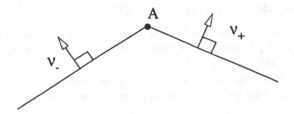

Figure 1.7.

If A is a vertex of the polygon $\partial\Omega$ in which ν is not well defined, the following two conditions will be prescribed:

$$(v_h.\nu_-)(A) \leq 0$$
$$(v_h.\nu_+)(A) \leq 0$$

where the meaning of ν_-, ν_+ is seen from Fig.1.7.

Let u_h be the solution to $\{\mathbf{K}_h, a, f\}$:

$$u_h \in \mathbf{K}_h : J(u_h) = \min_{v_h \in \mathbf{K}_h} J(v_h), \tag{1.278}$$

where J is the total potential energy functional given by (1.82).

As before, denote by $\mathcal{T}_S$ the isomorphism between $\mathbf{V}_h$ and $\mathbf{R}^{n(h)}$ ($\dim \mathbf{V}_h = n(h)$) and by $\mathcal{T}_S^{-1}$ its inverse. Then $\mathcal{U} \equiv \mathcal{T}_S^{-1}(\mathbf{K}_h)$ is given by

$$\mathcal{U} = \{\vec{\alpha} \in \mathbf{R}^{n(h)} \mid \mathbf{B}\vec{\alpha} \leq 0\},$$

where $\mathbf{B}$ is a rectangular matrix $m(h) \times n(h)$ with $m(h)$ being the number of the contact conditions. Any row of $\mathbf{B}$ contains at most two nonzero elements, namely the coordinates of ν.

The algebraic representation of (1.278) leads to the following *quadratic programming problem*:

$$\vec{\alpha}^* \in \mathcal{U} : \mathcal{J}(\vec{\alpha}^*) = \min_{\vec{\alpha} \in \mathcal{U}} \mathcal{J}(\vec{\alpha}), \tag{1.279}$$

where

$$\mathcal{J}(\vec{\alpha}) = \frac{1}{2}(\vec{\alpha}, \mathbf{A}\vec{\alpha}) - (\vec{F}, \vec{\alpha})$$

is the algebraic representation of J.

The rate of convergence of u_h to u depends on regularity assumptions imposed on the solution u. One of possible results is formulated in

Theorem 1.49 *Let the solution $u \in H^2(\Omega; \mathbf{R}^2) \cap K$ and let the normal component of the contact stress vector belong to $L^2(\Gamma_c)$. Then*

$$\|u - u_h\|_{1,\Omega} = \mathcal{O}(h^{\frac{3}{4}}), \quad h \to 0+.$$

For the proof we refer to (Haslinger et al., 1996).

This result can be improved by imposing additional assumptions on the behaviour of u on Γ_c in order to increase the rate of convergence to $\mathcal{O}(h)$ (see Hlaváček et al., 1988).

Let us consider the case when no regularity assumptions are at our disposal. Only, what we can prove in this case is the convergence itself:

$$\|u - u_h\|_{1,\Omega} \to 0, \quad h \to 0 + .$$

Since $\mathbf{K}_h$ is the internal approximation of $\mathbf{K}$ we have to verify (1.172), only. To this end we use the following density result:

Lemma 1.6 *Let* $\overline{\Gamma}_u \cap \overline{\Gamma}_c = \emptyset$ *and let there exist a finite number of points of* $\overline{\Gamma}_P \cap \overline{\Gamma}_c$, $\overline{\Gamma}_u \cap \overline{\Gamma}_P$. *Then the set*

$$\mathbf{K} \cap C(\overline{\Omega}; \mathbf{R}^2)$$

is dense in $\mathbf{K}$ *with respect to the* $H^1(\Omega; \mathbf{R}^2)$*-norm.*

This result can be obtained by adapting the proof from Hlaváček et al., 1988.

On the basis of this lemma we immediately obtain

Theorem 1.50 *Let all the assumptions of Lemma 1.6 be satisfied. Then for any regular family* $\{\mathcal{T}_h\}$ *of triangulations of* $\overline{\Omega}$ *we have*

$$\|u - u_h\|_{1,\Omega} \to 0, \quad h \to 0+,$$

where u, u_h *is the solution to* $\{\mathbf{K}, a, f\}$, $\{\mathbf{K}_h, a, f\}$, *respectively.*

Next we shall consider a slightly more complicated unilateral condition. Suppose that the body, represented by a polygonal domain Ω is unilaterally supported by a rigid foundation R with a curved boundary ∂R and touching Γ_c at a point C. Moreover let Γ_c be given by a straight line segment (see Fig.1.8). We introduce the local cartesian coordinate system (η, ξ) such that the η-axis coincides with Γ_c. Then the non-penetration condition can be expressed in the form

$$u.\xi \leq s \quad \text{in } [a, b],$$

where $[a, b]$ is a part of Γ_c (a zone of possible contact), containing C, s is a function whose graph is a part of ∂R over $[a, b]$ and ξ is the outward unit normal vector at C with respect to $\partial \Omega$.

The convex set $\mathbf{K} \subset H^1(\Omega; \mathbf{R}^2)$ is now defined by

$$\mathbf{K} = \{v \in H^1(\Omega; \mathbf{R}^2) \mid v = 0 \text{ on } \Gamma_u, \ v.\xi \leq s \text{ on } [a, b]\}. \tag{1.280}$$

Let $\{\mathcal{T}_h\}$, $h \to 0+$ be a regular family of triangulations of $\overline{\Omega}$, consistent with the decomposition of $\partial \Omega$ into Γ_u, Γ_P and Γ_c. Then the approximation of $\mathbf{K}$ is defined as follows:

$$\mathbf{K}_h = \{v_h \in \mathbf{V}_h \mid (v_h.\xi)(A) \leq s(A) \text{ for any} \tag{1.281}$$
$$\text{contact node } A \in \overline{\Gamma}_c \setminus \overline{\Gamma}_u\}.$$

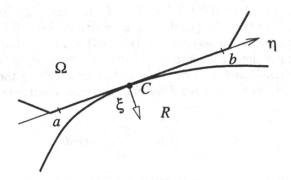

Figure 1.8.

Generally, $\mathbf{K}_h$ is an *external* approximation of $\mathbf{K}$ since the unilateral condition is prescribed only at the contact nodes. Denote by $\mathcal{U} \equiv \mathcal{T}_S^{-1}(\mathbf{K}_h)$, where $\mathcal{T}_S^{-1}$ has the same meaning as before. Then

$$\mathcal{U} = \{\vec{\alpha} \in \mathbf{R}^{n(h)} \mid \mathbf{B}\vec{\alpha} \le \vec{s}\},$$

with the same matrix $\mathbf{B}$ as before and $\vec{s} \in \mathbf{R}^{m(h)}$ being the vector whose components are equal to the values of s at the contact nodes A.

Let u, u_h be solutions to $\{K, a, f\}$, $\{K_h, a, f\}$ with $\mathbf{K}$, $\mathbf{K}_h$ defined by (1.280), (1.281), respectively.

To prove that u_h tends to u one has to use:

Lemma 1.7 *Let $s : [a, b] \to \mathbf{R}$ be a Lipschitz function in $[a, b]$ and let $\mathbf{K}$, $\mathbf{K}_h$ be defined by (1.280), (1.281), respectively. Then (1.172) and (1.173) hold true.*

For the proof see Haslinger and Neittaanmäki, 1996.

As a consequence of this lemma we have the following convergence result:

Theorem 1.51 *For any regular family $\{\mathcal{T}_h\}$ of triangulations of $\overline{\Omega}$ it holds that*

$$\|u - u_h\|_{1,\Omega} \to 0, \quad h \to 0+,$$

where u, u_h are solutions to $\{K, a, f\}, \{K_h, a, f\}$, respectively.

We end up this part by a finite element approximation of the parabolic equation (1.128):

$$\begin{cases} u'(t) + Au(t) = h(t) & \text{in } \Omega_T = \Omega \times (0, T), \\ u(t) = 0 & \text{on } \Gamma \times (0, T), \\ \frac{\partial u}{\partial \nu_A}(t) = g(t) & \text{on } \partial\Omega_T \setminus \overline{\Gamma} \times (0, T), \\ u(0) = u_0 & \text{in } \Omega, \end{cases}$$

in which A is defined by (1.54) satisfying (1.56) and (1.57), $\Omega \subset \mathbf{R}^n$ ($n \leq 3$), Γ is a nonempty open subset of $\partial\Omega$, $h \in L^2(0,T;L^2(\Omega))$ and $g \in L^2(0,T;L^2(\partial\Omega \setminus \overline{\Gamma}))$. The space V is given by (1.256) and $H = L^2(\Omega)$. Let $\{\mathcal{T}_h\}$ be a regular family of triangulations of a polygonal (polyhedral) domain $\overline{\Omega}$ that is consistent with the decomposition of $\partial\Omega$ into Γ and $\partial\Omega \setminus \overline{\Gamma}$. Further, let the space V_h be defined by (1.257) and k be the time step, characterizing the equidistant partition of $[0,T]$. The functional f defining the right hand side in (1.109) is of the form

$$\langle f(t), v \rangle = \int_\Omega h(t)v dx + \int_{\partial\Omega \setminus \overline{\Gamma}} g(t)v ds.$$

Clearly, $f \in L^2(0,T;V^*)$. Its approximation $\{f_k^i\}_{i=0}^m$ can be defined, for example, as follows:

$$f_k^i = \frac{1}{k} \int_{k(i-\frac{1}{2})}^{k(i+\frac{1}{2})} f(t)dt, \quad i = 1, ..., m(k) - 1, \tag{1.282}$$

$$f_k^0 = \frac{2}{k} \int_0^{\frac{1}{2}k} f(t)dt, \quad f_k^{m(k)} = \frac{2}{k} \int_{T-\frac{1}{2}k}^T f(t)dt.$$

Then it is easy to verify that (1.241) and (1.242) are satisfied.

Remark 1.26 *If h and g were more regular, e.g., elements of $C([0,T];L^2(\Omega))$ and $C([0,T];L^2(\partial\Omega \setminus \overline{\Gamma}))$, respectively, then instead of (1.282) one could use*

$$f_k^i = f(ik), \quad i = 0, ..., m(k) \tag{1.283}$$

as the approximation of f.

Finally, we assume that the initial condition $u_0 \in H^2(\Omega)$ and $\{u_{0h}\} = \{r_h u_0\}$ where $r_h u_0$ is the piecewise linear lagrange interpolate of u_0. Denote by $u_{h,k}^\theta$ and u the unique solutions of the θ-scheme and of the continuous problem (1.128), respectively. Then as a direct application of Theorem 1.44 we obtain

Theorem 1.52 *Let $V \cap H^2(\Omega)$ be dense in V. Then it holds:*

(i) If $\theta \in [\frac{1}{2}, 1]$ the θ-scheme is stable and convergent in the sense that $u_{h,k}^\theta$ tends weakly to u in $L^2(0,T;V)$ as $h, k \to 0+$.

(ii) If $\theta \in [0, \frac{1}{2})$ and the family of triangulations $\{\mathcal{T}_h\}$ is strongly regular then the θ-scheme is stable and convergent (in the above sense) under the condition $k/h^2 < C$, where $C > 0$ a constant large enough.

Proof: The density of $\{V_h\}$ in V (in the sense of (1.156)) has been already proven in Theorem 1.46. For the verification of (1.183) we use Theorem 1.45:

$$|u_0 - u_{0h}|_{0,\Omega} = |u_0 - r_h u_0|_{0,\Omega} \leq \hat{c}h^2 |u_0|_{2,\Omega} \to 0, \quad \text{as } h \to 0+.$$

If the approximation of f is defined by (1.282) then it is an easy exercise to verify (1.241) and (1.242) (see also Chapter 4). Then Theorem 1.44 (i) implies that the θ-scheme is stable and convergent for $\theta \in [\frac{1}{2}, 1]$.

The inverse inequality (1.243) is satisfied with $s(h) \equiv \hat{c}h^{-1}$ as follows from Lemma 1.5. Taking into account Remark 1.22 we see that the θ-scheme, $\theta \in [0, \frac{1}{2})$, is stable and convergent if

$$\frac{k}{h^2} < C \equiv \frac{\alpha}{2(1-\theta)M^2\hat{c}^2}.$$

$\square$

References

Adams, R. A. (1975). *Sobolev Spaces*. Academic Press, New York.

Aubin, J.-P. and Clarke, F. H. (1979). Shadow prices and duality for a class of optimal control problems. *SIAM J. Control Optimization*, 17:567–586.

Aubin, J.-P. and Ekeland, I. (1984). *Applied Nonlinear Analysis*. J. Wiley and Sons, New York.

Aubin, J.-P. and Frankowska, H. (1990). *Set-valued analysis*, volume 2 of *Systems & Control: Foundations & Applications*. Birkhäuser, Boston.

Axelsson, O. and Barker, V. A. (1984). *Finite Element Solution of Boundary Value Problems*. Academic Press, Orlando.

Barbu, V. (1993). *Analysis and control of nonlinear infinite-dimensional systems*, volume 190 of *Mathematics in Science and Engineering*. Academic Press, Boston.

Brézis, H. (1973). *Opérateurs Maximaux Monotones et Semigroupes de Contractions dans les Espaces de Hilbert*. North-Holland Publ. Co. Amsterdam and American Elsevier Publ. Co., New York.

Browder, F. E. and Hess, P. (1972). Nonlinear mappings of monotone type in Banach spaces. *J. Funct. Anal.*, 11:251–294.

Ciarlet, P. G. (1978). *The Finite Element Method for Elliptic Problems*. North Holland, Amsterdam, New York, Oxford.

Clarke, F. (1983). *Optimization and Nonsmooth Analysis*. J. Wiley, New York.

Doktor, P. (1973). On the density of smooth functions in certain subspaces of Sobolev spaces. *Comment. Math. Univ. Carolin.*, 14:609–622.

Duvaut, G. and Lions, J. L. (1976). *Inequalities in Mechanics and Physics*. Springer-Verlag, Berlin, Heidelberg, New York.

Ekeland, I. and Temam, R. (1976). *Convex Analysis and Variational Problems*. North-Holland, Amsterdam.

Fučik, S. and Kufner, A. (1980). *Nonlinear Differential Equations*. Studies in Applied Mechanics 2. Elsevier, Amsterdam, New York.

Glowinski, R. (1984). *Numerical Methods for Nonlinear Variational Problems*. Springer-Verlag, New York.

Glowinski, R., Lions, J. L., and Trémoliéres, R. (1981). *Numerical analysis of variational inequalities*, volume 8 of *Studies in Mathematics and its Applications*. North Holland, Amsterdam, New York.

Haslinger, J., Hlaváček, I., and Nečas, J. (1996). Numerical methods for unilateral problems in solid mechanics. In Ciarlet, P. G. and Lions, J. L., editors, *Handbook of Numerical Analysis*. North Holland.

Haslinger, J. and Neittaanmäki, P. (1996). *Finite Element Approximation for Optimal Shape, Material and Topology Design*. J. Wiley, second edition.

Hlaváček, I., Haslinger, J., Nečas, J., and Lovišek, J. (1988). *Numerical Solution of Variational Inequalities*. Springer Series in Applied Mathematical Sciences 66. Springer-Verlag, New York.

Kikuchi, N. and Oden, J. T. (1988). *Contact Problems in Elasticity: A Study of Variational Inequalities and Finite Element Methods*. SIAM Studies in Applied Mathematics 8. SIAM, Philadelphia.

Kufner, A., John, O., and Fučik, S. (1977). *Function Spaces*. Noordhoff International Publishing Leyden; Academia, Prague,.

Křižek, M. and Neittaanmäki, P. (1990). *Finite Element Approximation of Variational Problems and Applications*. Longman Scientific & Technical, Harlow.

Landes, R. and Mustonen, V. (1987). A strongly nonlinear parabolic intial value problem. *Ark. f. Mat.*, 25:29–40.

Lions, J. L. (1969). *Quelques Méthodes de résolution des problèmes aux limites non linéaires*. Dunod/Gauthier-Villairs, Paris.

Miettinen, M. (1996). A parabolic hemivariational inequality. *Nonlinear Analysis*, 26:725–734.

Moreau, J. J. (1967). *Fonctionnelles Convexes. Séminaire sur les équations aux dérivées partielles. Collège de France*. Paris.

Neittaanmäki, P. and Tiba, D. (1994). *Optimal control of nonlinear parabolic systems. Theory, algorithms, and applications.*, volume 179 of *Monographs and Textbooks in Pure and Applied Mathematics*. Marcel Dekker, New York.

Nečas, J. (1967). *Les Méthodes Directes en Théorie des Equations Elliptiques*. Masson, Paris.

Nečas, J. and Hlaváček, I. (1981). *Mathematical Theory of Elastic and Elasto-Plastic Bodies: An Introduction*. Elsevier, Amsterdam.

Rockafellar, R. T. (1969). *Convex Analysis*. Princeton Univ. Press, Princeton.

Thomée, V. (1984). *Galerkin finite element methods for parabolic problems*. Springer-Verlag, Berlin, Heidelberg, New York.

Trémoliéres, R. (1972). *Inéquations variationnelles: existence, approximations, résolution*. PhD thesis, Université de Paris VI.

Yosida, K. (1965). *Functional Analysis*. Springer-Verlag, Berlin.

Zeidler, E. (1990a). *Nonlinear functional analysis and its applications. II/A. Linear monotone operators*. Springer-Verlag, Berlin, New York.

Zeidler, E. (1990b). *Nonlinear functional analysis and its applications. II/B. Nonlinear monotone operators*. Springer-Verlag, Berlin, New York.

2 NONSMOOTH MECHANICS. CONVEX AND NONCONVEX PROBLEMS.

2.1 INTRODUCTION

Nonlinear, multivalued and possibly nonmonotone relations arise in several areas of mechanics. A multivalued or complete relation is a relation with complete vertical branches. Boundary laws of this kind connect boundary (or interface) quantities. A contact relation or a locking mechanism between boundary displacements and boundary tractions in elasticity is a representative example. Material constitutive relations with complete branches connect stress and strain tensors, or, in simplified theories, equivalent stress and strain quantities. A locking material or a perfectly plastic one is represented by such a relation. The question of nonmonotonicity is more complicated. One aspect concerns nonmonotonicity of a constitutive or a boundary law. Certainly, at a local microscopic level a nonmonotone relation corresponds to an unstable material or boundary law. Examples from damage or fracture mechanics may be presented. On a macroscopic level the complete mechanical behaviour of structural components can be described with such nonmonotone and possibly multivalued relations. A typical example of this kind is the delamination process of a composite structure, where local delaminations, crack propagation and interface or crack contact effects lead to a sawtooth overall load-displacement relation (see Panagiotopoulos and Baniotopoulos, 1984, Mistakidis and Stavroulakis, 1998, Li and Carlsson, 1999). The latter relation is adopted here as a constitutive law for the study of the structure at a macroscopic level. Another reason for nonmonotonicity is the large displacement or deformation effects.

Let us consider that there exists a convex deformation energy potential which is a function of some appropriate strain quantity. In a kinematically nonlinear mechanical theory the geometric compatibility relation, which connects strains with displacements of the structure, is nonlinear. Therefore the same potential energy, considered as a composite function of a convex function with a nonlinear relation, is, in general, nonconvex in the displacement variables.

In the case that the described mechanisms are conservative one may equivalently derive the previously mentioned mechanical relations from nonsmooth, generally nonconvex potentials. The term *superpotential*, initially proposed by Moreau for the convex case, has been adopted. It is clear that differentiation of a nondifferentiable function requires the use of nonsmooth analysis tools. For the convex case, the convex analysis subdifferential is an appropriate tool. A number of nonconvex generalizations have been proposed. The Clarke-Rockafellar generalized gradient has been extensively used by Panagiotopoulos in applications on mechanics (Panagiotopoulos, 1983, Panagiotopoulos, 1985, Panagiotopoulos, 1993). Other concepts, including the Demyanov-Rubinov quasidifferential, have also been investigated or are in the stage of development (Panagiotopoulos, 1988, Panagiotopoulos and Stavroulakis, 1992, Dem'yanov et al., 1996). In principle, to cope with nondifferentiability one has to enlarge the notion of the gradient and to adopt set-valued gradients. The relations of the mechanical problem result from minimality or, more general, stationarity conditions for the adopted superpotential. Instead of equations one has differential inclusions and inequalities, as one readily sees in the theory of nonlinear, constrained optimization. In the weak formulation, which is familiar to the computational mechanics community, instead of variational equations one has variational inequalities, for the convex case, or hemivariational inequalities, for the nonconvex one. The scientific discipline that uses nonsmooth analysis tools for the study of inequality problems in mechanics has been called, by Panagiotopoulos, *Nonsmooth Mechanics* (Moreau and Panagiotopoulos, 1988, Moreau et al., 1988).

Before entering into the details, let us summarize the link between nonsmooth analysis, optimization and nonsmooth mechanics. In fact, for example in elastostatics, from a potential energy minimization problem one derives the weak (variational) and, under certain assumptions, the strong (pointwise) description of the problem. For classical, smooth potentials, the necessary optimality condition (without side constraints) requires that the gradient of the potential energy is equal to zero at the optimum. This leads to a system of, in general, nonlinear equations. If one considers the equivalent requirement that the directional derivative of the potential function is equal to zero for all directions emanating from the solution (i.e., the equilibrium) point, one gets the weak formulation in the form of a variational equation. For historical reasons and since the most frequently used function is the potential energy function of a system written in terms of displacement variables and the gradient of this function plays the role of a stress or force vector, the above mentioned relation is called the *principle of virtual work*. In a dynamic analysis framework, where the potential is expressed in terms of velocities, the term *principle of virtual power* is used.

Analogously, the optimality condition for a convex nondifferentiable superpotential is expressed by a set–valued equation or a convex differential inclusion, where the set–valued generalization of the classical gradient, the subdifferential of convex analysis, appears. This set-valued equation is equivalent to a system of equations, inequalities and complementarity conditions which describe, in a pointwise way, the mechanical problem. Equivalently, the directional derivative of a nondifferentiable function for all directions emanating from the minimum must be greater or equal to zero. This requirement leads to a variational inequality problem. Unilateral contact problems are typical examples of structural analysis problems with kinematic inequality constraints which physically describe the no–penetration restriction of the unilateral contact mechanism.

For nonconvex, nondifferentiable potentials the substationarity relation (in the sense of Clarke-Rockafellar) takes the form of a hemivariational inequality. In certain cases one may work further, using for instance quasidifferentiability and difference convex optimization techniques, to decompose hemivariational inequalities into systems of variational inequalities, as it has been described elsewhere (see Panagiotopoulos, 1988, Panagiotopoulos, 1993, Dem'yanov et al., 1996, Mistakidis and Stavroulakis, 1998).

In this chapter simple mechanical models are used to demonstrate the previously outlined ideas. Emphasis is given on nonlinear elastostatics and on the more straightforward class of problems which concern elastic bodies with multivalued, monotone and nonmonotone boundary or interface laws. This is the case of contact and adhesion problems (possibly coupled with frictional effects). Extensions to nonlinear material constitutive laws and dynamic problems are given later in this chapter. It should be mentioned that the style of this chapter is more or less engineering oriented, without proofs and with a few concrete functional analysis definitions. Rigorous mathematical formulations and proofs are given in the other chapters of this book.

2.2 NONLINEAR ELASTOSTATICS

Simple models of nonlinearly elastic static problems will be used here to demonstrate the formulation of variational equations, inequalities and hemivariational inequalities. First, linear elastic structures with monotone and nonmonotone, multivalued, boundary and interface laws are considered. The classical relations of the problem are given. Then the superpotential formulation of several appropriate nonlinear boundary and interface relations are presented. If one uses subdifferentials for the mathematical description of, say, a boundary condition, one speaks also about a subdifferential boundary condition. Then, by using the virtual work relation for the structure together with the inequalities which are due to the nonlinear boundary relations, one formulates variational and hemivariational inequality problems and corresponding potential energy minimization or critical point problems. This section closes with an analogous derivation for nonlinear material laws.

Nonlinear boundary and interface laws

Description of the problem. Let $\Omega \subset \mathbf{R}^3$ be an open bounded subset occupied by a deformable body in its undeformed state. The boundary of Ω is denoted by $\partial\Omega$. The points $x \in \Omega, x = \{x_i\}, i = 1, 2, 3$, are referred to a Cartesian coordinate system. The linear (small displacement and small deformation) elastostatic analysis problem is described by *the equilibrium equation*:

$$\tau_{ij,j} + F_i = 0 \tag{2.1}$$

where τ is the stress tensor, F_i is the volume force vector, indices $i, j = 1, 2$ (resp. $= 1, 2, 3$) for two-(resp. three-)dimensional problems and the notation $\tau_{ij,j} \equiv \partial\tau_{ij}/\partial x_j$ is used. Moreover, one has *the strain-displacement compatibility* relation:

$$\varepsilon_{ij} = \frac{1}{2}(u_{i,j} + u_{j,i}), \tag{2.2}$$

where u is the displacement vector and $u_{i,j} \equiv \partial u_i/\partial x_j$. Further for a linearly elastic material one has *the constitutive law*:

$$\tau_{ij} = c_{ijhk}\varepsilon_{hk}, \tag{2.3}$$

where $c = \{c_{ijhk}\}, i, j, h, k = 1, 2, 3$, is the elasticity tensor which satisfies the well-known *symmetry* and *ellipticity properties*

$$\begin{cases} c_{ijhk} = c_{jihk} = c_{hkij} \\ \exists \alpha_0 > 0 : c_{ijhk}\varepsilon_{ij}\varepsilon_{hk} \geq \alpha_0\varepsilon_{ij}\varepsilon_{ij} \quad \text{a.e. in } \Omega \text{ and } \forall \varepsilon_{ij} = \varepsilon_{ji} \in \mathbf{R}. \end{cases} \tag{2.4}$$

Relations (2.1), (2.2), (2.3) hold pointwise for each $x \in \Omega$. If they are coupled with appropriate boundary conditions they fully describe boundary value problems of elastostatics.

Let us denote the bilinear form of linear elasticity by $a(\cdot, \cdot)$, where

$$a(u, v) = \int_\Omega c_{ijhk}\varepsilon_{ij}(u)\varepsilon_{hk}(v)dx. \tag{2.5}$$

On the assumption of small deformations one writes the virtual work relation:

$$\int_\Omega \tau_{ij}(u)\varepsilon_{ij}(v - u)dx = \int_\Omega F_i(v_i - u_i)dx + \int_{\partial\Omega} \tau_{ij}\nu_j(v_i - u_i)ds \quad \forall v \in \mathbf{V} \tag{2.6}$$

for $u \in \mathbf{V}$. Here $\mathbf{V}$ is equal to $H^1(\Omega; \mathbf{R}^3)$.

Relation (2.6) is obtained from the operator equations of the problem by applying the Green's formula (1.16), and is the expression of the principle of virtual work for the body when considered free, i.e., with no constraints on its boundary $\partial\Omega$. Thus, for the derivation of (2.6) one multiplies the equilibrium equation (2.1) by a virtual variation $v_i - u_i$ and then one integrates over Ω.

Then, on the assumption of appropriately smooth functions, one applies (1.16) by taking into account the strain-displacement relation (2.2).

If one considers contact and friction effects it may be advantageous to write, instead of (2.1), the relation:

$$\int_\Omega \tau_{ij}\varepsilon_{ij}(v-u)dx = \int_\Omega F_i(v_i-u_i)dx + \int_{\partial\Omega} T_\nu(v_\nu-u_\nu)ds \qquad (2.7)$$

$$+ \int_{\partial\Omega} T_{t_i}(v_{t_i}-u_{t_i})ds \qquad \forall v \in \mathbf{V}.$$

Here the last term (2.6) has been decomposed into the work of the normal and the work of the tangential to the boundary tractions. Relation (2.6) or (2.7) will be coupled with the boundary conditions in the sequel for the derivation of certain variational problems.

To give an idea of how this technique works in the classical case, let us assume that the support boundary conditions $T_\nu = 0$ and $u_{t_i} = 0$, $i = 1, 2, 3$, hold on $\partial\Omega$. Then (2.7) with (2.3) lead to the variational equation:

$$\begin{cases} \text{Find } u \in \mathbf{V}_0 \text{ such that} \\ a(u,v) = \int\limits_\Omega F_i v_i dx \quad \forall v \in \mathbf{V}_0, \end{cases} \qquad (2.8)$$

where $\mathbf{V}_0 = \{v \in H^1(\Omega; \mathbf{R}^3) \mid v_{t_i} = 0 \text{ on } \partial\Omega\}$.

Superpotential boundary and interface laws. Let $T = \{T_i\}$ be the stress vector on $\partial\Omega$, where $T_i = \tau_{ij}\nu_j$, $\tau = \{\tau_{ij}\}$ and $\nu = \{\nu_i\}$ is the outward unit normal vector on $\partial\Omega$. The vector T is decomposed into a normal component T_ν and a tangential component $T_t = (T_{t_i})$ with respect to $\partial\Omega$, where

$$T_\nu = \tau_{ij}\nu_j\nu_i \quad \text{and} \quad T_{t_i} = \tau_{ij}\nu_j - T_\nu\nu_i. \qquad (2.9)$$

Analogously to T_ν and T_t, u_ν and u_t denote the normal and the tangential components of the displacement vector u with respect to $\partial\Omega$. T_ν and u_ν are considered as positive if they are parallel to ν.

Maximal monotone operators $\beta_i : \mathbf{R} \to 2^\mathbf{R}$ are introduced, such that monotone, possibly multivalued boundary conditions can be expressed in the form

$$-T_i \in \beta_i(u_i), \quad i = 1, 2, 3. \qquad (2.10)$$

Then (see e.g. Panagiotopoulos, 1985 p. 57) convex, lower semicontinuous and proper functionals j_i on $\mathbf{R}$ may be determined, up to additive constants, such that

$$\beta_i = \partial j_i, \quad i = 1, 2, 3. \qquad (2.11)$$

Then (2.10) is written as

$$-T_i \in \partial j_i(u_i), \quad i = 1, 2, 3. \qquad (2.12)$$

This relation is a subdifferential boundary condition and is understood point-wisely, i.e., as a relation between $-T_i(x) \in \mathbf{R}$ and $u_i(x) \in \mathbf{R}$ at every point $x \in \partial\Omega$. The graph of β_i, referred to a Cartesian system Oxy, is a complete nondecreasing curve in $\mathbf{R}^2$ which is generally multivalued. This means that the graph of β_i may include segments parallel to both coordinate axes. Moreover j_i is a local superpotential, which after integration over the whole boundary leads to the global superpotential describing the energy contribution of the whole boundary law. From another point of view the boundary condition (2.10) may be considered as the material law of a fictive spring of zero length at x in the ith-direction.

Analogously to (2.10), (2.12) one defines boundary conditions of the form:

$$-T_\nu \in \beta_\nu(u_\nu) = \partial j_\nu(u_\nu), \quad -T_t \in \partial j_t(u_t). \qquad (2.13)$$

In dynamic mechanical problems, similar boundary conditions may be defined between T and the partial time derivative of the displacement u', or the velocity v.

Within the previously introduced general framework, the classical *support* boundary condition $u_i = 0$ can be put in the form (2.10) through the operator

$$\beta_i(u_i) = \begin{cases} \mathbf{R} & \text{if } u_i = 0 \\ \emptyset & \text{otherwise} \end{cases}, \qquad (2.14)$$

or through the functional $j_i(u_i) = \{0 \text{ if } u_i = 0 \text{ and } \infty \text{ otherwise}\}$. Analogously the *loaded* boundary condition $T_i = C_i$ is written in the form (2.10) or (2.12) with $\beta_i(u_i) = -C_i$ (C_i given) or $j_i(u_i) = -C_i u_i$ (no summation) for every $u_i \in \mathbf{R}$.

Unilateral contact relations between a boundary and a rigid support read as

$$\begin{aligned} &\text{if } u_\nu < u_0, \quad \text{then } T_\nu = 0; \\ &\text{if } u_\nu = u_0, \quad \text{then } T_\nu \leq 0, \end{aligned} \qquad (2.15)$$

where u_0 denotes an initial distance (gap) between the structure and the rigid support. Relations (2.15) may equivalently be expressed by the linear complementarity form:

$$T_\nu \leq 0, \quad u_\nu - u_0 \leq 0, \quad \text{and} \quad T_\nu(u_\nu - u_0) = 0 \quad \text{on } \partial\Omega. \qquad (2.16)$$

The respective operator β_ν reads:

$$\beta_\nu(u_\nu) = \begin{cases} 0 & \text{if } u_\nu < u_0, \\ [0, +\infty) & \text{if } u_\nu = u_0 \\ \emptyset & \text{if } u_\nu > u_0 \end{cases}, \qquad (2.17)$$

and the corresponding superpotential has the form

$$j_\nu(u_\nu) = \begin{cases} 0 & \text{if } u_\nu \leq u_0 \\ +\infty & \text{if } u_\nu > u_0. \end{cases} \qquad (2.18)$$

Note here that the previous unilateral contact law gives rise to the local variational inequality:

$$\begin{cases} \text{Find } u_\nu \in (-\infty, u_0] \text{ such that} \\ -T_\nu \left(u_\nu^* - u_\nu \right) \leq 0 \quad \forall u_\nu^* \in (-\infty, u_0]. \end{cases} \tag{2.19}$$

The last example of monotone laws given here is the *static Coulomb friction law* (we consider that $\Omega \subset \mathbf{R}^2$):

$$\text{if } |T_t| < \mu |T_\nu|, \text{ then } u_t = 0,$$

$$\text{if } |T_t| = \mu |T_\nu|, \text{ then there exists}$$

$$\lambda \geq 0 \text{ such that } u_t = -\lambda T_t. \tag{2.20}$$

The symbol $\mu = \mu(x) > 0$ denotes the coefficient of friction and $|\cdot|$ the usual absolute value. Relations (2.20) can be written in the form

$$-T_t \in \beta_t(u_t), \tag{2.21}$$

where

$$\beta_t(u_t) = \begin{cases} [-\mu |T_\nu|, +\mu |T_\nu|] & \text{if } u_t = 0 \\ \mu |T_\nu| & \text{if } u_t > 0 \\ -\mu |T_\nu| & \text{if } u_t < 0. \end{cases} \tag{2.22}$$

Let us assume further that $T_\nu = C_\nu$, where C_ν is given, and denote the quantity $\mu |C_\nu|$ by $(T_t)_0$. Then

$$\beta_t(u_t) = \partial j_t(u_t), \text{ where } j_t(u_t) = (T_t)_0 |u_t|, \tag{2.23}$$

(see Fig.2.1).

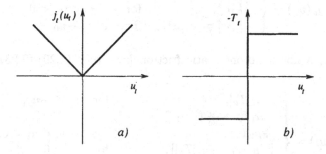

Figure 2.1.   Simplified static Coulomb friction law and the corresponding convex superpotential.

Recall that the initial Coulomb's law of friction involves the tangential velocity u'_t. Nevertheless, the study of the simplified static relation (2.23) makes sense, since it arises, after time discretization within each time step of a time-marching algorithm.

The subdifferential simplified static Coulomb friction model (2.23) gives rise to the local variational inequality

$$j_t(u_t^*) - j_t(u_t) \geq -T_t(u_t^* - u_t), \quad \forall u_t^* \in \mathbf{R}. \tag{2.24}$$

Nonmonotone, possibly multivalued laws are described analogously by relation (2.10), where β_i is, in this case, nonmonotone. Filling-in the vertical branches and integrating the multifunction β_i one constructs a nonconvex, generally nondifferentiable function j_i on $\mathbf{R}$. It is obvious that for writing of an analogous to (2.12), one needs an appropriate differentiation tool which takes care of both nonconvexity and nondifferentiability issues. The generalized subdifferential in the sense of Clarke, denoted by $\bar{\partial}$ is used here (see Definition 1.5). Thus one writes the nonmonotone law in the form:

$$-T \in \bar{\partial} j(u), \tag{2.25}$$

where j is a locally Lipschitz superpotential function. Instead of (2.25) one may also consider the boundary laws

$$-T_\nu \in \bar{\partial} j_\nu(u_\nu), \quad - T_t \in \bar{\partial} j_t(u_t). \tag{2.26}$$

For example, a *delamination* law can be expressed in the form (see Fig. 2.2):

$$-T_\nu \in \begin{cases} k_1 u_\nu, & \text{for} \quad u_\nu \geq 0, \\ k_2 u_\nu, & \text{for} \quad -u_0 < u_\nu < 0, \\ [-k_2 u_0, 0], & \text{for} \quad u_\nu = -u_0, \\ 0, & \text{for} \quad u_\nu < -u_0, \end{cases} \tag{2.27}$$

where k_1, k_2 are given positive constants. Relation (2.27) can be written in the form of the first law in (2.26) by means of the nonconvex superpotential:

$$j_\nu(u_\nu) = \begin{cases} \frac{1}{2} k_1 u_\nu^2, & \text{for} \quad u_\nu \geq 0, \\ \frac{1}{2} k_2 u_\nu^2, & \text{for} \quad -u_0 < u_\nu < 0, \\ \frac{1}{2} k_2 u_0^2 = \text{const.}, & \text{for} \quad u_\nu < -u_0. \end{cases} \tag{2.28}$$

Analogously, a nonmonotone static friction law (cf. (2.20)-(2.23)) may be considered:

$$-T_t \in \beta_t(u_t) = \begin{cases} -\mu_2 |T_\nu|, & \text{for} \quad u_t \leq -u_0, \\ -\mu_1 T_\nu + \frac{u_t}{u_0} (-\mu_1 T_\nu \\ \quad +\mu_2 T_\nu), & \text{for} \quad -u_0 < u_t < 0, \\ [-\mu_1 |T_\nu|, +\mu_1 |T_\nu|], & \text{for} \quad u_t = 0, \\ \mu_1 T_\nu - \frac{u_t}{u_0} (\mu_1 T_\nu - \mu_2 T_\nu), & \text{for} \quad 0 < u_t < u_0, \\ \mu_2 |T_\nu|, & \text{for} \quad u_0 < u_t. \end{cases} \tag{2.29}$$

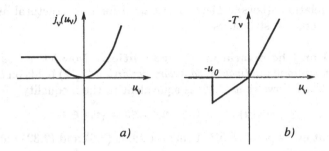

Figure 2.2.   Delamination law and the corresponding nonconvex superpotential.

Model (2.29), with $\mu_2 < \mu_1$, is a nonmonotone approximation of a stiction type friction law, with a different (higher) static friction coefficient μ_1 than the dynamic friction coefficient μ_2 (cf. Pfeiffer and Glocker, 1996). In this case, as previously (cf., (2.22), (2.23)), assuming that the contact traction is kept constant $T_\nu = C_\nu$ and $\Omega \subset \mathbf{R}^2$, one may construct the nonconvex superpotential $j_t(\cdot)$, by integrating β in (2.29):

$$j_t(u_t) = \int_0^{u_t} \beta(s)ds \qquad (2.30)$$

and (2.29) can be expressed in the form $-T_t \in \bar{\partial}j_t(u_t)$ (see Fig. 2.3).

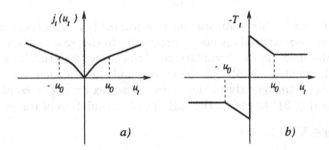

Figure 2.3.   Simplified nonmonotone friction law and the corresponding nonconvex super-potential.

Finally, relation (2.25) is by definition equivalent to the local hemivariational inequality

$$j^\circ(u; v - u) \geq T_i(v_i - u_i) \quad \forall v = \{v_i\} \in \mathbf{R}^3, \qquad (2.31)$$

where j° is the generalized directional derivative of j° (see Definition 1.4).

Note here that if u is replaced by the relative displacement $[u]$ between adjacent sides of a given interface and T is interpretted as an interface traction, then all aforementioned boundary conditions can be used for the description of interface relations (laws). They can be used for the structural analysis of multibody or cracked structures.

Variational and hemivariational inequalities. Now let us assume that the general *monotone multivalued boundary condition* (2.12) holds on $\partial\Omega$. From Definition 1.2 it follows that (2.12) is equivalent to the inequality

$$j(v) - j(u) \geq -T_i(v_i - u_i) \quad \forall v = \{v_i\} \in \mathbf{R}^3, \tag{2.32}$$

which holds at every point of $\partial\Omega$. Using relations (2.6) and (2.32) one gets the variational inequality:

$$\begin{cases} \text{Find } u \in \mathbf{V} \text{ such that} \\ a(u, v - u) + \displaystyle\int_{\partial\Omega} (j(v) - j(u)) \, ds \\ \geq \displaystyle\int_{\Omega} F_i(v_i - u_i) dx \quad \forall v \in \mathbf{V} \end{cases} \tag{2.33}$$

assuming that the integral over $\partial\Omega$ exists. Furthermore, by using the convex functional

$$J(v) = \begin{cases} \int_{\partial\Omega} j(v) ds, & \text{if } j(v) \in L^1(\partial\Omega) \\ +\infty, & \text{otherwise,} \end{cases} \tag{2.34}$$

one may also write the variational inequality:

$$\begin{cases} \text{Find } u \in \mathbf{V} \text{ such that} \\ a(u, v - u) + J(v) - J(u) \geq \displaystyle\int_{\Omega} F_i(v_i - u_i) dx \quad \forall v \in \mathbf{V}. \end{cases} \tag{2.35}$$

Let us consider further *nonmonotone multivalued boundary laws* and the corresponding hemivariational inequality problems. In this case the basic building element is the definition of boundary conditions and material laws based on Clarke subdifferential (2.25) and the corresponding local hemivariational inequality (2.31). Analogously to the previous convex case, one combines relations (2.6) and (2.31) to obtain the variational formulation of the problem:

$$\begin{cases} \text{Find } u \in \mathbf{V} \text{ such that} \\ a(u, v - u) + \displaystyle\int_{\partial\Omega} j^\circ(u; v - u) ds \geq \displaystyle\int_{\Omega} F_i(v_i - u_i) dx \quad \forall v \in \mathbf{V} \end{cases} \tag{2.36}$$

assuming the integral over $\partial\Omega$ exists. In the case that one considers separately the mechanical behaviour in the normal and the tangential to the boundary direction (i.e., relations (2.26)), one proceeds analogously and writes the following variational problem:

$$\begin{cases} \text{Find } u \in \mathbf{V} \text{ such that} \\ a(u, v - u) + \displaystyle\int_{\partial\Omega} j_\nu^\circ(u_\nu; v_\nu - u_\nu) ds + \displaystyle\int_{\partial\Omega} j_t^\circ(u_t; v_t - u_t) ds \\ \geq \displaystyle\int_{\Omega} F_i(v_i - u_i) dx \quad \forall v \in \mathbf{V}. \end{cases} \tag{2.37}$$

The last type of variational expressions which involve $j^\circ(\cdot;\cdot)$ or $j^\circ_\nu(\cdot;\cdot)$ and $j^\circ_t(\cdot;\cdot)$ has been introduced and studied in mechanics by P.D. Panagiotopoulos, who named them *hemivariational inequalities* (see, among others, Panagiotopoulos, 1985, Panagiotopoulos, 1993, Naniewicz and Panagiotopoulos, 1995, Motreanu and Panagiotopoulos, 1998). Note that in the more general case in which j or j_ν and j_t are not locally Lipschitz the generalized directional derivatives $j^\circ(\cdot;\cdot)$ in (2.36) and $j^\circ_\nu(\cdot;\cdot), j^\circ_t(\cdot;\cdot)$ in (2.37) have to be replaced by more general objects (see Panagiotopoulos, 1985). In this book we restrict ourselves to the case of locally Lipschitz functionals.

One should recall here that solutions of variational problems, like the variational equations, and the systems of variational inequalities or the hemivariational inequalities derived previously, satisfy the operator equations of the problem, e.g., the equilibrium equation, and the boundary conditions of the problem in a weak sense. This means, roughly speaking, that these relations are satisfied in an integral form, in the body or on the boundary of the structure respectively, where the integral may be also seen as a weighted average of the integrands and is defined by the adopted functional framework. This question is connected with requirements that must be posed on the finite element interpolation (element choice) and on the numerical accuracy of the solutions they produce. Details on this point can be found in Duvaut and Lions, 1972, Panagiotopoulos, 1985, for variational inequality problems and in Panagiotopoulos, 1985, Panagiotopoulos, 1993, for hemivariational inequality problems.

Potential energy and critical point formulation. Let us consider the potential energy functional of the previously described elastostatic analysis problems:

$$\Pi(v) = \frac{1}{2}a(v,v) + J(v) - (\mathcal{F},v), \quad v \in \mathbf{V}. \tag{2.38}$$

Here J is the convex (respectively nonconvex) potential energy contribution of the monotone (resp. nonmonotone) boundary laws (see also (2.34) and $\mathcal{F}$ represents external applied forces.

For the convex case, the structural analysis problem can be written as a minimization problem for the potential energy Π. Taking into account the possible nonsmoothness of this functional, one writes the first order necessary and sufficient optimality conditions in the form of the variational inclusion (set-valued equation) (see Theorem 1.23):

$$\begin{cases} \text{Find } u \in \mathbf{V} \text{ such that} \\ 0 \in \partial\Pi(u). \end{cases} \tag{2.39}$$

Here the subdifferential ∂ in the sense of convex analysis has been used. Moreover, for simplicity, it has been assumed that all constraints of the problem (e.g. boundary support conditions) are included in Π through some penalty function (cf. (2.14), or that the corresponding variables have been eliminated from the description of the problem (i.e., they do not appear in space $\mathbf{V}$).

In the nonconvex case one can only write substationarity problems, i.e., critical point problems, in the sense of Clarke-Rockafellar (see Theorem 1.24):

$$\begin{cases} \text{Find } u \in \mathbf{V} \text{ such that} \\ 0 \in \bar{\delta}\Pi(u). \end{cases} \tag{2.40}$$

The bundle type algorithms of Chapter 5 solve directly (2.39), for the convex case, and (2.40), for the nonconvex case.

Nonlinear material laws

Certain classes of nonlinear material laws can be extracted from potential, strain energy density functions. This approach has been widely adopted in the construction of material models, which are valid for arbitrary large deformations (hyperelastic materials). The adoption of nondifferentiable, convex and nonconvex potentials extends this approach and allows to consider material laws with complete vertical or horizontal branches (locking or perfect plasticity effects). Here some simple examples of these laws will be given. They relate the stress tensor $\tau = \{\tau_{ij}\}$ and strain tensor $\varepsilon = \{\varepsilon_{ij}\}$, or their time derivatives, in a small deformation theory. More details can be found in Panagiotopoulos, 1993.

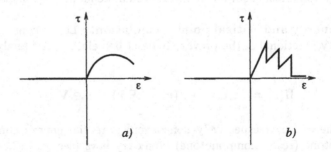

Figure 2.4.   Onedimensional nonmonotone stress-strain laws.

Description of the problem and variational formulation.   One considers the elastostatic boundary value problem defined by (2.1), (2.2) and coupled with a nonlinear superpotential material law. For monotone laws we take the superpotential constitutive law:

$$\tau \in \partial w(\varepsilon), \tag{2.41}$$

where w is a tensor-valued, convex, lower semicontinuous and proper strain energy density function defined on $\mathbf{R}^6$.

For nonmonotone laws one uses, analogously, the relation:

$$\tau \in \bar{\partial} w(\varepsilon), \tag{2.42}$$

where w is a locally Lipschitz, generally nonconvex tensor-valued strain energy function.

In analogy to the discussion of the previous part concerning the elastostatic problem with nonlinear boundary conditions, one has the equivalent local variational inequality of (2.41):

$$w(\varepsilon^*) - w(\varepsilon) \geq \tau_{ij}(\varepsilon_{ij}^* - \varepsilon_{ij}), \quad \forall \varepsilon^* \in \mathbf{R}^6, \qquad (2.43)$$

and the local hemivariational inequality of (2.42):

$$w^0(\varepsilon^* - \varepsilon) \geq \tau_{ij}(\varepsilon_{ij}^* - \varepsilon_{ij}), \quad \forall \varepsilon^* \in \mathbf{R}^6. \qquad (2.44)$$

Relations (2.43) and (2.44) hold for $\varepsilon \in \mathbf{R}^6$.

Finally one combines the virtual work relation (2.6) with the appropriate inequality (2.43) or (2.44) to obtain the variational formulation of the studied elastostatic problem with a nonlinearity expressed by a monotone or nonmonotone, multivalued relation. Obviously one defines the corresponding global forms or $L^1(\Omega)$-extensions of the above superpotential laws according to (2.34) (with $\partial\Omega$ replaced by Ω). Examples of material laws which can be written in the form of (2.41) and (2.42) are given below.

Superpotential material constitutive laws. Monotone material laws which can be derived from convex, generally nondifferentiable potentials are discussed first.

For the description of *elastic ideally locking* materials one may consider a function w defined on $\mathbf{R}^6$ by:

$$w(\varepsilon) = w_0(\varepsilon) + I_K(\varepsilon), \quad \varepsilon = \{\varepsilon_{ij}\}, \qquad (2.45)$$

where w_0 is a continuously differentiable convex function and I_K is the indicator function of the convex closed subset of $\mathbf{R}^6$

$$K = \{\varepsilon \mid Q(\varepsilon) \leq 0\}, \qquad (2.46)$$

where Q is a convex, continuously differentiable function in $\mathbf{R}^6$ such that $0 \in K$. The superpotential relation (2.41) reads in this case as follows:

$$\tau_{ij} \in [\partial w(\varepsilon)]_{ij} = \frac{\partial w_0(\varepsilon)}{\partial \varepsilon_{ij}} + \partial I_K(\varepsilon), \qquad (2.47)$$

or

$$\tau_{ij} = \frac{\partial w_0(\varepsilon)}{\partial \varepsilon_{ij}} + \bar{\tau}_{ij}, \qquad (2.48)$$

where $\bar{\tau} = \{\bar{\tau}_{ij}\}$ is an element of the outward normal cone to K at the point $\varepsilon \in K$, which is characterized by means of the following variational inequality (see Definition 1.3):

$$\bar{\tau}_{ij}(\varepsilon_{ij}^* - \varepsilon_{ij}) \leq 0 \quad \forall \varepsilon^* \in K. \qquad (2.49)$$

Using the method of lagrange multipliers λ to describe the latter locking stress contribution, one may rewrite relation (2.48) in the equivalent form

$$\tau_{ij} = \frac{\partial w_0(\varepsilon)}{\partial \varepsilon_{ij}} + \lambda \frac{\partial Q(\varepsilon)}{\partial \varepsilon_{ij}}, \quad \lambda \geq 0, \quad \lambda Q(\varepsilon) = 0, \quad Q(\varepsilon) \leq 0. \qquad (2.50)$$

Accordingly, if $Q(\varepsilon) < 0$, then $\lambda = 0$, and the material behaves like a nonlinear elastic material. If $Q(\varepsilon) = 0$ then no finite or infinite increment of the stresses can cause an increase in the value of the function $Q(\varepsilon)$. This is called an ideal-locking effect. One possible form for $Q(\varepsilon)$ gives the Prager criterion:

$$Q(\varepsilon) = \frac{1}{2}\varepsilon_{ij}^D \varepsilon_{ij}^D - k^2, \qquad (2.51)$$

where $\varepsilon^D = \{\varepsilon_{ij}^D\}$ is the strain deviator, and k is an appropriate material constant.

A simplified version of relations (2.47) or (2.50) would be to consider a linearly elastic contribution with $w_0(\varepsilon) = \frac{1}{2}c_{ijhk}\varepsilon_{ij}\varepsilon_{hk}$, thus $\partial w_0(\varepsilon)/\partial \varepsilon_{ij} = c_{ijhk}\varepsilon_{hk}$ (cf. (2.3)).

One may generalize the previous model by considering a *nonconvex locking criterion*. Let L be a closed subset of the strain space $\mathbf{R}^6$ and let I_L be the indicator function of L. Assuming linear elastic part, then the relation

$$\tau - c\varepsilon \in \bar{\partial} I_L(\varepsilon) \qquad (2.52)$$

generalizes (2.47) for a locking criterion defined by a closed but generally *nonconvex surface* in the strain space. Here we denote by $\bar{\partial}$ also the generalized gradient of the extended valued (possibly) nonconvex functional $I_L : \mathbf{R}^6 \to [0, \infty]$ (see Remark 1.7).

One may also consider that the nonlinear elastic part of the previous material law is given by a nonmonotone relation. Thus, it may be derived from a nonconvex strain energy density $\bar{w}_0(\varepsilon)$. In this case the material law reads (cf. (2.42)):

$$\tau \in \bar{\partial} w(\varepsilon) + \bar{\partial} I_L(\varepsilon) \qquad (2.53)$$

This relation describes the behaviour of a material obeying both the nonmonotone multivalued law $\tau \in \bar{\partial} w(\varepsilon)$ and the locking criterion defined by the closed set L made arbitrarily large.

Once again it should be emphasized that nonmonotone material laws may be seen as macroscopic relations appropriate for the description of a material behaviour, including local instabilities, fractures etc., at a macroscopic, phenomenological way. Analogously, nonconvex locking effects arise, for instance, in the macroscopic mechanical description of fiber-reinforced composite materials.

More general relations can be considered if one replaces the stress or the strain quantities in (2.41) or (2.42) by their time derivatives and then considers the resulting time dependent (rate) material law. In this case one should work with strain and displacement rates in the geometric compatibility relation (2.2). Further one writes a virtual power relation, by replacing v and u in (2.6) with their time derivatives and proceeds further in an analogous way.

Generalizations

A small number of other applications of multivalued relations in statics and dynamics, as well as in related modeling problems are outlined here. Some related information, including references to original publications, is given at the end of this chapter.

Monotone and nonmonotone boundary conditions in *plate theory* can be written. Let w be the vertical displacement of the plate, ν the normal to the boundary. The boundary laws connect the bending moment M_ν with the boundary rotation $\partial w / \partial \nu$:

$$M_\nu \in \beta_1 \left(\frac{\partial w}{\partial \nu} \right) = \bar{\partial} j_1 \left(\frac{\partial w}{\partial \nu} \right), \qquad (2.54)$$

and the total shearing force K_ν with the boundary displacement w:

$$-K_\nu \in \beta_2(w) = \bar{\partial} j_2(w). \qquad (2.55)$$

In *delamination* analysis of *composite* plates one considers analogous relations between the interlaminar tractions f and the corresponding relative displacements $[w] = w_1 - w_2$:

$$f \in \bar{\partial} j_3([w]), \qquad (2.56)$$

where, for example, a plate with two layers has been considered and $w_i, i = 1, 2$ denotes the displacement of the layer i. Theoretical studies of hemivariational and variational-hemivariational inequality problems arising in composite plates have been presented, among others, in Panagiotopoulos and Stavroulakis, 1988, Panagiotopoulos and Stavroulakis, 1990, Panagiotopoulos, 1993.

Analogous relations can be considered in *semipermeability* problems connected, for instance, with fluid mechanics applications (Panagiotopoulos, 1985, Haslinger et al., 1993).

In dynamics (without impact phenomena) one considers boundary value problems on the space-time domain $\Omega \times (0, T)$, where the time ranges over the time interval $[0, \infty)$. Then, in the place of static equilibrium relation (2.1), one has:

$$\tau_{ij,j} + F_i = \rho u_i'' \quad \text{in} \quad \Omega \times (0, T), \qquad (2.57)$$

where u_i'' is the acceleration vector, and $\rho = \rho(x), x \in \Omega$ is the density of the body. Without entering into details one can say that one uses a virtual power relation (analogous to (2.6)), monotone and nonmonotone material or boundary laws as previously (in both time and space variables). The variational or hemivariational inequality problem which may be formulated in this way have analogous form to the static ones (cf., e.g., (2.33), (2.36)) with the additional inertial terms. For a boundary relation

$$-T \in \bar{\partial} j(u'), \qquad (2.58)$$

one gets the hyperbolic hemivariational inequality:

$$
\begin{cases}
\text{Find } u : [0, T] \to V \text{ with } u'(t) \in V \text{ and } u''(t) \in [L^2(\Omega)]^3 \text{ such that} \\
(\rho u'', v - u') + a(u, v - u') + \int_{\partial\Omega} j^0(u', v - u') dx \\
\geq \int_{\Omega} F_i(v_i - u') dx, \quad \forall v \in V \quad + \text{ initial conditions..}
\end{cases}
\tag{2.59}
$$

The theoretical study of essentially nonsmooth and possibly nonconvex problems in dynamics, including impact effects, is still an open problem. First attempts in this direction are included in Panagiotopoulos and Liolios, 1989, Panagiotopoulos, 1995, Goeleven, 1997, Panagiotopoulos and Glocker., 1998. See also the mathematical theory of the dynamic hemivariational inequalities in Goeleven et al., 1999, Miettinen and Panagiotopoulos, 1999, Pop and Panagiotopoulos, 1998.

2.3 LITERATURE REVIEW

The link between convex analysis and variational problems has been discussed by Duvaut and Lions, 1972, Ekeland and Temam, 1976. The extension to nonconvex problems has been proposed and studied in Panagiotopoulos, 1985, Panagiotopoulos, 1993. Existence results for hemivariational or variational-hemivariational inequality problems and concrete applications in delamination and interface problems have been considered, among others, in Panagiotopoulos, 1985, Panagiotopoulos, 1993, Naniewicz and Panagiotopoulos, 1995, Panagiotopoulos and Koltsakis, 1987, Motreanu and Panagiotopoulos, 1998.

Noncoercive problems have been addressed in Goeleven and Mentagui, 1995, Goeleven and Théra, 1995, Goeleven et al., 1996, Adly et al., 1996.

Buckling, bifurcation and eigenvalue problems for hemivariational inequalities have been considered in Motreanu and Panagiotopoulos, 1995, Goeleven, 1997,

References

Adly, S., Goeleven, D., and Théra, M. (1996). Recession mappings and noncoercive variational inequalities. *Nonlinear Analysis Theory Methods and Applications*, 26(9):1573–1604.

Dem'yanov, V. F., Stavroulakis, G. E., Polyakova, L. N., and Panagiotopoulos, P. D. (1996). *Quasidifferentiability and nonsmooth modelling in mechanics, engineering and economics*. Kluwer Academic, Dordrecht.

Duvaut, G. and Lions, J. L. (1972). *Les inéquations en méchanique et en physique*. Dunod, Paris.

Ekeland, I. and Temam, R. (1976). *Convex analysis and variational problems*. North-Holland, Amsterdam.

Goeleven, D. (1997). A bifurcation theory for nonconvex unilateral laminated plate problem formulated as a hemivariational inequality involving a potential operator. *Zeitschrift für Angewandte Mathematik und Mechanik (ZAMM)*, 77(1):45–51.

Goeleven, D. and Mentagui, D. (1995). Well-posed hemivariational inequalities. *Numerical Functional Analysis and Optimization*, 16(7–8):909–921.

Goeleven, D., Miettinen, M., and Panagiotopoulos, P. D. (1999). Dynamic hemivariational inequalities and their applications. *to appear in J. Opt. Theory Appl.*

Goeleven, D., Stavroulakis, G. E., and Panagiotopoulos, P. D. (1996). Solvability theory for a class of hemivariational inequalities involving copositive plus matrices. Applications in robotics. *Mathematical Programming Ser. A*, 75(3):441–465.

Goeleven, D. and Théra, M. (1995). Semicoercive variational hemivariational inequalities. *Journal of Global Optimization*, 6:367–381.

Haslinger, J., Baniotopoulos, C. C., and Panagiotopoulos, P. (1993). A boundary multivalued integral "equation" approach to the semipermeability problem. *Applications of Mathematics*, 38:39–60.

Li, X. and Carlsson, L. A. (1999). The tilted sandwich debond (TSD) specimen for face/core interface fracture characterization. *Journal of Sandwich Structures and Materials*, 1:60–75.

Miettinen, M. and Panagiotopoulos, P. D. (1999). On parabolic hemivariational inequalities and applications. *Nonlinear Analysis*, 35:885–915.

Mistakidis, E. S. and Stavroulakis, G. E. (1998). *Nonconvex optimization in mechanics. Smooth and nonsmooth algorithms, heuristics and engineering applications by the F.E.M.* Kluwer Academic Publisher, Dordrecht, Boston, London.

Moreau, J. and Panagiotopoulos, P. D., editors (1988). *Nonsmooth mechanics and applications*, volume 302 of *CISM Lect. Notes*, New York. Springer.

Moreau, J. J., Panagiotopoulos, P. D., and Strang, G., editors (1988). *Topics in nonsmooth mechanics*, Basel-Boston. Birkhäuser.

Motreanu, D. and Panagiotopoulos, P. (1998). *Minimax Theorems and Qualitative Properties of the Solutions of Hemivariational Inequalities*. Kluwer Academic Publisher, Dordrecht, Boston, London.

Motreanu, D. and Panagiotopoulos, P. D. (1995). An eigenvalue problem for a hemivariational inequality involving a nonlinear compact operator. *Set Valued Analysis*, 3.

Naniewicz, Z. and Panagiotopoulos, P. D. (1995). *Mathematical theory of hemivariational inequalities and applications*. Marcel Dekker, New York.

Panagiotopoulos, P. D. (1983). Nonconvex energy functions. Hemivariational inequalities and substationary principles. *Acta Mechanica*, 42:160–183.

Panagiotopoulos, P. D. (1985). *Inequality problems in mechanics and applications. Convex and nonconvex energy functions.* Birkhäuser, Basel, Boston, Stuttgart.

Panagiotopoulos, P. D. (1988). Nonconvex superpotentials and hemivariational inequalities. Quasidifferentiability in mechanics. In Moreau, J. J. and Panagiotopoulos, P. D., editors, *Nonsmooth Mechanics and Applications*, number 302 in CISM Lect. Notes, New York.

Panagiotopoulos, P. D. (1993). *Hemivariational inequalities. Applications in mechanics and engineering.* Springer, Berlin, Heidelberg, New York.

Panagiotopoulos, P. D. (1995). Variational principles for contact problems including impact phenomena. In Raous, M., Jean, M., and Moreau, J., editors, *Contact Mechanics*, pages 431–440. Plenum.

Panagiotopoulos, P. D. and Baniotopoulos, C. C. (1984). A hemivariational inequality and substationarity approach to the interface problem. Theory and prospects of applications. *Engineering Analysis*, 1:20–31.

Panagiotopoulos, P. D. and Glocker., C. (1998). Analytical mechanics. addendum i: Inequality constraints with elastic impacts. the convex case. *Zeitschrift für angewandte Mathematik und Mechanik (ZAMM)*, 78(4):219–229.

Panagiotopoulos, P. D. and Koltsakis, E. (1987). Interlayer slip and delamination effect. *Proc. Canadian Soc. Mech. Eng.*, 11:43–52.

Panagiotopoulos, P. D. and Liolios, A. (1989). On the dynamic of inelastic shocks. a new approach. In *Greek-German Seminar on Structural Dynamics and Earthquake Engineering*, pages 12–18. Hellenic Society of Theoretical and Applied Mechanics.

Panagiotopoulos, P. D. and Stavroulakis, G. E. (1988). Variational-hemivariational inequality approach to the laminated plate theory under subdifferential boundary conditions. *Quarterly of Applied Mathematics*, pages 409–430.

Panagiotopoulos, P. D. and Stavroulakis, G. E. (1990). The delamination effect in laminated von Karman plates under unilateral boundary conditions. A variational - hemivariational inequality approach. *Journal of Elasticity*, 23:69–96.

Panagiotopoulos, P. D. and Stavroulakis, G. E. (1992). New types of variational principles based on the notion of quasidifferentiability. *Acta Mechanica*, 94:171–194.

Pfeiffer, F. and Glocker, C. (1996). *Multibody dynamics with unilateral contacts*. John Wiley, New York.

Pop, G. and Panagiotopoulos, P. D. (1998). On a type of hyperbolic variational-hemivariational inequalities. *preprint*.

11 Finite Element Approximation of Hemivariational Inequalities

3 APPROXIMATION OF ELLIPTIC HEMIVARIATIONAL INEQUALITIES

From the previous chapter we know that there exist many important problems in mechanics in which constitutive laws are expressed by means of nonmonotone, possibly multivalued relations (nonmonotone multivalued stress-strain or reaction-displacement relations,e.g). The resulting mathematical model leads to an inclusion type problem involving multivalued nonmonotone mappings or to a substationary type problem for a nonsmooth, nonconvex superpotential expressed in terms of calculus of variation. It is the aim of this chapter to give a detailed study of a discretization of such a type of problems including the convergence analysis. Here we follow closely Miettinen and Haslinger, 1995, Miettinen and Haslinger, 1997.

We start this chapter by an abstract formulation of a class of static hemivariational inequalities of *scalar type*. Wording "scalar" means that a nonmonotone law relates two *scalar physical* quantities. At the end of this chapter we extend the analysis to a *vector case*, as well.

Let $V \subset H^1(\Omega; \mathbf{R}^d)$, $d \geq 1$, be a space of (*vector*) functions, defined in a bounded domain $\Omega \subset \mathbf{R}^n$ with the Lipschitz boundary $\partial\Omega$ and let V^* stand for its dual with a duality pairing denoted by $\langle \cdot, \cdot \rangle$. Throughout this chapter we shall suppose that

$$V \cap C^\infty(\overline{\Omega}; \mathbf{R}^d) \text{ is dense in } V. \tag{3.1}$$

Let $a : V \times V \to \mathbf{R}$ be a bounded, V-elliptic bilinear form and $f \in V^*$ be given. In order to define a nonmonotone relation, we first introduce a real valued function $b : \omega \times \mathbf{R} \to \mathbf{R}$. Next, we shall deal with two types of ω's:

(i) ω is *a subdomain* of Ω with the Lipschitz boundary $\partial\omega$;

(ii) $\omega \subset \partial\Omega$ is a nonempty, open portion (in $\partial\Omega$).

In what follows we shall suppose that b satisfies the following assumptions:

$$(x, \xi) \mapsto b(x, \xi) \text{ is measurable in } \omega \times \mathbf{R}; \qquad (3.2)$$
$$x \mapsto b(x, \xi) \text{ is continuous in } \omega \text{ for a.a. } \xi \in \mathbf{R}; \qquad (3.3)$$

$$\forall r > 0 \quad \exists c \equiv c(r) > 0 \text{ such that} \qquad (3.4)$$
$$|b(x, \xi)| \le c \quad \forall x \in \omega \text{ and a.a. } |\xi| \le r;$$

$$\exists \bar{\xi} > 0 \text{ such that} \qquad (3.5)$$
$$\operatorname*{ess\,sup}_{\xi \in (-\infty, -\bar{\xi})} \sup_{x \in \omega} b(x, \xi) \le 0 \le \operatorname*{ess\,inf}_{\xi \in (\bar{\xi}, \infty)} \inf_{x \in \omega} b(x, \xi).$$

Additional assumptions on b will appear at the moment, when it will be necessary.

With any b, satisfying (3.2)-(3.5) the *multifunction* $\hat{b} : \omega \times \mathbf{R} \to 2^{\mathbf{R}}$ will be associated. Below we describe its construction:

For any $\varepsilon > 0$ we first define two auxiliary functions $\underline{b}_\varepsilon, \bar{b}_\varepsilon : \omega \times \mathbf{R} \to \mathbf{R}$ by

$$\underline{b}_\varepsilon(x, \xi) = \operatorname*{ess\,inf}_{|\tau - \xi| \le \varepsilon} b(x, \tau), \quad \bar{b}_\varepsilon(x, \xi) = \operatorname*{ess\,sup}_{|\tau - \xi| \le \varepsilon} b(x, \tau). \qquad (3.6)$$

Because of their monotonicity, the following limits exist:

$$\underline{b}(x, \xi) = \lim_{\varepsilon \searrow 0+} \underline{b}_\varepsilon(x, \xi), \quad \bar{b}(x, \xi) = \lim_{\varepsilon \searrow 0+} \bar{b}_\varepsilon(x, \xi). \qquad (3.7)$$

Having $\underline{b}, \bar{b}$ at our disposal, we define

$$\hat{b}(x, \xi) = [\underline{b}(x, \xi), \bar{b}(x, \xi)], \quad (x, \xi) \in \omega \times \mathbf{R}. \qquad (3.8)$$

Roughly speaking, $\hat{b}$ results from the generally discontinuous function b by "filling in the gaps".

Further, let Z be a space of *real* valued functions, defined in ω and $\Pi : V \to Z$ be a linear continuous mapping satisfying

$$y \in V \cap C(\overline{\Omega}; \mathbf{R}^d) \implies \Pi y \in L^\infty(\omega). \qquad (3.9)$$

Other assumptions on Π will be specified later. Finally let Y be another space of real valued functions, defined in ω and which is in duality with Z.

Finally, let Y be another space of *real valued* functions, defined in ω. By $\langle \cdot, \cdot \rangle_{Y \times Z}$ we denote a pairing between Y and Z.

We start with

Definition 3.1 *A pair of functions $(u, \Xi) \in V \times Y$ is said to be a solution of a hemivariational inequality of scalar type iff*

$$\begin{cases} a(u,v) + \langle \Xi, \Pi v \rangle_{Y \times Z} = \langle f, v \rangle & \forall v \in V; \\ \Xi(x) \in \hat{b}(x, (\Pi u)(x)) & \text{for a.a. } x \in \omega. \end{cases} \tag{P}$$

Remark 3.1 *(Some comments to the definition of (P)). The choice of V as a subspace of $H^1(\Omega; \mathbf{R}^d)$ reflects the fact that hemivariational inequalities of the second order are treated.*

The choice of Y and Z strongly depends on the behaviour of the function b. In the next sections we shall show that $Y \equiv L^1(\omega) \cap V^*$ when (3.4) is satisfied. Since $Z \supset L^\infty(\omega)$ as follows from (3.9), the pairing between Y and Z will be understood in the following sense:

$$\langle \Xi, \Pi v \rangle_{Y \times Z} = \int_\omega \Xi \Pi v \, d\mu \quad \text{if } v \in V \cap C(\overline{\Omega}; \mathbf{R}^d),$$

where the integral over ω is either the volume integral, if ω is a subdomain of Ω or the surface integral, when $\omega \subset \partial\Omega$. If the mapping $\varphi : v \mapsto \langle \Xi, \Pi v \rangle_{Y \times Z}$, $v \in V \cap C(\overline{\Omega}; \mathbf{R}^d)$, will be continuous in V, then due to (3.1) one can extend φ to the whole V and Ξ will be identified with an element of V^*. Stronger assumptions on the growth of b, presented later, permit us to take $Y = L^q(\omega)$ for some $q > 1$. In this case $Z = L^{q'}(\omega)$, where $\frac{1}{q} + \frac{1}{q'} = 1$ and the duality between Y and Z will be realized by the integral over ω for any $v \in V$.

The presence of the mapping Π in the definition of (P) is due to the fact that elements of V are in general vector valued *functions, while (P) is a scalar hemivariational inequality. Here are possible examples of Π:*

- *if $d = 1$, i.e., V is the space of real valued functions then typically $\Pi y = y|_\omega$;*

- *if $d > 1$ and $y = (y_1, ..., y_d) \in V$ then $\Pi y = y_j|_\omega$ for some $j \in \{1, ..., d\}$ provided that ω is a subdomain of Ω;*

- *if $d > 1$ and $\omega \subset \partial\Omega$ then*

$$\Pi y = y.\nu|_\omega,$$

i.e., Π associates with y its normal component $y.\nu$ on ω.

Let us apply the previous abstract setting to two examples of hemivariational inequalities. We start with the one-dimensional case, presented in Introduction.

Example 3.1 We have:

$$V = \{v \in H^2(I) \mid v(0) = v'(0) = 0\}, \quad I = (0, l),$$

$$Y = Z = \mathbf{R}, \quad \omega = \{l\}, \quad \Pi v = v(l),$$

$$a(u, v) = \int_I u''v'' dx, \quad \langle f, v \rangle = \int_I fv dx, \quad f \in L^2(I)$$

$$\hat{b}(x, \xi) \equiv \hat{b}(\xi) = \begin{cases} 0, & \text{if } \xi \in (-\infty, 0) \cup (a, \infty), \ a > 0, \\ k\xi, & \text{if } \xi \in (0, a), \\ [0, ka], & \text{if } \xi = a. \end{cases}$$

Since $H^2(I) \hookrightarrow C^1(I)$, the mapping Π is continuous from V to Z and the corresponding hemivariational inequality takes the form:

$$\begin{cases} \text{Find } (u, \Xi) \in V \times \mathbf{R} \text{ such that} \\ a(u, v) + \Xi v(l) = \langle f, v \rangle \quad \forall v \in V \\ \Xi \in \hat{b}(u(l)). \end{cases}$$

Example 3.2 Let us consider a plane, elastic body Ω subject to body forces F and surface tractions P, which is in a bilateral contact with a rigid foundation, and obeying a nonmonotone simplified friction law on a part Γ_c. The boundary $\partial\Omega$ is divided into three open disjoint sets Γ_u, Γ_P and Γ_c (see Fig.3.1).

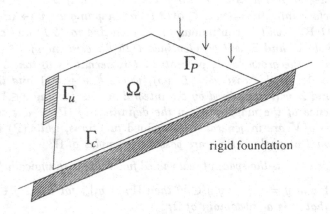

Figure 3.1.

We look for a displacement field $u = (u_1, u_2)$ satisfying the following equations and the boundary conditions (the meaning of the symbols has been introduced in Section 1.3):

$$-\quad \tau_{ij,j}(u) + F_i = 0 \quad \text{in } \Omega, \quad i = 1, 2;$$

$$-\quad \tau_{ij}(u) = c_{ijkl}\varepsilon_{kl}(u), \quad \varepsilon_{kl}(u) = \frac{1}{2}\left(\frac{\partial u_k}{\partial x_l} + \frac{\partial u_l}{\partial x_k}\right);$$

$$-\quad u_i = 0 \quad \text{on } \Gamma_u, \quad i = 1, 2;$$

$$-\quad \tau_{ij}\nu_j = P_i \quad \text{on } \Gamma_P, \quad i = 1, 2;$$

$$-\quad u_\nu = 0, \quad \tau_{ij}\nu_j t_i + \Xi = 0 \quad \text{on } \Gamma_c;$$

$$-\quad \Xi(x) \in \hat{b}(u_t(x)) \quad \text{on } \Gamma_c.$$

The symbol $\hat{b}$ stands for a multifunction, describing the nonmonotone friction law and satisfying (3.2)-(3.5). Then

$$V = \{v \in H^1(\Omega; \mathbf{R}^2) \mid v = 0 \text{ on } \Gamma_u, \ v_\nu = 0 \text{ on } \Gamma_c\},$$

$$Y = L^1(\Gamma_c) \cap V^*, \quad Z = L^2(\Gamma_c), \quad \omega = \Gamma_c, \quad \Pi v = v_t|_{\Gamma_c},$$

$$a(u, v) = \int_\Omega c_{ijkl}\varepsilon_{ij}(u)\varepsilon_{kl}(v)dx,$$

$$\langle f, v \rangle = \int_\Omega F_i v_i dx + \int_{\Gamma_P} P_i v_i ds, \quad F \in L^2(\Omega; \mathbf{R}^2), \ P \in L^2(\Gamma_P; \mathbf{R}^2).$$

From the Rellich's theorem it follows that Π is continuous from V into Z and (3.9) clearly holds.

The corresponding hemivariational inequality, describing the equilibrium state of Ω reads as follows:

$$\begin{cases} \text{Find } (u, \Xi) \in V \times L^1(\Gamma_c) \cap V^* \text{ such that} \\ a(u, v) + \langle \Xi, v_t \rangle_{Y \times Z} = \langle f, v \rangle \quad \forall v \in V; \\ \Xi(x) \in \hat{b}(u_t(x)) \quad \text{a.e. on } \Gamma_c, \end{cases}$$

where the duality between $L^1(\Gamma_c) \cap V^*$ and Z is understood in the sense of Remark 3.1.

In the subsequent parts of this chapter we shall introduce and analyse the *full approximation* of (P), i.e., we shall approximate both components u and Ξ of the solution by appropriate u_h and Ξ_h, respectively, lying in their own finite dimensional spaces. Besides of that we shall approximate also the bilinear form a and the linear term f (by using a numerical integration, e.g.). Under appropriate assumptions on approximated data we shall show that approximate solutions are close on subsequences to solutions of (P) in a suitable topology. As a byproduct, the existence result for (P) will be recovered.

3.1   AUXILIARY RESULTS

We start by presentation of auxiliary results, which will be used in subsequent parts.

Let $b : \omega \times \mathbf{R} \to \mathbf{R}$ be a function satisfying (3.2)-(3.5). Then it is readily seen that there exists $\rho > 0$ such that

$$\begin{cases} b(x, \xi) \geq 0 & \forall x \in \omega \quad \text{and a.a. } \xi \in (\bar{\xi}, \infty); \\ b(x, \xi) \leq 0 & \forall x \in \omega \quad \text{and a.a. } \xi \in (-\infty, -\bar{\xi}); \\ |b(x, \xi)| \leq \rho & \forall x \in \omega \quad \text{and a.a. } \xi \in (-\bar{\xi}, \bar{\xi}). \end{cases} \quad (3.10)$$

Let $\beta \in C_0^\infty([-1, 1])$ be such that $\beta \geq 0$ in $[-1, 1]$, $\int_{-1}^1 \beta(\tau)d\tau = 1$ and define the *regularization* of b as follows:

$$\begin{aligned} b^\kappa(x, \xi) &= \frac{1}{\kappa} \int_{-\infty}^\infty b(x, \xi - \tau)\beta\left(\frac{\tau}{\kappa}\right)d\tau \quad (3.11) \\ &= -\frac{1}{\kappa} \int_{-\infty}^\infty b(x, \tau)\beta\left(\frac{\xi - \tau}{\kappa}\right)d\tau \end{aligned}$$

with $\kappa \to 0+$. Properties of b^κ are summarized in

Lemma 3.1 *Let $b : \omega \times \mathbf{R} \to \mathbf{R}$ satisfy (3.2)-(3.5). Then for any $\kappa > 0$ it holds that*

$$-\text{the function } b^\kappa \in C(\omega \times \mathbf{R}); \tag{3.12}$$

- there exist two numbers $\hat{\rho}, \hat{\xi} > 0$ independent of $\kappa > 0$ such that

$$\begin{cases} b^\kappa(x,\xi) \geq 0 & \forall x \in \omega \quad \forall \xi \geq \hat{\xi}; \\ b^\kappa(x,\xi) \leq 0 & \forall x \in \omega \quad \forall \xi \leq -\hat{\xi}; \\ |b^\kappa(x,\xi)| \leq \hat{\rho} & \forall x \in \omega \quad \forall |\xi| \leq \hat{\xi}. \end{cases} \tag{3.13}$$

Proof: (3.12) results from the well-known theorem on the continuity of the Lebesgue integral with respect to a parameter and from (3.3).

From the definition of β we see that

$$b^\kappa(x,\xi) = \frac{1}{\kappa} \int_{-\kappa}^{\kappa} b(x,\xi - \tau)\beta\left(\frac{\tau}{\kappa}\right)d\tau \tag{3.14}$$

$$\leq \underset{|\tau| \leq \kappa}{\text{ess sup}} b(x,\xi - \tau) \equiv \overline{b}_\kappa(x,\xi)$$

and, similarly:

$$b^\kappa(x,\xi) \geq \underset{|\tau| \leq \kappa}{\text{ess inf}} b(x,\xi - \tau) \equiv \underline{b}_\kappa(x,\xi). \tag{3.15}$$

From (3.14) and (3.15) we easily arrive at (3.13).   □

Remark 3.2 *A result similar to (3.13) (with possibly different $\hat{\rho}, \hat{\xi}$) holds true for $\underline{b}_\kappa, \overline{b}_\kappa, \underline{b}, \overline{b},$ as well.*

In the sequel also the continuity of the mappings

$$x \mapsto \underline{b}_\varepsilon(x,\xi), \quad x \mapsto \overline{b}_\varepsilon(x,\xi) \qquad \forall \xi \in \mathbf{R}$$

will be needed. In order to guarantee this property we shall suppose that the mapping $x \mapsto b(x,\xi)$, $x \in \omega$, is *uniformly continuous* in the following sense:

$$\begin{cases} \text{given } (x,\xi) \in \omega \times \mathbf{R} \text{ and } \delta > 0 \ \exists \varepsilon_0 > 0 \\ \text{and } \gamma \equiv \gamma(x,\xi,\delta) \text{ such that} \\ \text{for a.a. } \xi' \in (\xi - \varepsilon_0, \xi + \varepsilon_0) \text{ and} \\ \text{any } x' \in (x - \gamma, x + \gamma) : |b(x,\xi') - b(x',\xi')| < \delta. \end{cases} \tag{3.16}$$

Then it is easy to prove

Lemma 3.2 *Let (3.16) be satisfied. Then the functions*

$$x \mapsto \underline{b}_\varepsilon(x,\xi), \quad x \mapsto \overline{b}_\varepsilon(x,\xi)$$

are continuous in ω for any $\xi \in \mathbf{R}$ and any $\varepsilon > 0$.

Another important result needed in what follows is

Proposition 3.1 *Let $\{\eta_k\}$, $\{\xi_k\}$ be such that $\eta_k \in \hat{b}(x, \xi_k)$ for any k (keeping x fixed) and $\eta_k \to \eta$, $\xi_k \to \xi$. Then $\eta \in \hat{b}(x, \xi)$.*

Proof: Since $\eta_k \in \hat{b}(x, \xi_k)$, then it holds that

$$\underline{b}_\varepsilon(x, \xi_k) \le \eta_k \le \overline{b}_\varepsilon(x, \xi_k) \tag{3.17}$$

for any $\varepsilon > 0$. Let $\varepsilon > 0$ be fixed. Then for k sufficiently large we have

$$\overline{b}_\varepsilon(x, \xi_k) \le \overline{b}_{2\varepsilon}(x, \xi)$$
$$\underline{b}_\varepsilon(x, \xi_k) \ge \underline{b}_{2\varepsilon}(x, \xi).$$

From this and (3.17) we obtain

$$\underline{b}_{2\varepsilon}(x, \xi) \le \eta_k \le \overline{b}_{2\varepsilon}(x, \xi). \tag{3.18}$$

Passing to the limit with $k \to \infty$, then with $\varepsilon \to 0+$ we arrive at the assertion. $\square$

Lemma 3.3 *The function*

$$(x, \xi) \mapsto \overline{b}_\varepsilon(x, \xi), \quad (x, \xi) \in \omega \times \mathbf{R},$$

is lower semicontinuous on $\omega \times \mathbf{R}$, and, the function

$$(x, \xi) \mapsto \underline{b}_\varepsilon(x, \xi) \quad (x, \xi) \in \omega \times \mathbf{R},$$

is upper semicontinuous on $\omega \times \mathbf{R}$.

Proof: We prove only the first part, since the second one can be proven in the same way. Let $(x_0, \xi_0) \in \omega \times \mathbf{R}$ be given and $\{(x_k, \xi_k)\}$ be a sequence tending to (x_0, ξ_0). Let $\delta > 0$ be given. From the definition of $\overline{b}_\varepsilon$ it follows that there exists a subset $B_\delta \subset (\xi_0 - (\varepsilon - \gamma_\delta), \xi_0 + (\varepsilon - \gamma_\delta))$, $0 < \gamma_\delta < \varepsilon$, such that meas $B_\delta > 0$ and

$$b(x_0, \xi) \ge \overline{b}_\varepsilon(x_0, \xi_0) - \delta \quad \forall \xi \in B_\delta. \tag{3.19}$$

Using the fact that $x_k \to x_0$ we have

$$\liminf_{k \to \infty} \int_{B_\delta} b(x_k, \xi) d\xi \ge \int_{B_\delta} \liminf_{k \to \infty} b(x_k, \xi) d\xi$$
$$\overset{(3.3)}{=} \int_{B_\delta} b(x_0, \xi) d\xi \overset{(3.19)}{\ge} \text{meas } B_\delta (\overline{b}_\varepsilon(x_0, \xi_0) - \delta).$$

Hence there exists $k_0 \in \mathbf{N}$ such that $\forall k \ge k_0 \ \exists B_k \subset B_\delta$, meas $B_k > 0$, and

$$b(x_k, \xi) \ge \overline{b}_\varepsilon(x_0, \xi_0) - 2\delta \quad \forall \xi \in B_k. \tag{3.20}$$

Let us choose $k_1 \in \mathbf{N}$ such that $|\xi_0 - \xi_k| \leq \gamma_\delta$ as $k \geq k_1$. If $k \geq \max(k_0, k_1)$, then $B_k \subset (\xi_k - \varepsilon, \xi_k + \varepsilon)$ and

$$\overline{b}_\varepsilon(x_k, \xi_k) \geq \overline{b}_\varepsilon(x_0, \xi_0) - 2\delta$$

making use of (3.20). Letting $\delta \to 0+$, we obtain

$$\liminf_{k \to \infty} \overline{b}_\varepsilon(x_k, \xi_k) \geq \overline{b}_\varepsilon(x_0, \xi_0).$$

$\square$

3.2 DISCRETIZATION

In this section we define an appropriate discretization of (P) and prove the existence of its solutions. As mentioned at the beginning of this chapter, we shall approximate simultaneously *both* components of the solution. For this reason we shall introduce two types of finite element spaces, one for the approximation of u and the other approximating Ξ.

Let $h > 0$ be a *discretization parameter* and $\mathcal{D}_h, \mathcal{T}_h$ be partitions of $\overline{\Omega}$ and $\overline{\omega}$, respectively, whose norms do not exceed h. With any $\mathcal{D}_h$ and $\mathcal{T}_h$ two finite element spaces $V_h \subset V$ and $Y_h \subset Y$ will be associated:

$$V_h = \{v_h \in C(\overline{\Omega}; \mathbf{R}^d) \mid v_h|_T \in (P_k(T))^d \ \forall T \in \mathcal{D}_h\} \cap V \quad (3.21)$$

$$Y_h = \{\mu_h \in L^\infty(\omega) \mid \mu_h|_K \in P_0(K) \ \forall K \in \mathcal{T}_h\}. \quad (3.22)$$

The fact that Y_h is the space of *piecewise constant* functions over $\mathcal{T}_h$ will be important in what follows.

Remark 3.3 *The partition $\mathcal{D}_h$ characterizing V_h will be a classical one made of triangles, tetrahedrons,... and satisfying (1.246)-(1.249). On the contrary, the partition $\mathcal{T}_h$ used when defining Y_h will use elements, whose shapes can be more complicated (quadilaterals, polygons,...).*

Next we shall suppose that any $K_i \in \mathcal{T}_h$ is *a closed subset* of $\overline{\omega}$ with $\mathrm{int}_\omega K_i \neq \emptyset \ \forall i$,

$$\overline{\omega} = \cup_{i=1}^m K_i$$

and $\mathrm{int}_\omega K_i \cap \mathrm{int}_\omega K_j = \emptyset$ for $i \neq j$, where the symbol $\mathrm{int}_\omega K$ stands for the interior of K in ω.

Denote by $W_h \subset Z$ the image of V_h with respect to Π:

$$W_h = \Pi(V_h) \quad (3.23)$$

and define a *linear mapping P_h* from W_h into Y_h.

Now, in each $K_i \in \mathcal{T}_h$ exactly one point $x_h^i \in K_i$ will be selected. The values of functions from Y_h at $\{x_h^i\}$ will be interpreted as the degrees of freedom, i.e., if $\mu_h \in Y_h$ then there exist $c_i \in \mathbf{R}$, $i = 1, ..., m$, such that

$$\mu_h = \sum_{i=1}^m c_i \mathcal{X}_{\mathrm{int}_\omega K_i}(x)$$

and $c_i = \mu_h(x_h^i)$. The symbol $\mathcal{X}_\theta$ stands for the characteristic function of a set θ.

Next we shall consider a *sequence* of parameters h tending to zero and the respective families of $\{\mathcal{D}_h\}$, $\{\mathcal{T}_h\}$, $\{V_h\}$ and $\{Y_h\}$.

As we have already mentioned, the bilinear form a and the linear term f can be also approximated. To this end we introduce a sequence $\{a_h\}$, where each $a_h \in \{a_h\}$ is a bilinear form defined in $V_h \times V_h$ and satisfying:

(*the uniform boundedness*):

$$\exists \tilde{M} > 0 : |a_h(y_h, z_h)| \leq \tilde{M}\|y_h\|\|z_h\| \quad \forall y_h, z_h \in V_h, \ \forall h \to 0+; \quad (3.24)$$

(*the uniform V_h-ellipticity*):

$$\exists \tilde{\alpha} > 0 : a_h(y_h, y_h) \geq \tilde{\alpha}\|y_h\|^2 \quad \forall y_h \in V_h, \ \forall h \to 0+ . \quad (3.25)$$

Similarly, a sequence $\{f_h\}$, $f_h : V_h \to \mathbf{R}$ will approximate $f \in V^*$. We suppose that

(*the uniform boundedness*):

$$\exists \tilde{\beta} > 0 : |\langle f_h, z_h \rangle_h| \leq \tilde{\beta}\|z_h\| \quad \forall z_h \in V_h, \ \forall h \to 0+, \quad (3.26)$$

where $\langle \cdot, \cdot \rangle_h$ denotes a duality pairing between V_h and V_h^*.

Remark 3.4 *The uniform boundedness of a_h, f_h and the uniform V_h-ellipticity of a_h with respect to h are required to get estimates which do not depend on h.*

The *discretization* of (P) now reads as follows:

$$\left\{ \begin{array}{l} \text{Find } (u_h, \Xi_h) \in V_h \times Y_h \text{ such that} \\ a_h(u_h, v_h) + \displaystyle\int_\omega \Xi_h P_h(\Pi v_h) d\mu = \langle f_h, v_h \rangle_h \quad \forall v_h \in V_h \\ \Xi_h(x) \in \hat{b}(\displaystyle\sum_{i=1}^m \mathcal{X}_{\mathrm{int}_\omega K_i}(x) x_h^i, P_h(\Pi u_h)(x)) \quad \text{for a.a. } x \in \omega. \end{array} \right. \quad (\mathrm{P})_h$$

The integral over ω will be interpreted either as the volume integral if $\mathrm{int}_\Omega \omega \neq \emptyset$ or the surface integral if $\omega \subset \partial\Omega$.

Remark 3.5 *Since both, Ξ_h and $P_h(\Pi u_h)$ are piecewise constant in ω (elements of Y_h), the last inclusion in $(\mathrm{P})_h$ is equivalent to the following m inclusions at the points x_h^i, $i = 1, ..., m$:*

$$\Xi_h(x_h^i) \in \hat{b}(x_h^i, P_h(\Pi u_h)(x_h^i)) \quad \forall i = 1, ..., m.$$

If higher order polynomials were used for the construction of Y_h, such an equivalence would be no longer true.

In what follows we prove the existence of solutions to $(P)_h$. To this end we introduce the following *regularized* version of $(P)_h$:

$$
\begin{cases}
\text{Find } u_h^\kappa \in V_h \text{ such that} \\
a_h(u_h^\kappa, v_h) + \displaystyle\int_\omega b^\kappa(\sum_{i=1}^m \mathcal{X}_{\text{int}_\omega K_i}(x) x_h^i, P_h(\Pi u_h^\kappa)(x)) \\
P_h(\Pi v_h) d\mu = \langle f_h, v_h \rangle_h \quad \forall v_h \in V_h,
\end{cases}
\qquad (P)_h^\kappa
$$

where b^κ is the regularization of b defined by (3.11) and $\kappa \to 0+$.

We start by proving the existence of solutions to $(P)_h^\kappa$.

Lemma 3.4 *Let all the assumptions concerning of a_h, f_h, b and P_h be satisfied. Then $(P)_h^\kappa$ has at least one solution u_h^κ for any $h, \kappa > 0$, which is bounded uniformly with respect to $h, \kappa > 0$:*

$$
\exists r > 0 : \|u_h^\kappa\| \le r \quad \forall h, \kappa > 0.
$$

Proof: Define a mapping $T_h^\kappa : V_h \to V_h^*$ by

$$
\langle T_h^\kappa y_h, z_h \rangle_h \equiv a_h(y_h, z_h) + \int_\omega b^\kappa(\sum_{i=1}^m \mathcal{X}_{\text{int}_\omega K_i}(x) x_h^i, \qquad (3.27)
$$

$$
P_h(\Pi y_h)(x)) P_h(\Pi z_h)(x) d\mu - \langle f_h, z_h \rangle_h.
$$

This mapping is continuous in V_h. Indeed, for any $y_h, \bar{y}_h \in V_h$ it holds:

$$
\|T_h^\kappa y_h - T_h^\kappa \bar{y}_h\|_{*,h} \equiv \sup_{\|z_h\| \le 1} \langle T_h^\kappa y_h - T_h^\kappa \bar{y}_h, z_h \rangle_h
$$

$$
\le \tilde{M}\|y_h - \bar{y}_h\| + c \int_\omega |b^\kappa(\sum_{i=1}^m \mathcal{X}_{\text{int}_\omega K_i}(x) x_h^i, P_h(\Pi y_h)(x))
$$

$$
- b^\kappa(\sum_{i=1}^m \mathcal{X}_{\text{int}_\omega K_i}(x) x_h^i, P_h(\Pi \bar{y}_h)(x))| d\mu
$$

as follows from (3.24) and the fact that P_h and Π are linear and V_h is finite dimensional. If $\bar{y}_h \to y_h$ then $T_h^\kappa \bar{y}_h \to T_h^\kappa y_h$ making use of (3.12).

Next we show that there exists a number $r > 0$ independent of h and κ such that

$$
\langle T_h^\kappa z_h, z_h \rangle_h > 0 \quad \forall \|z_h\| = r, \ \forall h, \kappa > 0. \qquad (3.28)
$$

The first and the third term on the right hand side of (3.27) can be estimated from below:

$$
a_h(z_h, z_h) - \langle f_h, z_h \rangle_h \ge \tilde{\alpha}\|z_h\|^2 - \tilde{\beta}\|z_h\| \qquad (3.29)
$$

as follows from (3.25) and (3.26). Further:

$$
\int_\Omega b^\kappa(\sum_{i=1}^m \mathcal{X}_{\text{int}_\omega K_i}(x) x_h^i, P_h(\Pi z_h)(x)) P_h(\Pi z_h)(x) d\mu
$$

$$
= \int_{|P_h(\Pi z_h)| \le \hat{\xi}} + \int_{|P_h(\Pi z_h)| > \hat{\xi}},
$$

where $\hat{\xi}$ is the same as in Lemma 3.1. The second integral on the right hand side is nonnegative, as follows from $(3.13)_{1,2}$, while the first one is bounded from below:

$$\int_{|P_h(\Pi z_h)| \leq \hat{\xi}} \geq -\hat{\rho}\hat{\xi} \operatorname{meas} \omega \tag{3.30}$$

making use of $(3.13)_3$. From (3.29) and (3.30) we finally obtain:

$$\langle T_h^\kappa z_h, z_h \rangle_h \geq \tilde{\alpha}\|z_h\|^2 - \tilde{\beta}\|z_h\| - \hat{\rho}\hat{\xi} \operatorname{meas} \omega.$$

Since $\tilde{\alpha}, \tilde{\beta}, \hat{\rho}, \hat{\xi}$ are independent of κ and h, one can find $r > 0$ such that (3.28) holds. The existence of $u_h^\kappa \in V_h$ solving $T_h^\kappa u_h^\kappa = 0$ and such that $\|u_h^\kappa\| \leq r$ $\forall h, \kappa > 0$ now follows from Theorem 1.26. $\qquad\square$

The existence of solutions to $(P)_h$ will be now proven by letting $\kappa \to 0+$ in $(P)_h^\kappa$. Indeed, we have:

Theorem 3.1 *Let all the assumptions, concerning of a_h, f_h, b and P_h be satisfied. Then for any $h > 0$ there exists a solution (u_h, Ξ_h) of $(P)_h$ such that $\|u_h\| \leq r$, where r does not depend on h.*

Proof: From Lemma 3.4 we know that $(P)_h^\kappa$ has a solution u_h^κ satisfying $\|u_h^\kappa\| \leq r$ for any $\kappa > 0$. Thus there exist : a subsequence of $\{u_h^\kappa\}$ and a function $u_h \in V_h$ such that[*]

$$u_h^\kappa \to u_h \quad \text{as } \kappa \to 0+ \tag{3.31}$$

and $\|u_h\| \leq r$ $\forall h > 0$. From (3.4) we see that also the sequence $\{b^\kappa(x_h^i, P_h(\Pi u_h^\kappa)(x_h^i))\}$ is bounded so that a function $\Xi_h \in Y_h$ exists and

$$b^\kappa(x_h^i, P_h(\Pi u_h^\kappa)(x_h^i)) \to \Xi_h(x_h^i), \quad \kappa \to 0+ \tag{3.32}$$

for all $i = 1, ..., m$. Next we show that the pair $(u_h, \Xi_h) \in V_h \times Y_h$ is a solution to $(P)_h$.

Letting $\kappa \to 0+$ in $(P)_h^\kappa$ and using (3.31) and (3.32) we arrive at

$$a_h(u_h, v_h) + \int_\omega \Xi_h P_h(\Pi v_h) d\mu = \langle f_h, v_h \rangle_h \quad \forall v_h \in V_h.$$

It remains to verify that $\Xi_h(x_h^i) \in \hat{b}(x_h^i, P_h(\Pi u_h)(x_h^i))$ $\forall i = 1, ..., m$ (see Remark 3.5). Since both P_h and Π are linear and the spaces are finite dimensional, it follows from (3.31) that

$$P_h(\Pi u_h^\kappa) \to P_h(\Pi u_h) \quad \text{in } L^\infty(\omega)$$

as $\kappa \to 0+$, i.e., for any $\varepsilon > 0$ there exists $\kappa_0 > 0$ such that

$$|P_h(\Pi u_h^\kappa)(x) - P_h(\Pi u_h)(x)| < \frac{\varepsilon}{2} \tag{3.33}$$

---

[*]In what follows we shall denote subsequences by the same symbols as the original sequences.

holds for any $x \in \text{int}_\omega K_i$, $i = 1, ..., m$ and $\kappa \leq \kappa_0$. Taking $\kappa \leq \min(\kappa_0, \frac{\varepsilon}{2})$ we have

$$b^\kappa(x_h^i, P_h(\Pi u_h^\kappa)(x_h^i)) \overset{(3.14)}{\leq} \bar{b}_\kappa(x_h^i, P_h(\Pi u_h^\kappa)(x_h^i))$$

$$\leq \bar{b}_{\frac{\varepsilon}{2}}(x_h^i, P_h(\Pi u_h^\kappa)(x_h^i)) \overset{(3.33)}{\leq} \bar{b}_\varepsilon(x_h^i, P_h(\Pi u_h)(x_h^i)).$$

Here we also used the monotonicity of $\bar{b}_\varepsilon$ with respect to ε. Letting first $\kappa \to 0+$ and then $\varepsilon \to 0+$ we arrive at $\Xi_h(x_h^i) \leq \bar{b}(x_h^i, P_h(\Pi u_h)(x_h^i))$ $\forall i = 1, ..., m$. The inequality $\Xi_h(x_h^i) \geq \underline{b}(x_h^i, P_h(\Pi u_h)(x_h^i))$ can be proven in a similar way.   □

Remark 3.6 *The fact that Y_h is the space of piecewise constant functions over $\mathcal{T}_h$ is very important. Indeed, if it is so then the mapping*

$$x \mapsto b^\kappa(\sum_{i=1}^m \mathcal{X}_{\text{int}_\omega K_i}(x)x_h^i, P_h(\Pi u_h)(x)), \quad x \in \omega$$

is piecewise constant over $\mathcal{T}_h$, as well. This, together with (3.32) yield

$$\int_\omega b^\kappa(\sum_{i=1}^m \mathcal{X}_{\text{int}_\omega K_i}(x)x_h^i, P_h(\Pi u_h)(x))P_h(\Pi z_h)(x)d\mu$$

$$\to \int_\omega \Xi_h P_h(\Pi z_h)d\mu, \quad \kappa \to 0+.$$

The solution (u_h, Ξ_h) of $(P)_h$ is not unique, in general. Let us examine the uniqueness of Ξ_h at the moment, when u_h is already at our disposal. This problem is of a practical importance. Numerical methods which will be used for the realization of $(P)_h$ enable us to find u_h, only. Provided that Ξ_h is unique (for such u_h), one can recover it from the equation in $(P)_h$.

We can formulate the following simple result, guaranteeing the uniqueness of Ξ_h knowing u_h.

Theorem 3.2 *Let the mapping P_h maps W_h onto Y_h. If $(u_h, \Xi_h) \in V_h \times Y_h$ is a solution to $(P)_h$, then the second component Ξ_h is unique, keeping u_h fixed.*

Proof: Let (u_h, Ξ_h), $(u_h, \bar{\Xi}_h)$ be two solutions to $(P)_h$. Then the difference $(\Xi_h - \bar{\Xi}_h) \in Y_h$ satisfies

$$\int_\omega (\Xi_h - \bar{\Xi}_h)P_h(\Pi v_h)d\mu = 0 \quad \forall v_h \in V_h. \tag{3.34}$$

By assumption there exists $\bar{v}_h \in V_h$ such that $P_h(\Pi \bar{v}_h) = \Xi_h - \bar{\Xi}_h$. From this and (3.34) we see that $\Xi_h = \bar{\Xi}_h$ in ω.   □

Remark 3.7 *Let $\dim W_h = \dim Y_h$. Then P_h maps W_h onto Y_h iff from $P_h w_h = 0$, $w_h \in W_h$, we have that $w_h = 0$.*

3.3   CONVERGENCE ANALYSIS

This section will be devoted to the study of the relation between solutions to $(P)_h$ and (P), when the discretization parameter $h \to 0+$. We shall show that the corresponding solutions are close on subsequences in the weak topology of $V \times L^1(\omega)$. Stronger assumptions on b will enable us to improve the results.

We shall need additional assumptions on the approximating data. In what follows, we shall suppose that:

$$\forall v \in V \cap C^\infty(\overline{\Omega}; \mathbf{R}^d) \; \exists \{v_h\}, \; v_h \in V_h, \text{ such that} \qquad (3.35)$$
$$v_h \to v \quad \text{in the } V \text{ and } C(\overline{\Omega}; \mathbf{R}^d)\text{-norms.}$$

Further we suppose that there is a rule which with any $V_h \in \{V_h\}$ associates the unique $Y_h \in \{Y_h\}$ (we write $V_h \rightsquigarrow Y_h$) and the cartesian product $V_h \times Y_h$ is then used in $(P)_h$. The family $\{V_h \times Y_h\}$, where $V_h \rightsquigarrow Y_h$, has to be such that the mappings $P_h : W_h \to Y_h$ introduced in Section 3.2 satisfy (recall that $W_h = \Pi(V_h)$):

$$y_h \rightharpoonup y \text{ in } V, \; y_h \in V_h \implies \exists a \text{ subsequence of } \{y_h\} \text{ such that} \qquad (3.36)$$
$$P_h(\Pi y_h)(x) \to \Pi y(x) \text{ for a.a. } x \in \omega \text{ as } h \to 0+;$$

$$y_h \to y \in V \cap C^\infty(\overline{\Omega}; \mathbf{R}^d) \text{ in } C(\overline{\Omega}; \mathbf{R}^d), \; y_h \in V_h \qquad (3.37)$$
$$\implies P_h(\Pi y_h) \to \Pi y \quad \text{in } L^\infty(\omega), \text{ as } h \to 0+.$$

Finally, we shall need the following assumptions relating $\{a_h\}$ to a and $\{f_h\}$ to f:

$$y_h \rightharpoonup y, \; z_h \to z \quad \text{in } V, \; y_h, z_h \in V_h \implies a_h(y_h, z_h) \to a(y, z); \qquad (3.38)$$

$$z_h \to z \quad \text{in } V, \; z_h \in V_h \implies \langle f_h, z_h \rangle_h \to \langle f, z \rangle \text{ as } h \to 0+. \qquad (3.39)$$

Then we prove

Theorem 3.3 *Let all the assumptions concerning of $\{V_h\}$, $\{a_h\}$, $\{f_h\}$, $\{P_h\}$ and b be satisfied. Let $\{(u_h, \Xi_h)\}$ be a sequence of solutions to $(P)_h$ with $\{u_h\}$ being bounded in V independently of h. Then there exist : a subsequence of $\{(u_h, \Xi_h)\}$ and an element $(u, \Xi) \in V \times L^1(\omega) \cap V^*$ such that*

$$\begin{cases} u_h \rightharpoonup u & \text{in } V, \\ \Xi_h \rightharpoonup \Xi & \text{in } L^1(\omega), \; h \to 0+. \end{cases} \qquad (3.40)$$

Moreover, (u, Ξ) is a solution to (P). Furthermore, any cluster point of $\{(u_h, \Xi_h)\}$ in the sense of (3.40) is a solution to (P).

Proof: The sequence $\{u_h\}$ being bounded in V one can find a subsequence of $\{u_h\}$ and $u \in V$ such that $(3.40)_1$ holds. In the next step we shall show that $\{\Xi_h\}$ is weakly compact in $L^1(\omega)$ by using Theorem 1.4.

First we estimate the integral $\int_\omega |\Xi_h P_h(\Pi u_h)| d\mu$ as follows (see Remark 3.2):

$$\int_\omega |\Xi_h P_h(\Pi u_h)| d\mu = \int_{|P_h(\Pi u_h)|>\hat\xi} |\Xi_h P_h(\Pi u_h)| d\mu \qquad (3.41)$$

$$+ \int_{|P_h(\Pi u_h)|\leq\hat\xi} |\Xi_h P_h(\Pi u_h)| d\mu$$

$\leq$ (the first integrand is nonnegative $\Longrightarrow$)

$$\leq \int_\omega \Xi_h P_h(\Pi u_h) d\mu + 2\int_{|P_h(\Pi u_h)|\leq\hat\xi} |\Xi_h P_h(\Pi u_h)| d\mu$$

$\leq$ (the definition of $(P)_h \Longrightarrow$)

$$\leq -a_h(u_h,u_h) + \langle f_h,u_h\rangle_h + 2\hat\rho\hat\xi \operatorname{meas}\omega$$

$$\overset{(3.26)}{\leq} \tilde\beta\|u_h\| + 2\hat\rho\hat\xi \operatorname{meas}\omega \leq C$$

holds for any $h > 0$. From (3.41) we see that for any $\gamma > 0$ given, there exists $q_0 > 0$ such that

$$\frac{1}{q_0}\int_\omega |\Xi_h P_h(\Pi u_h)| d\mu \leq \frac{\gamma}{2} \quad \forall h > 0. \qquad (3.42)$$

From (3.4) the existence of $\delta > 0$ such that

$$\delta \underset{|P_h(\Pi u_h)(x)|\leq q_0}{\operatorname{ess\,sup}} |\Xi_h(x)| \leq \frac{\gamma}{2} \quad \forall h > 0 \qquad (3.43)$$

follows. Let $\omega_0 \subset \omega$ be such that $\operatorname{meas}\omega_0 < \delta$. Then from (3.42),(3.43) and the inequality

$$|\Xi_h(x)| \leq \frac{1}{q_0}|\Xi_h P_h(\Pi u_h)| + \underset{|P_h(\Pi u_h)(x)|\leq q_0}{\operatorname{ess\,sup}} |\Xi_h(x)|$$

valid in ω we see that the integrals $\int_{\omega_0} |\Xi_h| d\mu$ are bounded by γ for any $h > 0$ whenever $\operatorname{meas}\omega_0 < \delta$. From Theorem 1.4 the existence of a subsequence of $\{\Xi_h\}$ and of a function $\Xi \in L^1(\omega)$ satisfying $(3.40)_2$ follows.

Now, we prove that (u,Ξ) solves (P). Let $\bar z \in V \cap C^\infty(\overline\Omega; \mathbf{R}^d)$ be fixed. Then there exists a sequence $\{\bar z_h\}$, $\bar z_h \in V_h$ such that $\bar z_h \to \bar z$ in V and $C(\overline\Omega; \mathbf{R}^d)$ (see (3.35)) and from (3.37):

$$P_h(\Pi\bar z_h) \to \Pi\bar z \quad \text{in } L^\infty(\omega). \qquad (3.44)$$

From the definition of $(P)_h$ it follows that

$$a_h(u_h,\bar z_h) + \int_\omega \Xi_h P_h(\Pi\bar z_h) d\mu = \langle f_h,\bar z_h\rangle_h.$$

Passing here to the limit with $h \to 0+$, using (3.38)-(3.40) and (3.44) we get that

$$a(u,\bar z) + \int_\omega \Xi\Pi\bar z d\mu = \langle f,\bar z\rangle \quad \forall \bar z \in V \cap C^\infty(\overline\Omega; \mathbf{R}^d).$$

From this we see that there exists a constant $c > 0$ such that

$$\left| \int_\omega \Xi \Pi \bar{z} d\mu \right| \le c \|\bar{z}\| \quad \forall \bar{z} \in V \cap C^\infty(\overline{\Omega}; \mathbf{R}^d).$$

Since $V \cap C^\infty(\overline{\Omega}; \mathbf{R}^d)$ is dense in V (see (3.1)), the functional

$$z \mapsto \int_\omega \Xi \, \Pi z d\mu, \quad z \in V \cap C^\infty(\overline{\Omega}; \mathbf{R}^d)$$

can be extended in a unique way to the whole V. Thus Ξ can be identified with the element of V^*.

To complete the proof it remains to verify that $\Xi(x) \in \hat{b}(x, \Pi u(x))$ for a.a. $x \in \omega$. From (3.36) and Theorem 1.5 we may assume that $\{P_h(\Pi u_h)\}$ converges to Πu uniformly up to small sets: given $\varepsilon > 0$, $\delta > 0$ there exist: a set $\omega_0 \subset \omega$, meas $\omega_0 < \delta$ and $h_1 > 0$ such that

$$|P_h(\Pi u_h)(x) - \Pi u(x)| < \frac{\varepsilon}{2} \quad \text{for all } x \in \omega \setminus \omega_0, \, \forall h \le h_1. \tag{3.45}$$

Let $h \le \min(h_1, \frac{\varepsilon}{2})$. Then the definition of $(P)_h$, the monotonicity of $\varepsilon \mapsto \bar{b}_\varepsilon$ and (3.45) yield:

$$\Xi_h(x) \quad \le \quad \bar{b}(\sum_{i=1}^m \mathcal{X}_{\text{int}_\omega K_i}(x) x_h^i, P_h(\Pi u_h)(x)) \tag{3.46}$$

$$\le \quad \bar{b}_{\frac{\varepsilon}{2}}(\sum_{i=1}^m \mathcal{X}_{\text{int}_\omega K_i}(x) x_h^i, P_h(\Pi u_h)(x))$$

$$\le \quad \bar{b}_\varepsilon(\sum_{i=1}^m \mathcal{X}_{\text{int}_\omega K_i}(x) x_h^i, \Pi u(x)) \quad \text{for a.a. } x \in \omega \setminus \omega_0.$$

Let $\Phi \in L^\infty(\omega)$, $\Phi \ge 0$ in ω be given. Then from $(3.40)_2$, (3.46) and Lemma 3.2 we find out that

$$\int_{\omega \setminus \omega_0} \Xi \Phi d\mu = \lim_{h \to 0+} \int_{\omega \setminus \omega_0} \Xi_h \Phi d\mu$$

$$\overset{(3.46)}{\le} \quad \limsup_{h \to 0+} \int_{\omega \setminus \omega_0} \bar{b}_\varepsilon(\sum_{i=1}^m \mathcal{X}_{\text{int}_\omega K_i}(x) x_h^i, \Pi u(x)) \Phi(x) d\mu$$

$$\overset{\text{(Theorem 1.6)}}{\le} \quad \int_{\omega \setminus \omega_0} \limsup_{h \to 0+} \bar{b}_\varepsilon(\sum_{i=1}^m \mathcal{X}_{\text{int}_\omega K_i}(x) x_h^i, \Pi u(x)) \Phi(x) d\mu$$

$$= \int_{\omega \setminus \omega_0} \bar{b}_\varepsilon(x, \Pi u(x)) \Phi(x) d\mu$$

holds for any $\varepsilon > 0$ sufficiently small.

Since $V_h \subset C(\overline{\Omega}; \mathbf{R}^d)$ and (3.45) holds we see that Πu is bounded in $\omega \setminus \omega_0$ so that

$$\int_{\omega \setminus \omega_0} \Xi \Phi d\mu \leq \limsup_{\varepsilon \to 0+} \int_{\omega \setminus \omega_0} \overline{b}_\varepsilon(x, \Pi u(x)) \Phi(x) d\mu$$

$$\overset{\text{Th. 1.6}}{\leq} \int_{\omega \setminus \omega_0} \limsup_{\varepsilon \to 0+} \overline{b}_\varepsilon(x, \Pi u(x)) \Phi(x) d\mu = \int_{\omega \setminus \omega_0} \overline{b}(x, \Pi u(x)) \Phi(x) d\mu.$$

Similarly, we prove that

$$\int_{\omega \setminus \omega_0} \Xi \Phi d\mu \geq \int_{\omega \setminus \omega_0} \underline{b}(x, \Pi u(x)) \Phi(x) d\mu.$$

Hence $\Xi(x) \in \hat{b}(x, \Pi u(x))$ for a.a. $x \in \omega \setminus \omega_0$. If $\operatorname{meas} \omega_0 = \delta \to 0+$ we finally obtain that $\Xi(x) \in \hat{b}(x, \Pi u(x))$ for a.a. $x \in \omega$. At the same time we proved that any cluster point of $\{(u_h, \Xi_h)\}$ in the sense of (3.40) solves (P). $\square$

Remark 3.8 *Let us comment the assumptions, under which Theorem 3.3 holds:*

(i) (3.35) is a standard result of the finite element approximation theory;

(ii) (3.36) will be the consequence of the following (stronger) property of the mapping P_h:

$$y_h \rightharpoonup y \quad in \ V, \ y_h \in V_h \implies \|P_h(\Pi y_h) - \Pi y\|_{L^2(\omega)} \to 0, \quad h \to 0+;$$

(iii) when a_h and f_h arise from a and f, respectively, by using the numerical integration, assumptions (3.25),(3.26),(3.38) and (3.39) will be satisfied when an appropriate quadrature formula will be used (see Ciarlet, 1978).

In what follows we improve the convergence results of Theorem 3.3. We shall formulate additional assumptions, under which:

(j) $\Xi \in L^{q'}(\omega)$ for some $q' > 1$;

(jj) the weak convergence of u_h to u is replaced by the strong one in V.

Instead of (3.4) we shall suppose that b satisfies the following growth condition:

$$\begin{cases} \exists \text{ constants } c_1, c_2 > 0 \text{ such that} \\ |b(x, \xi)| \leq c_1 + c_2 |\xi|^{\frac{q}{q'}} \quad \text{for a.a. } x \in \omega, \ \xi \in \mathbf{R}, \end{cases} \tag{3.47}$$

where $1/q + 1/q' = 1$, $1 \leq q < q^*$ if $n > 2$ or $q \in [1, \infty)$ if $n = 2$. The number q^* is equal to $2n/(n-2)$ if $\operatorname{int}_\Omega \omega \neq \emptyset$ or to $2(n-1)/(n-2)$ if $\omega \subset \partial\Omega$. Clearly $\underline{b}$ and $\overline{b}$ satisfy (3.47), as well.

Let the mapping Π from the definition of (P) be *continuous* from V into $L^q(\omega)$:

$$\exists c > 0 : \|\Pi v\|_{L^q(\omega)} \leq c\|v\| \quad \forall v \in V. \tag{3.48}$$

Instead of (3.36) we shall require the following stronger assumption satisfied by $P_h : W_h \to Y_h$:

$$y_h \rightharpoonup y \quad \text{in } V, \ y_h \in V_h \implies \exists s \geq q \text{ such that} \qquad (3.49)$$
$$\|P_h(\Pi y_h) - \Pi y\|_{L^s(\omega)} \to 0, \quad \text{as } h \to 0+ .$$

If (3.47)-(3.49) are satisfied, the sequence $\{\Xi_h\}$ is bounded in $L^{q'}(\omega)$. Indeed:

$$\int_\omega |\Xi_h|^{q'} d\mu \overset{(3.47)}{\leq} \int_\omega \left(c_1 + c_2 |P_h(\Pi u_h)|^{\frac{q}{q'}} \right)^{q'} d\mu$$

$$\leq C \int_\omega \left(1 + |P_h(\Pi u_h)|^q \right) d\mu.$$

Therefore

$$\limsup_{h \to 0+} \int_\omega |\Xi_h|^{q'} d\mu \overset{(3.49)}{\leq} C(1 + \|\Pi u\|^q_{L^q(\omega)}) \overset{(3.48)}{\leq} C.$$

Thus one can find a subsequence of $\{\Xi_h\}$ and a function $\Xi \in L^{q'}(\omega)$ such that

$$\Xi_h \rightharpoonup \Xi \quad \text{in } L^{q'}(\omega). \qquad (3.50)$$

At the same time we may assume that $u_h \rightharpoonup u$ in V. Then it is very easy to show that the pair (u, Ξ) is a solution to (P). Indeed, let $\bar{v} \in V$ be given. Then there exists a sequence $\{\bar{v}_h\}$, $\bar{v}_h \in V_h$ such that $\bar{v}_h \to \bar{v}$ in V (see (3.35)) and at the same time

$$P_h(\Pi \bar{v}_h) \to \Pi \bar{v} \quad \text{in } L^q(\omega) \qquad (3.51)$$

as follows from (3.49). The definition of $(P)_h$ yields:

$$a_h(u_h, \bar{v}_h) + \int_\omega \Xi_h P_h(\Pi \bar{v}_h) d\mu = \langle f_h, \bar{v}_h \rangle_h.$$

Passing to the limit with $h \to 0+$ we obtain that

$$a(u, \bar{v}) + \int_\omega \Xi \Pi \bar{v} d\mu = \langle f, \bar{v} \rangle$$

making use of (3.38), (3.39), (3.50) and (3.51). The inclusion $\Xi(x) \in \hat{b}(x, \Pi u(x))$ a.e. in ω can be verified in the same way as we did in Theorem 3.3.

Remark 3.9 *Imposing the additional assumptions on the data, we increased the regularity of Ξ. The duality pairing $\langle \Xi, \Pi v \rangle_{Y \times Z}$ is now realized by $\int_\omega \Xi \Pi v d\mu$ for any $v \in V$, $Y = L^{q'}(\omega)$, $Z = L^q(\omega)$, $1/q + 1/q' = 1$. Let us observe also that (3.37), as well as Dunford-Pettis theorem, were not used, since the weak compactness of $\{\Xi_h\}$ in $L^{q'}(\omega)$ follows from its boundedness in $L^{q'}(\omega)$, $q' > 1$. Moreover the density of $\{V_h\}$ only in the norm of V is needed.*

Now, we are able to improve the convergence result of Theorem 3.3. To this end we introduce the following stronger assumptions, relating $\{a_h\}$ to a and $\{f_h\}$ to f. Instead of (3.38),(3.39) we shall suppose:

$$y_h \to y, \; z_h \to z \; \text{ in } V, \; y_h, z_h \in V_h \implies \tag{3.52}$$
$$a_h(y_h, z_h) \to a(y, z) \; \& \; a_h(z_h, y_h) \to a(z, y);$$

$$z_h \rightharpoonup z \; \text{ in } V, \; z_h \in V_h \implies \langle f_h, z_h \rangle_h \to \langle f, z \rangle \text{ as } h \to 0+. \tag{3.53}$$

Then we have:

Theorem 3.4 *Let (3.24)-(3.26),(3.35), (3.47)-(3.49),(3.52) and (3.53) be satisfied. Let $\{(u_h, \Xi_h)\}$ be a sequence of solutions to $(P)_h$ with $\{u_h\}$ being bounded in V independently of h. Then there exist : a subsequence of $\{(u_h, \Xi_h)\}$ and a pair $(u, \Xi) \in V \times L^{q'}(\omega)$ with $q' > 1$ from (3.47) such that*

$$\begin{cases} u_h \to u & \text{in } V; \\ \Xi_h \rightharpoonup \Xi & \text{in } L^{q'}(\omega), \; h \to 0+. \end{cases} \tag{3.54}$$

Moreover (u, Ξ) solves (P). Furthermore any cluster point of $\{(u_h, \Xi_h)\}$ in the sense of (3.54) is a solution to (P).

Proof: It remains to prove $(3.54)_1$. Let $\{\bar{v}_h\}$, $\bar{v}_h \in V_h$ be such that (see (3.35) and (3.49))

$$\bar{v}_h \to u \; \text{ in } V \text{ and } P_h(\Pi \bar{v}_h) \to \Pi u \; \text{ in } L^s(\omega), \; s \geq q. \tag{3.55}$$

Then from (3.25) it follows that

$$\tilde{\alpha} \|u_h - \bar{v}_h\|^2 \leq a_h(u_h - \bar{v}_h, u_h - \bar{v}_h) \tag{3.56}$$
$$= a_h(u_h, u_h - \bar{v}_h) - a_h(\bar{v}_h, u_h - \bar{v}_h) \to 0 \quad \text{as } h \to 0+.$$

Indeed, the second term on the right hand side of (3.56) tends to zero, as follows from (3.52) and (3.55) using that $u_h - \bar{v}_h \rightharpoonup 0$ in V. The first term can be estimated as follows:

$$|a_h(u_h, u_h - \bar{v}_h)| \overset{(P)_h}{=} \left| \int_\omega \Xi_h(P_h(\Pi u_h) - P_h(\Pi \bar{v}_h)) d\mu - \langle f_h, u_h - \bar{v}_h \rangle_h \right|$$
$$\leq \|\Xi_h\|_{L^{q'}(\omega)} \|P_h(\Pi u_h) - P_h(\Pi \bar{v}_h)\|_{L^q(\omega)} + |\langle f_h, u_h - \bar{v}_h \rangle_h| \to 0+$$

as follows from (3.53), the boundedness of $\{\|\Xi_h\|_{L^{q'}(\omega)}\}$ and the fact that $P_h(\Pi u_h) - P_h(\Pi \bar{v}_h) \to 0$ in $L^q(\omega)$ (see (3.49),(3.55)). The strong convergence of an appropriate subsequence of $\{u_h\}$ follows from (3.56) and the triangle inequality

$$\|u - u_h\| \leq \|u - \bar{v}_h\| + \|\bar{v}_h - u_h\|.$$

$\square$

3.4   CONSTRUCTION OF FINITE ELEMENT SPACES AND INTERPOLATION OPERATORS

In the previous sections we formulated abstract assumptions, under which approximate solutions are close on subsequences to a solution of (P). We have seen that the choice of V_h, Y_h and of the mapping P_h plays the crucial role in the convergence analysis. In the present section we shall specify their construction and verify all the assumptions, guaranteeing the convergence.

Throughout this section we shall suppose that Ω is a *plane polygonal domain* and that $\{\mathcal{D}_h\}$, $h \to 0+$ is a *regular system of triangulations* of $\overline{\Omega}$ in the sense of Definition 1.8.

We start with the case $V = H^1(\Omega; \mathbf{R}^d)$, i.e., no homogenous Dirichlet boundary conditions on the boundary $\partial\Omega$ are prescribed. By V_h we denote the space of all *continuous piecewise linear functions* over $\mathcal{D}_h$:

$$V_h = \{v_h \in C(\overline{\Omega}; \mathbf{R}^d) \mid v_h|_T \in (P_1(T))^d \ \forall T \in \mathcal{D}_h\}. \tag{3.57}$$

As far as Y_h is concerned, its definition will depend on ω. Next we shall discuss two cases: *(i)* $\omega = \Omega$, *(ii)* $\omega = \partial\Omega$ with straightforward modifications when ω is a proper part of Ω or $\partial\Omega$ (see examples presented later on).

CASE $\omega = \Omega$

Type I.   The partition $\mathcal{T}_h$ characterizing Y_h is made of triangles and quadrilaterals, constructed as follows: if $T_1, T_2 \in \mathcal{D}_h$ are two adjacent triangles such that their common edge $T' \equiv T_1 \cap T_2$ is not a part of $\partial\Omega$ then $K_i \in \mathcal{T}_h$ is the *quadrilateral* whose sides joint the barycentres of T_1, T_2 to their vertices lying on the common edge T' (see Fig.3.2).

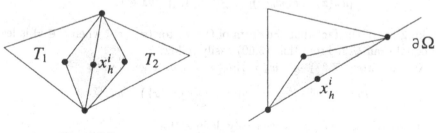

Figure 3.2.

Figure 3.3.

If $T \in \mathcal{D}_h$ is a boundary triangle whose edge T' is a part of $\partial\Omega$, then $K_i \in \mathcal{T}_h$ is the *triangle* defined by the vertices of T on $\partial\Omega$ and by the barycentre of T (see Fig.3.3). The midpoints of edges of all $T \in \mathcal{D}_h$ serve as the degrees of freedom of functions from Y_h. Then Y_h is the space of all piecewise constant functions over $\mathcal{T}_h$, whose construction has been just described.

Let the mapping Π transforming vector valued functions $y \in V$, $y = (y_1, ..., y_d)$, into scalar ones be defined as follows:

$$\Pi y = y_j \quad \text{for some } j \in \{1, ..., d\}. \tag{3.58}$$

Then $W_h \equiv \Pi(V_h)$ is given by

$$W_h = \{w_h \in C(\overline{\Omega}) \mid w_h|_T \in P_1(T) \, \forall T \in \mathcal{D}_h\},$$

i.e., W_h is the space of all continuous piecewise linear *scalar functions* over $\mathcal{D}_h$.

The mapping $P_h : W_h \to Y_h$ is defined as the *piecewise constant Lagrange interpolation* of $w_h \in W_h$ using the values of w_h at $\{x_h^i\}$ as the degrees of freedom:

$$P_h w_h = \sum_{i=1}^{m} w_h(x_h^i) \mathcal{X}_{\text{int}_\Omega K_i}(x). \tag{3.59}$$

Now we prove that P_h, defined by (3.59), satisfy (3.36) and (3.37). We start with

Lemma 3.5 *It holds that:*

$$\|P_h w_h - w_h\|_{L^p(\Omega)} \le \frac{h}{2}\|\nabla w_h\|_{(L^p(\Omega))^n} \quad \forall w_h \in W_h, \, \forall p \in [1, \infty] \tag{3.60}$$

and also (3.37) is satisfied.

Proof: Let $T \in \mathcal{D}_h$, $K \in \mathcal{T}_h$ and take an arbitrary point $x \in \Theta \equiv K \cap T$. Since $w_h \in W_h$ is linear in T and $w_h(x_h^i) = P_h w_h(x)$ for all $x \in \text{int}\,\Theta$ we have:

$$\begin{aligned}
w_h(x) &= w_h(x_h^i) + (x - x_h^i).\nabla w_h(x) \\
&= P_h w_h(x) + (x - x_h^i).\nabla w_h(x).
\end{aligned}$$

Hence

$$|w_h(x) - P_h w_h(x)| \le \frac{h}{2}|\nabla w_h(x)| \quad \forall x \in \Theta$$

making use of the fact that the norm of the vector $(x - x_h^i)$ when $x \in \Theta$ is less or equal than $h/2$. From this, (3.60) easily follows.

Now we prove (3.37). Set $w_h = \Pi y_h$, $w = \Pi y$. Then

$$\|P_h w_h - w\|_{L^\infty(\Omega)} = \max_{i=1,...,m} \max_{x \in K_i} |w_h(x_h^i) - w(x)|.$$

From the triangle inequality we easily deduce that

$$\max_{x \in K_i} |w_h(x_h^i) - w(x)| \le \|w_h - w\|_{C(\overline{\Omega})} + \max_{x \in K_i} |w(x_h^i) - w(x)|. \tag{3.61}$$

The first term on the right hand side of (3.61) tends to zero, because of the assumptions. Now, let $\varepsilon > 0$ be given. Since w is uniformly continuous in $\overline{\Omega}$, there exists $h_0 > 0$ such that for any $K_i \in \mathcal{T}_h$ with $\text{diam}\, K_i \le h_0$ one has:

$$\max_{x \in K_i} |w(x_h^i) - w(x)| < \varepsilon \quad \forall i = 1, ..., m,$$

i.e., the second term tends to zero, as well. From this and (3.61), the property (3.37) follows. $\square$

Consequence 3.1 *Let V_h, Y_h, P_h and Π be the same as above. Then (3.49) holds with $s = 2$. Indeed: from $y_h \rightharpoonup y$ in V and the Rellich's theorem it follows that $w_h \to w$ in $L^2(\Omega)$, where $w_h = \Pi y_h$, $w = \Pi y$. Moreover $\{\|w_h\|_1\}$ is bounded. The triangle inequality and (3.60) (with $p = 2$) yield:*

$$\|P_h w_h - w\|_{0,\Omega} \le \|P_h w_h - w_h\|_{0,\Omega} + \|w_h - w\|_{0,\Omega} \qquad (3.62)$$
$$\le ch + \|w_h - w\|_{0,\Omega} \to 0 \quad as \ h \to 0 + .$$

Also (3.36) is a direct consequence of (3.62).

Remark 3.10 *P_h is one-to-one mapping from W_h into Y_h. Indeed:*

$$P_h w_h = 0, \ w_h \in W_h \iff w_h(x_h^i) = 0 \ \ \forall i = 1, ..., m.$$

A piecewise linear function being equal to zero at the midpoint of any edge of any $T \in \mathcal{D}_h$ is identically equal to zero in Ω. On the other hand, since the number of the edges is greater than the number of the triangles from $\mathcal{D}_h$, we have that $\dim Y_h > \dim W_h$ and P_h does not map W_h onto Y_h.

Type II. The only difference between the previous and the present type consists in the definition of $\mathcal{T}_h$. The system $\{x_h^i\}$ coincides with the set of all the nodes of the triangulation $\mathcal{D}_h$. Let x_h^i be an *inner* node of $\mathcal{D}_h$. Then the corresponding K_i, containing x_h^i is obtained by joining the barycentres of all $T \in \mathcal{D}_h$ sharing x_h^i as a common vertex to the midpoints of edges of T issuing from x_h^i (with the adequate modification when $x_h^i \in \partial\Omega$, see Fig.3.4).

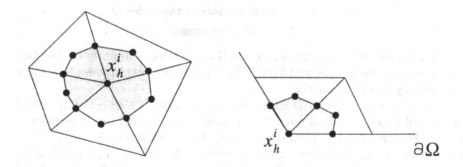

Figure 3.4.

The mapping P_h is still defined by (3.59) but using Type II elements K_i. It is readily seen that Lemma 3.5 is still valid with the following modification of (3.60) (see Glowinski, 1984):

$$\|P_h w_h - w_h\|_{L^p(\Omega)} \le \frac{2}{3}h\|\nabla w_h\|_{(L^p(\Omega))^n} \quad \forall w_h \in W_h, \ \forall p \in [1,\infty]. \quad (3.63)$$

Remark 3.11 *In contrast to the previous case we see that* $\dim W_h = \dim Y_h$ *and* P_h *is one-to-one mapping of* W_h *onto* Y_h *(see Remark 3.7).*

Now let us pass to the case when

$$V = \{v \in H^1(\Omega; \mathbf{R}^d) \mid v = 0 \text{ on } \Gamma_u\},$$

where $\Gamma_u \subseteq \partial\Omega$ is a nonempty, open part in $\partial\Omega$, i.e., the homogenous Dirichlet boundary condition is prescribed on Γ_u. Let $\{\mathcal{D}_h\}$ be a regular family of partitions of $\overline{\Omega}$, which is *consistent* with the decomposition of $\partial\Omega$ into Γ_u and $\partial\Omega \setminus \overline{\Gamma}_u$. Then one has to modify the definition of V_h in the following sense:

$$V_h = \{v_h \in C(\overline{\Omega}; \mathbf{R}^d) \mid v_h|_T \in (P_1(T))^d \quad \forall T \in \mathcal{D}_h, \qquad (3.64)$$
$$v_h = 0 \text{ on } \Gamma_u\}.$$

Let Π be given by (3.58). Then

$$W_h \equiv \Pi(V_h) = \{w_h \in C(\overline{\Omega}) \mid w_h|_T \in P_1(T) \quad \forall T \in \mathcal{D}_h,$$
$$w_h = 0 \text{ on } \Gamma_u\}.$$

If Type I or Type II elements are used when constructing $\mathcal{T}_h$, then $\dim Y_h$ is always greater than $\dim W_h$. Indeed:

$$\dim W_h \quad = \quad \text{the total number of the nodes of } \mathcal{D}_h$$
$$- \text{ the number of the nodes of } \mathcal{D}_h \text{ on } \overline{\Gamma}_u,$$

while

$$\dim Y_h \quad = \quad \text{the number of all edges of}$$
$$T \in \mathcal{D}_h \text{ (Type I elements)}$$

or

$$\dim Y_h \quad = \quad \text{the total number of the nodes of}$$
$$\mathcal{D}_h \text{ (Type II elements).}$$

As we shall see later, the case when $\dim W_h = \dim Y_h$ is important. For this reason we shall modify the partition $\mathcal{T}_h$ for Type II elements in order to achieve the equality $\dim Y_h = \dim W_h$.

Let $\mathcal{T}_h$ be a partition of $\overline{\Omega}$ constructed from Type II elements introduced above and let $K_1 \in \mathcal{T}_h$ be a boundary element whose at least one edge belongs to Γ_u. Then K_1 shares a common side with an inner element $K_2 \in \mathcal{T}_h$. From K_1 and K_2 we create a new boundary element $K \equiv K_1 \cup K_2$ (see Fig.3.5). Thus our new partition of $\overline{\Omega}$, still denoted by $\mathcal{T}_h$, will contain all such modified boundary elements (along Γ_u), completed by the classical Type II elements introduced above. Using this new $\mathcal{T}_h$ in the definition of Y_h and P_h given by (3.59) we see that $\dim Y_h = \dim W_h$. It is also readily seen that the error estimate similar to (3.63) again holds true:

$$\|P_h w_h - w_h\|_{L^p(\Omega)} \leq ch\|\nabla w_h\|_{(L^p(\Omega))^n} \quad \forall w_h \in W_h, \forall p \in [1, \infty]$$

with a constant $c > 0$, which does not depend on h. Also (3.37) remains valid.

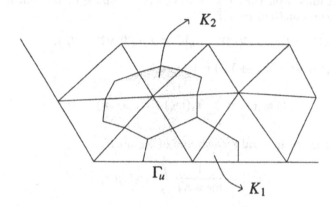

Figure 3.5.

CASE $\omega = \partial\Omega$

Any triangulation $\mathcal{D}_h$ of $\overline{\Omega}$, used for the construction of V_h defines a partition $\tilde{\mathcal{T}}_h \equiv \mathcal{D}_h|_{\partial\Omega}$ of $\partial\Omega$ made of edges T' of the boundary triangles $T \in \mathcal{D}_h$. Now, we present two types of Y_h constructed on $\partial\Omega$ and define $P_h : W_h \to Y_h$ satisfying the required assumptions.

We start again with the case when $V = H^1(\Omega; \mathbf{R}^d)$.

Type III.   We take $\mathcal{T}_h \equiv \tilde{\mathcal{T}}_h$, i.e., any element $K_i \in \mathcal{T}_h$ is represented by an edge $T' \subset \partial\Omega$ of a boundary triangle $T \in \mathcal{D}_h$ (see Fig.3.6). The system $\{x_h^i\}$ is given by all the midpoints of $K_i \in \mathcal{T}_h$.

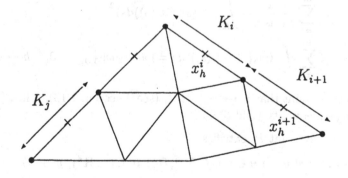

Figure 3.6.

First, let the mapping Π be defined by:

$$\Pi y = y_j|_{\partial\Omega} \quad \text{for some } j \in \{1, ..., d\}, \; y = (y_1, ..., y_d). \tag{3.65}$$

Then it is readily seen that $W_h \equiv \Pi(V_h)$ is the space of *all continuous* and *piecewise linear functions* over $\tilde{\mathcal{T}}_h$:

$$W_h = \{w_h \in C(\partial\Omega) \mid w_h|_{T'} \in P_1(T') \; \forall T' \in \tilde{\mathcal{T}}_h\}. \tag{3.66}$$

Let the mapping $P_h : W_h \to Y_h$ be defined by

$$P_h w_h(x) = \sum_{i=1}^{m} \pi_i(w_h) \mathcal{X}_{\text{int}_{\partial\Omega} K_i}(x), \tag{3.67}$$

where $\pi_i(w_h)$ is the integral mean value of w_h on K_i:

$$\pi_i(w_h) = \frac{1}{\text{meas}\,K_i} \int_{K_i} w_h(s)ds,$$

i.e., $P_h(w_h)$ is the $L^2(\partial\Omega)$-projection of $w_h \in W_h$ on the space Y_h. We prove:

Lemma 3.6 *Let $y_h \rightharpoonup y$ in V, $y_h \in V_h$ and let P_h be given by (3.67). Then*

$$\|P_h(\Pi y_h) - \Pi y\|_{0,\partial\Omega} \to 0, \quad h \to 0+, \tag{3.68}$$

i.e., (3.49) is satisfied with $s = 2$.

Proof: Let us denote $w_h = \Pi y_h$, $w = \Pi y$. Since the space of piecewise constant functions is dense in $L^2(\partial\Omega)$ and $w \in L^2(\partial\Omega)$, we have that

$$\|P_h w - w\|_{0,\partial\Omega} \to 0, \quad h \to 0+. \tag{3.69}$$

From the Hölder's inequality we obtain:

$$\|P_h w_h - P_h w\|_{0,\partial\Omega}^2 = \int_{\partial\Omega} \left(\sum_{i=1}^{m} \pi_i(w_h - w)\mathcal{X}_{\text{int}_{\partial\Omega} K_i}(s)\right)^2 ds$$

$$= \sum_{i=1}^{m} \frac{1}{\text{meas}\,K_i}\left(\int_{K_i}(w_h(s) - w(s))ds\right)^2$$

$$\leq \sum_{i=1}^{m} \int_{K_i}(w_h(s) - w(s))^2 ds = \|w_h - w\|_{0,\partial\Omega}^2 \to 0, \quad h \to 0+$$

taking into account that $V \hookrightarrow\hookrightarrow L^2(\partial\Omega)$ (see Theorem 1.13). This, (3.69) and the triangle inequality yield (3.68). $\qquad\square$

Now we prove that P_h satisfies (3.37).

Lemma 3.7 *Let $y_h \to y \in V \cap C^\infty(\overline{\Omega}; \mathbf{R}^d)$ in $C(\overline{\Omega}; \mathbf{R}^d)$, $y_h \in V_h$. Then*

$$P_h(\Pi y_h) \to \Pi y \quad \text{in } L^\infty(\partial\Omega), \quad h \to 0+. \tag{3.70}$$

Proof: Again denote $w_h = \Pi y_h$, $w = \Pi y$. Then

$$\|P_h w_h - w\|_{L^\infty(\partial\Omega)} = \max_i \max_{x \in K_i} |\pi_i(w_h) - w(x)|$$

$$= \max_i \max_{x \in K_i} |w_h(x_h^i) - w(x)|$$

using that $\pi_i(w_h) = w_h(x_h^i)$ for any linear function w_h. Since $w = y_j|_{\partial\Omega}$ is uniformly continuous on $\partial\Omega$ and $w_h \to w$ in $C(\partial\Omega)$, we can proceed exactly in the same way as in Lemma 3.5 when proving (3.37).   □

Next we shall consider another example of the mapping Π, which plays an important role in unilateral boundary value problems, namely:

$$\Pi y = y.\nu \quad \text{on } \partial\Omega, \tag{3.71}$$

associating with $y \in V \equiv H^1(\Omega; \mathbf{R}^d)$ its normal component $y.\nu$ on $\partial\Omega$. Since Ω is a polygonal domain, the unit normal vector ν is *piecewise constant* with a discontinuity located at the vertices of Ω. Let Σ be a system of all straight (closed) sides Γ of $\partial\Omega$. Then it is easy to see that

$$W_h \equiv \Pi(V_h) = \{w_h \in L^\infty(\partial\Omega) \mid w_h|_\Gamma \in C(\Gamma) \ \forall\Gamma \in \Sigma, \tag{3.72}$$
$$w_h|_{T'} \in P_1(T') \ \forall T' \in \tilde{\mathcal{T}}_h\}.$$

The mapping $P_h : W_h \to Y_h$ is still defined by (3.67). As before (see Lemma 3.6) one can prove that

$$\|P_h(y_h.\nu) - y.\nu\|_{0,\partial\Omega} \to 0, \quad h \to 0+$$

provided that $y_h \rightharpoonup y$ in V, i.e., (3.49) is satisfied with $s = 2$. Also (3.37) remains valid. Indeed, from $y_h \to y \in V \cap C^\infty(\overline{\Omega}; \mathbf{R}^d)$ in $C(\overline{\Omega}; \mathbf{R}^d)$ it follows that $y_h.\nu \to y.\nu$ in $C(\Gamma)$ for any $\Gamma \in \Sigma$. Further:

$$\|P_h(y_h.\nu) - y.\nu\|_{L^\infty(\partial\Omega)} = \max_{\Gamma\in\Sigma} \max_{x\in K_i\subset\Gamma} |(y_h.\nu)(x_h^i) - (y.\nu)(x)| \tag{3.73}$$
$$\to 0 \quad \text{as } h \to 0+$$

using the fact that $y.\nu$ is uniformly continuous on any $\Gamma \in \Sigma$.

Remark 3.12 *When Π is defined by (3.71) we see that $\dim W_h = \dim Y_h +$ the number of the vertices of Ω. It is readily seen that P_h maps W_h onto Y_h. If Π is given by (3.65) then $\dim W_h = \dim Y_h$ but this time P_h does not map W_h onto Y_h, since P_h is not one-to-one, in general.*

Type IV.   The system $\{x_h^i\}$ will be given by all the nodes of $\mathcal{D}_h$ lying on $\partial\Omega$. Let Σ have the same meaning as above. We describe how to construct $\mathcal{T}_h$. If x_h^{i+1} is not a vertex of Ω, then K_{i+1} is the segment joining the midpoints of edges $T' \subset \partial\Omega$ of $T \in \mathcal{D}_h$ sharing x_h^{i+1} as a common point. If x_h^i is a vertex of $\partial\Omega$, then K_i is a half of the edge $T' \subset \partial\Omega$ of $T \in \mathcal{D}_h$ issuing from x_h^i (see Fig.3.7), i.e., any such x_h^i is contained in two K_i's.

Now we restrict ourselves to the case when Π is defined by (3.71) and P_h by (3.67) using Type IV elements. Then W_h is given by (3.72). It is readily seen that Lemma 3.6 is still valid. Also (3.37) remains true. Indeed:

$$P_h(y_h.\nu)(x) = \sum_{i=1}^m \pi_i(y_h.\nu)\mathcal{X}_{\text{int}_{\partial\Omega} K_i}(x)$$
$$= \sum_{i=1}^m (y_h.\nu)(\bar{x}_h^i)\mathcal{X}_{\text{int}_{\partial\Omega} K_i}(x),$$

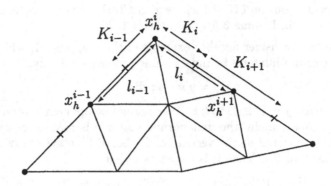

Figure 3.7.

where $\bar{x}_h^i$ is a point in K_i as follows from the integral mean value theorem applied to the function $y_h.\nu|_{K_i}$. Then we proceed exactly in the same way as in Lemma 3.7 using that $y.\nu$ is uniformly continuous on $\Gamma \in \Sigma$, provided that $y \in C^\infty(\overline{\Omega}; \mathbf{R}^d)$.

Remark 3.13 *When Ω is polygonal and Π is defined by (3.71), we have to exclude such partitions T_h made of elements K_i containing a vertex of Ω in their interior (see Fig.3.8), since $y_h.\nu$ has a jump at x_h^i and the integral mean value theorem cannot be applied. On the other hand, when Π is given by (3.65), elements K_i "going around" x_h^i are allowed since W_h contains functions continuous on the whole $\partial\Omega$.*

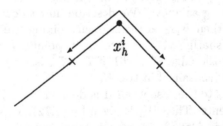

Figure 3.8.

Remark 3.14 *Let Π be given by (3.71) and W_h by (3.72). Then $\dim W_h = \dim Y_h$. Let us show that $P_h : W_h \to Y_h$ defined by (3.67) with Type IV elements* *is* one-to-one mapping W_h onto Y_h. *To this end it is sufficient to show that*

$$P_h w_h = 0, \ w_h \in W_h \implies w_h \equiv 0 \quad on \ \partial\Omega. \tag{3.74}$$

Indeed, (3.74) is equivalent to the system of linear algebraic equations formulated on any $\Gamma \in \Sigma$ for the nodal values of w_h at $x_h^i \in \Gamma$ with the following tridiagonal matrix:

$$
\mathbf{B} = \begin{pmatrix}
3l_1 & l_1 & 0 & \cdots\cdots\cdots & 0 \\
l_1 & 3(l_1 + l_2) & l_2 & 0 \ \cdots\cdots\cdots & 0 \\
0 & l_2 & 3(l_2 + l_3) & l_3 \ \cdots\cdots\cdots & 0 \\
\cdots\cdots\cdots\cdots\cdots\cdots\cdots\cdots\cdots\cdots\cdots \\
0 & 0 & 0 & 0 \ \cdots \ 0 \ l_{m'} & 3l_{m'}
\end{pmatrix}, \tag{3.75}
$$

where m' is the number of the nodes $x_h^i \in \Gamma$ and $l_i = |x_h^{i+1} - x_h^i|$. We see that $\mathbf{B}$ is a strictly diagonally dominant matrix, thus regular and (3.74) holds.

Let the partition T_h of $\partial\Omega$ be defined by Type IV elements and Π be given by (3.71). We change the definition of P_h. Instead of the $L^2(\partial\Omega)$-projection of W_h on Y_h we consider P_h to be the *piecewise constant lagrange interpolation operator* at $\{x_h^i\}$. If $w_h \in W_h$, where W_h is given by (3.72) then

$$P_h w_h(x) = \sum_i w_h(x_h^i) \mathcal{X}_{\mathrm{int}_{\partial\Omega} K_i}(x). \tag{3.76}$$

From Glowinski et al., 1981 we know that for this type of the lagrange interpolation operator on the boundary, (3.49) is again satisfied with $s = 2$. It is readily seen that P_h is one-to-one mapping of W_h onto Y_h.

Now we shall show how to treat the case when the homogenous Dirichlet boundary condition is prescribed on a nonempty, open part Γ_u of $\partial\Omega$:

$$V = \{v \in H^1(\Omega; \mathbf{R}^d) \mid v = 0 \ on \ \Gamma_u\}.$$

Let $\{\mathcal{D}_h\}$, $h \to 0+$, be a regular family of partitions of $\overline{\Omega}$, which is consistent with the decomposition of $\partial\Omega$ into Γ_u and $\partial\Omega \setminus \overline{\Gamma}_u$. The space V_h is defined by (3.64). Consider the mapping Π given by (3.71). Then

$$
\begin{aligned}
W_h \equiv \Pi(V_h) = \{w_h \in L^\infty(\partial\Omega) \mid w_h|_\Gamma \in C(\Gamma) \ \forall \Gamma \in \Sigma, \\
w_h|_{T'} \in P_1(T') \ \forall T' \in \tilde{T}_h, \ w_h = 0 \ on \ \overline{\Gamma}_u\},
\end{aligned}
$$

where the symbols Γ and Σ have the same meaning as in (3.72). The space W_h contains piecewise linear functions over $\tilde{T}_h \equiv \mathcal{D}_h|_{\partial\Omega}$ with possible discontinuities at the vertices of $\partial\Omega$ and vanishing at all the nodes of $\mathcal{D}_h$ on $\overline{\Gamma}_u$. It is easy to see that

$$
\begin{aligned}
\dim W_h \ = \ & \text{the number of the nodes of } T_h \text{ on } \partial\Omega \setminus \overline{\Gamma}_u \\
& + \text{the number of the vertices of } \Omega \text{ on } \partial\Omega \setminus \overline{\Gamma}_u.
\end{aligned}
$$

In order to construct the space Y_h of piecewise constant functions and to define the corresponding mapping $P_h : W_h \to Y_h$ by (3.67) we use Type IV elements which will give us the partition of $\partial\Omega \setminus \overline{\Gamma}_u$. Due to the consistency of $\mathcal{D}_h$ with the decomposition of $\partial\Omega$, any boundary node of Γ_u in $\partial\Omega$ is a node x_h^i of $\mathcal{D}_h$. The respective $K_i \in \mathcal{T}_h$ containing such a node will be represented by a half segment (see Fig.3.9).

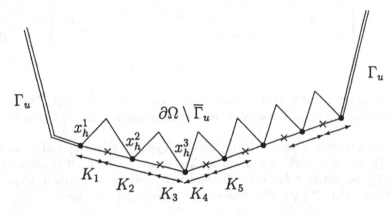

Figure 3.9.

Using such a partition $\mathcal{T}_h$ of $\partial\Omega \setminus \Gamma_u$ we have that

$$\dim Y_h = \dim W_h + \text{ the number of the boundary}$$
$$\text{points of } \Gamma_u \text{ in } \partial\Omega.$$

For the partition of $\partial\Omega \setminus \Gamma_u$ illustrated by Fig.3.9 we have $\dim Y_h = \dim W_h + 2$. Since the dimension of Y_h is greater than the dimension of W_h, the mapping $P_h : W_h \to Y_h$ can not be onto. To achieve the equality of dimensions of W_h and Y_h we have to modify the partition $\mathcal{T}_h$ of $\partial\Omega \setminus \Gamma_u$ as follows: any element $K_i \in \mathcal{T}_h$ containing a boundary node $x_h^i \in \overline{\Gamma}_u$ will be appended to its neighbour which is placed already in the interior of $\partial\Omega \setminus \overline{\Gamma}_u$, the union of both define a new boundary element K (in the situation depicted in Fig.3.9, this is, for example, the element $K = K_1 \cup K_2$) and the respective x_h^i will be dropped from $\{x_h^i\}$. After such modifications of $\mathcal{T}_h$ we see that $\dim W_h = \dim Y_h$. It is easy to verify that Lemmas 3.6 and 3.7 are still valid. It remains to show that the mapping P_h (given by (3.67)) is one-to-one mapping W_h onto Y_h. To this end it is sufficient to show that

$$P_h w_h = 0, \ w_h \in W_h \implies w_h = 0 \quad \text{on } \partial\Omega \setminus \overline{\Gamma}_u. \tag{3.77}$$

If $\Gamma \in \Sigma$ is a straight side of $\partial\Omega$ having the empty intersection with $\overline{\Gamma}_u$ then $w_h|_\Gamma = 0$ on $\overline{\Gamma}$ as follows from Remark 3.14. If Γ is a straight part of $\partial\Omega$ having common points with $\overline{\Gamma}_u$, then (3.77) leads to a new system of linear algebraic equations for the nodal values of w_h at $x_h^i \in \overline{\Gamma} \setminus \overline{\Gamma}_u$ (do not forget that $w_h(x_h^i) = 0$ if $x_h^i \in \overline{\Gamma} \cap \overline{\Gamma}_u$) with the matrix $\mathbf{B}$, given by (3.75) with the following modification of the first (or the last row):

$$l_1 + \frac{3}{4}l_2, \ \frac{l_2}{4}, \ 0, \ \ldots, \ 0 \quad \text{(the first row)}$$

$$0, \ \ldots, \ 0, \ \frac{1}{4}l_{n'-1}, \ l_{n'} + \frac{3}{4}l_{n'-1} \quad \text{(the last row)}.$$

The remaining rows of $\mathbf{B}$ are the same as in (3.75). The resulting matrix is again strictly diagonally dominant, thus regular. From this we conclude that $w_h = 0$ on $\overline{\Gamma} \setminus \overline{\Gamma}_u$.

If the mapping P_h is defined by (3.76), then P_h is one-to-one mapping W_h onto Y_h and (3.49) is again satisfied with $s = 2$.

Next we shall illustrate how the previous abstract results can be used in a concrete example. We shall restrict ourselves to Example 3.2 presented at the beginning of this chapter.

Suppose that Ω is a *plane, polygonal* domain whose boundary $\partial\Omega$ is decomposed into three open disjoint parts Γ_u, Γ_P and Γ_c, where different boundary conditions are prescribed. For the sake of simplicity we suppose that Γ_c, being in the bilateral contact with the rigid foundation (see Fig.3.1) is straight and the body Ω is rotated in such a way that Γ_c is placed in the x_1-axis. Then the conditions on Γ_c take the following simpler form:

$$\begin{cases} u_2(x) = 0, \quad -\sigma_{12}(x) + \Xi(x) = 0 \\ \Xi(x) \in \hat{b}(u_1(x)) \end{cases} \quad \text{for a.a. } x \in \Gamma_c. \tag{3.78}$$

We do not specify the choice of $\hat{b}$. Only what we suppose is that $\hat{b}$ is derived from a measurable function $b : \mathbf{R} \to \mathbf{R}$ satisfying (3.4) and (3.5) (let us notice that b does not depend on $x \in \Gamma_c$).

We define

$$V = \{v \in H^1(\Omega; \mathbf{R}^2) \mid v = 0 \text{ on } \Gamma_u, \ v_2 = 0 \text{ on } \Gamma_c\},$$
$$Y = L^1(\Gamma_c) \cap V^*, \quad \Pi v = v_1|_{\Gamma_c}, \quad \omega = \Gamma_c, \quad Z = L^2(\Gamma_c).$$

Let a and f be the same as in Example 3.2.

Then the corresponding hemivariational inequality reads as follows:

$$\begin{cases} \text{Find } (u, \Xi) \in V \times L^1(\Gamma_c) \cap V^* \text{ such that} \\ a(u, v) + \langle \Xi, v_1 \rangle_{Y \times Z} = \langle f, v \rangle \quad \forall v \in (v_1, v_2) \in V \\ \Xi(x) \in \hat{b}(u_1(x)) \quad \text{for a.a. } x \in \Gamma_c. \end{cases} \tag{3.79}$$

The duality $\langle \cdot, \cdot \rangle_{Y \times Z}$ is understood in the sense of Remark 3.1.

Let us pass to the approximation of (3.79). Let $\{\mathcal{D}_h\}$, $h \to 0+$, be a regular system of triangulations of $\overline{\Omega}$ consistent with the decomposition of $\partial\Omega$ and define:

$$V_h = \{v_h = (v_{h1}, v_{h2}) \in C(\overline{\Omega}; \mathbf{R}^2) \mid v_h|_T \in (P_1(T))^2 \ \forall T \in \mathcal{D}_h,$$
$$v_h = 0 \text{ on } \Gamma_u, \ v_{h2} = 0 \text{ on } \Gamma_c\}$$

and

$$Y_h = \{\mu_h \in L^\infty(\Gamma_c) \mid \mu_h|_{K_i} \in P_0(K_i) \; \forall K_i \in \mathcal{T}_h\},$$

where *the partition* $\mathcal{T}_h$ *of* Γ_c (only of Γ_c !) is constructed by using Type IV elements with the eventual modification, when $\overline{\Gamma}_u \cap \overline{\Gamma}_c \neq \emptyset$. The mapping P_h will be given by (3.67) or (3.76).

Now we shall verify all the assumptions guaranteeing the convergence of a subsequence of $\{(u_h, \Xi_h)\}$, being solutions to (P)$_h$.

If $V \cap C^\infty(\overline{\Omega}; \mathbf{R}^2)$ is dense in V (which is true for "reasonable" partitions of $\partial\Omega$ into Γ_u, Γ_P and Γ_c) then the system $\{V_h\}$ is dense in $V \cap C^\infty(\overline{\Omega}; \mathbf{R}^2)$ in the V- and $C(\overline{\Omega}; \mathbf{R}^2)$-norm, as follows from the classical approximation theory. Now, let us consider the case when the body Ω is made of a *homogenous* material, implying that the elasticity coefficients c_{ijkl} are *constant* in Ω. Then the bilinear form a can be evaluated exactly on V_h (the integrand is piecewise constant over $\mathcal{D}_h$). Thus no numerical integration is necessary, i.e., $a_h \equiv a$ for any $h > 0$ so that (3.24), (3.25) and (3.52) are automatically satisfied. For the approximation of the linear term $f \in V^*$ given by

$$\langle f, v \rangle \equiv \int_\Omega F_i v_i dx + \int_{\Gamma_P} P_i v_i ds$$

we use a simple rule for the numerical evaluation of integrals. To this end we shall suppose that $F \in C(\overline{\Omega}; \mathbf{R}^2)$ and $P \in C(\overline{\Gamma}_P; \mathbf{R}^2)$. The volume integral over a single element $T \in \mathcal{D}_h$ will be approximated by

$$\int_T \varphi(x)dx \approx \operatorname{meas} T \varphi(Q_T), \quad \varphi \in C(T), \tag{3.80}$$

where Q_T is the barycentre of T, while the surface integral over an edge $T' \subset \Gamma_P$ of $T \in \mathcal{D}_h$ is evaluated as follows:

$$\int_{T'} \varphi(s)ds \approx \operatorname{length} T' \varphi(Q_{T'}), \varphi \in C(T'), \tag{3.81}$$

where $Q_{T'}$ is the midpoint of T'. The resulting approximation $f_h \in V_h^*$ of $f \in V^*$ generated by (3.80) and (3.81) now reads as follows:

$$\langle f_h, z_h \rangle_h \equiv \sum_{T \in \mathcal{D}_h} \operatorname{meas} T (F_i z_{ih})(Q_T) \tag{3.82}$$

$$+ \sum_{T' \subset \Gamma_P} \operatorname{length} T' (P_i z_{ih})(Q_{T'}).$$

First we prove that (3.26) is satisfied. Indeed, since $z_{ih}|_T \in P_1(T)$ we have:

$$|z_{ih}(Q_T)| = \frac{1}{\operatorname{meas} T} \left| \int_T z_{ih}(x)dx \right| \tag{3.83}$$

$$\leq (\operatorname{meas} T)^{-\frac{1}{2}} \|z_{ih}\|_{0,T}$$

and similarly

$$|z_{ih}(Q_{T'})| = \frac{1}{\text{length } T'} \left| \int_{T'} z_{ih}(s)ds \right| \tag{3.84}$$

$$\leq (\text{length } T')^{-\frac{1}{2}} \|z_{ih}\|_{0,T'}.$$

Substituting (3.83) and (3.84) into (3.82) and using the fact that F, P are elements of $C(\overline{\Omega}; \mathbf{R}^2)$ and $C(\overline{\Gamma}_P; \mathbf{R}^2)$, respectively, we arrive at

$$|\langle f_h, z_h \rangle_h| \leq C\Big(\sum_{T \in \mathcal{D}_h} (\text{meas } T)^{\frac{1}{2}} (\|z_{1h}\|_{0,T} + \|z_{2h}\|_{0,T})$$

$$+ \sum_{T' \subset \Gamma_P} (\text{length } T')^{\frac{1}{2}} (\|z_{ih}\|_{0,T'} + \|z_{2h}\|_{0,T'}) \Big)$$

$$\leq C\Big((\text{meas } \Omega)^{\frac{1}{2}} \|z_h\|_{0,\Omega} + (\text{length } \Gamma_P)^{\frac{1}{2}} \|z_h\|_{0,\Gamma_P} \Big)$$

$$\leq \tilde{\beta} \|z_h\|_{1,\Omega} \quad \forall z_h \in V_h,$$

with a constant $\tilde{\beta} > 0$ *independent* of $h > 0$.

Now we verify (3.53). Let $z_h \rightharpoonup z$ in V, $z_h \in V_h$. Then

$$\langle f_h, z_h \rangle_h - \langle f, z \rangle = \{\langle f_h, z_h \rangle_h - \langle f, z_h \rangle\} \tag{3.85}$$

$$+ \{\langle f, z_h \rangle - \langle f, z \rangle\}.$$

The second term on the right hand side of (3.85) tends to zero as $h \to 0+$. Let us analyse the first term. We have

$$\int_T F_i z_{ih} dx - \text{meas } T(F_i z_{ih})(Q_T) \tag{3.86}$$

$$= \int_T (F_i(x) - F_i(Q_T))z_{ih}(x)dx$$

$$\leq C \max_{x \in T} \|F(x) - F(Q_T)\|(\text{meas } T)^{\frac{1}{2}} \|z_h\|_{0,T},$$

where $C > 0$ is an absolute constant. Since F is uniformly continuous in $\overline{\Omega}$, then for any $\varepsilon > 0$ one can find $h_0 \equiv h_0(\varepsilon) > 0$ such that

$$\max_{x \in T} \|F(x) - F(Q_T)\| \leq \varepsilon$$

holds for any $T \in \mathcal{D}_h$ satisfying diam $T \leq h_0$. Therefore (3.86) can be estimated by the term

$$\varepsilon C(\text{meas } T)^{\frac{1}{2}} \|z_h\|_{0,T} \quad \forall T \in \mathcal{D}_h, \text{ diam } T \leq h_0. \tag{3.87}$$

Similarly

$$\int_{T'} P_i z_{ih} ds - \text{length } T'(P_i z_{ih})(Q_{T'})$$

can be estimated by

$$\varepsilon C (\text{length}\, T')^{\frac{1}{2}} \|z_h\|_{0,T'}.$$

From this and (3.87) we arrive at the following estimate

$$|\langle f_h, z_h \rangle_h - \langle f, z_h \rangle| \leq \varepsilon C \{ (\text{meas}\,\Omega)^{\frac{1}{2}} \|z_h\|_{0,\Omega}$$
$$+ (\text{length}\, \Gamma_P)^{\frac{1}{2}} \|z_h\|_{1,\Omega} \} \leq \varepsilon C$$

which holds for any $h \leq h_0$ making use of the boundedness of $\{\|z_h\|_{1,\Omega}\}$. Hence (3.53) is verified.

Thus all the assumptions of Theorem 3.3 are fulfilled. Consequently, from any sequence $\{(u_h, \Xi_h)\}$ of approximate solutions such that $\{\|u_h\|_1\}$ is bounded one can find a subsequence such that

$$\begin{cases} u_h \rightharpoonup u & \text{in } V \\ \Xi_h \rightharpoonup \Xi & \text{in } L^1(\Gamma_c), \ h \to 0+ \end{cases} \tag{3.88}$$

and (u, Ξ) is a solution of (P).

The previous convergence result can be improved. Since (3.48) is satisfied for any $q \in (1, \infty)$ and (3.49) holds with with $s = q = 2$, we see that if b is such that (see (3.47) with $q = q' = 2$):

$$|b(\xi)| \leq c_1 + c_2|\xi|, \quad \forall \xi \in \mathbf{R},$$

where c_1, c_2 are positive constants, then $\Xi \in L^2(\Gamma_c)$ and

$$\begin{cases} u_h \to u & \text{in } V \\ \Xi_h \rightharpoonup \Xi & \text{in } L^2(\Gamma_c), \quad h \to 0+, \end{cases} \tag{3.89}$$

as follows from Theorem 3.4.

3.5   ALGEBRAIC REPRESENTATION

Let the discretization parameter $h > 0$ be *fixed*. For the simplicity of notations we drop the symbol h in what follows, so that instead of $V_h, Y_h, u_h,...$, we shall write $V, Y, u,...$ bearing in mind that the spaces are finite dimensional.

Suppose that $\dim V = n$, $\dim W = p$ and $\dim Y = m$ (recall that $W = \Pi(V)$). Then V, W and Y can be identified with $\mathbf{R}^n, \mathbf{R}^p$ and $\mathbf{R}^m$, respectively. Let (u, Ξ) be a solution to a (*discrete*) hemivariational inequality:

$$\begin{cases} a(u, v) + \displaystyle\int_\omega \Xi P(\Pi v) d\mu = \langle f, v \rangle & \forall v \in V \\ \Xi(x^i) \in \hat{b}(x^i, P(\Pi u)(x^i)) & \forall i = 1, ..., m. \end{cases} \tag{P}$$

Next we derive the algebraic form of (P). To this end, let the mappings $\Pi|_{V_h}$ and P be represented by $(p \times n)$, $(m \times p)$ matrices Π and $\mathbf{P}$, respectively. Since the integrand is piecewise constant over ω, we have:

$$\int_\omega \Xi P(\Pi v) d\mu = \sum_{i=1}^m c_i \Xi_i (\mathbf{P}(\mathbf{\Pi}\vec{v}))_i,$$

where $\vec{\Xi} = (\Xi_1, ..., \Xi_m)$ is the vector representing $\Xi \in Y$, $c_i = \text{meas } K_i$, $K_i \in \mathcal{T}_h$ and $\vec{v}$ is the vector of the nodal values of $v \in V$. Setting $\Xi_i := c_i \Xi_i$ we obtain the following *algebraic* representation of (P):

$$\begin{cases} \text{Find } (\vec{u}, \vec{\Xi}) \in \mathbf{R}^n \times \mathbf{R}^m \text{ such that} \\ (\mathbf{A}\vec{u}, \vec{v})_{\mathbf{R}^n} + (\vec{\Xi}, \mathbf{P}(\Pi\vec{v}))_{\mathbf{R}^m} = (\vec{f}, \vec{v})_{\mathbf{R}^n} \quad \forall \vec{v} \in \mathbf{R}^n \\ \Xi_i \in c_i \hat{b}(x^i, (\mathbf{P}(\Pi\vec{u})_i) \quad \forall i = 1, ..., m. \end{cases} \quad (\vec{\text{P}})$$

Here $\mathbf{A}$ is the standard stiffness matrix and $\vec{f}$ is the load vector, whose components are given by $\langle f, \varphi_j \rangle$, where $\{\varphi_j\}_{j=1}^n$ is the basis of V. For the sake of simplicity of notations we write $\Lambda = \mathbf{P}\Pi \in \mathcal{L}(\mathbf{R}^n; \mathbf{R}^m)$ in what follows.

To solve numerically the discrete hemivariational inequality we shall not use the formulation (P) (or ($\vec{\text{P}}$)) directly. Instead of that we find the corresponding superpotential $\mathcal{L}$ and we shall seek for its substationary points (see Definition 1.7). The reason is simple: methods for finding substationary points exist, on the contrary, no mathematically justified methods for solving (P) directly are available at present.

Below we describe how to construct the superpotential, whose substationary points are related to solutions of ($\vec{\text{P}}$). *From now on, we suppose that the bilinear form a is symmetric.*

Let b be a function, satisfying (3.2)-(3.5) and let $v \in V$ be given. We define the locally Lipschitz functional $\Phi : V \to \mathbf{R}$ by

$$\Phi(v) = \int_\omega \int_0^{P(\Pi v)(x)} b(x, t) dt d\mu.$$

Now, we shall *approximate* Φ by using the following integration formulae for the evaluation of the outer integral over ω:

$$\begin{aligned} \Phi(v) &\approx \sum_{i=1}^m c_i \int_0^{P(\Pi v)(x^i)} b(x^i, t) dt \\ &= \sum_{i=1}^m c_i \int_0^{(\Lambda\vec{v})_i} b(x^i, t) dt \equiv \Psi(\vec{v}), \end{aligned}$$

where $c_i = \text{meas } K_i$ and $x^i \in K_i$ are the same as in the definition of (P).

The superpotential related to the algebraic hemivariational inequality is now given by

$$\mathcal{L}(\vec{v}) = \frac{1}{2}(\mathbf{A}\vec{v}, \vec{v})_{\mathbf{R}^n} - (\vec{f}, \vec{v})_{\mathbf{R}^n} + \Psi(\vec{v}). \quad (3.90)$$

We prove

Theorem 3.5 *For every substationary point $\vec{u}$ of $\mathcal{L}$ there exists $\vec{\Xi} \in \mathbf{R}^m$ such that $\Lambda^T\vec{\Xi} \in \bar{\partial}\Psi(\vec{u})$ and the pair $(\vec{u}, \vec{\Xi})$ solves ($\vec{\text{P}}$).*

Proof: Define the function $\hat{\Psi} : \mathbf{R}^m \to \mathbf{R}$ by

$$\hat{\Psi}(\vec{\theta}) = \sum_{i=1}^m c_i \int_0^{\theta_i} b(x^i, t) dt \equiv \sum_{i=1}^m \hat{\Psi}_i(\theta_i), \quad (3.91)$$

$\vec{\theta} = (\theta_1, ..., \theta_m)$, i.e., $\hat{\Psi}$ is the sum of m functions, each of them depending on one (its own) variable θ_i. Observe that

$$\Psi(\vec{v}) = \hat{\Psi}(\Lambda \vec{v}) = \sum_{i=1}^{m} \hat{\Psi}_i((\Lambda \vec{v})_i).$$

If $\vec{s} \in \bar{\partial}\hat{\Psi}(\vec{\theta})$ for some $\vec{\theta} \in \mathbf{R}^m$, $\vec{s} = (s_1, ..., s_m)$, then $s_i \in \bar{\partial}\hat{\Psi}_i(\theta_i)$ as follows from Proposition 1.10. From the definition of the generalized directional derivative we have that

$$
\begin{aligned}
\hat{\Psi}_i^\circ(\theta_i; v) &= \limsup_{\substack{z \to 0 \\ t \to 0+}} \frac{\hat{\Psi}_i(\theta_i + z + tv) - \hat{\Psi}_i(\theta_i + z)}{t} \\
&= \limsup_{\substack{z \to 0 \\ t \to 0+}} \frac{c_i}{t} \int_{\theta_i+z}^{\theta_i+z+tv} b(x^i, t) dt.
\end{aligned}
\tag{3.92}
$$

Since both, z and t tend to zero, we see that for any $\varepsilon > 0$ it holds that $[\theta_i + z, \theta_i + z + tv] \subset [\theta_i - \varepsilon, \theta_i + \varepsilon]$ provided that z and t are small enough. Thus

$$\underline{b}_\varepsilon(x^i, \theta_i) \le b(x^i, t) \le \bar{b}_\varepsilon(x^i, \theta_i) \tag{3.93}$$

for a.a. $t \in [\theta_i + z, \theta_i + z + tv]$.

Let $v > 0$. Then (3.92) and (3.93) yield:

$$\hat{\Psi}_i^\circ(\theta_i, v) \le c_i \bar{b}_\varepsilon(x^i, \theta_i) v \quad \forall \varepsilon > 0.$$

Passing to the limit with $\varepsilon \to 0+$ we have

$$\hat{\Psi}_i^\circ(\theta_i, v) \le c_i \bar{b}(x^i, \theta_i) v \quad \forall v > 0. \tag{3.94}$$

In a similar way we prove that

$$\hat{\Psi}_i^\circ(\theta_i, v) \le c_i \underline{b}(x^i, \theta_i) v \quad \forall v \le 0. \tag{3.95}$$

From (3.94), (3.95) and the definition of the generalized gradient it follows that any element $s_i \in \bar{\partial}\hat{\Psi}_i(\theta_i)$ satisfies

$$s_i \in c_i \hat{b}(x^i, \theta_i) \quad \forall i = 1, ..., m. \tag{3.96}$$

Since $\Psi(\vec{v}) = \hat{\Psi}(\Lambda \vec{v})$ and Λ is smooth, Theorem 1.22 says that $\bar{\partial}\Psi(\vec{v}) \subset \Lambda^T \bar{\partial}\hat{\Psi}(\Lambda \vec{v})$. In other words: for any $\vec{s}' \in \bar{\partial}\Psi(\vec{v})$ there exists $\vec{s} \in \bar{\partial}\hat{\Psi}(\Lambda \vec{v})$ such that $\vec{s}' = \Lambda^T \vec{s}$.

Now we are ready to prove the statement of the theorem.

Let $\vec{u} \in \mathbf{R}^n$ be a substationary point of $\mathcal{L}$:

$$0 \in \bar{\partial}\mathcal{L}(\vec{u}) = \mathbf{A}\vec{u} - \vec{f} + \bar{\partial}\Psi(\vec{u}). \tag{3.97}$$

The last equality follows from Proposition 1.10. Then (3.97) is equivalent to say that there exists $\vec{\Xi}' \in \mathbf{R}^n$ such that

$$\begin{cases} (A\vec{u}, \vec{v})_{\mathbf{R}^n} + (\vec{\Xi}', \vec{v})_{\mathbf{R}^n} = (\vec{f}, \vec{v})_{\mathbf{R}^n} & \forall \vec{v} \in \mathbf{R}^n \\ \vec{\Xi}' \in \bar{\partial}\Psi(\vec{u}). \end{cases} \tag{3.98}$$

From what it has been said above, there exists $\vec{\Xi} \in \mathbf{R}^m$ such that $\vec{\Xi}' = \Lambda^T\vec{\Xi}$ and $\vec{\Xi} \in \bar{\partial}\hat{\Psi}(\Lambda\vec{u})$. Substituting $\vec{\Xi}'$ into $(3.98)_1$ we obtain that

$$(A\vec{u}, \vec{v})_{\mathbf{R}^n} + (\vec{\Xi}, \Lambda\vec{v})_{\mathbf{R}^m} = (\vec{f}, \vec{v})_{\mathbf{R}^n} \quad \forall \vec{v} \in \mathbf{R}^n.$$

From (3.96) we have that $\Xi_i \in c_i\hat{b}(x^i, (\Lambda\vec{u})_i)$ $\forall i = 1, ..., m$. $\square$

From the last theorem we see that all substationary points of $\mathcal{L}$ correspond to the first component of a solution to $(\vec{P})$. On the other hand it could happen that the set of all solutions to $(\vec{P})$ is larger, i.e., there are solutions to $(\vec{P})$ which cannot be interpreted as substationary points of $\mathcal{L}$. A natural question arises, namely, under which conditions both formulations are equivalent. The answer is given in

Theorem 3.6 *Let for any $i = 1, ..., m$ there exist one-sided limits $b(x^i, \xi\pm)$ for any $\xi \in \mathbf{R}$ and let the mapping P maps $\mathbf{R}^p$ onto $\mathbf{R}^m$. If $(\vec{u}, \vec{\Xi}) \in \mathbf{R}^n \times \mathbf{R}^m$ is a solution to $(\vec{P})$ then $\vec{u}$ is a substationary point of $\mathcal{L}$ and $\Lambda^T\vec{\Xi} \in \bar{\partial}\Psi(\vec{u})$.*

Proof: Since the mapping $\hat{\Psi} : \mathbf{R}^m \to \mathbf{R}$ defined by (3.91) is the sum of m functions $\hat{\Psi}_i$ depending only on one, its own variable θ_i, we have:

$$\left(\bar{\partial}\hat{\Psi}(\vec{\theta})\right)_i = \bar{\partial}\hat{\Psi}_i(\theta_i) \quad \forall i = 1, ..., m. \tag{3.99}$$

In the proof of Theorem 3.5 we have shown (see (3.96)) that $\bar{\partial}\hat{\Psi}_i(\theta_i) \subset c_i\hat{b}(x^i, \theta_i)$. If the one-sided limits $b(x^i, \xi\pm)$ exist for any $\xi \in \mathbf{R}$ then (see Chang, 1981):

$$\bar{\partial}\hat{\Psi}_i(\theta_i) = c_i\hat{b}(x^i, \theta_i), \quad \forall i = 1, ..., m. \tag{3.100}$$

Since $\Psi(\vec{v}) = \hat{\Psi}(\Lambda\vec{v})$ and Λ maps $\mathbf{R}^n$ onto $\mathbf{R}^m$ (do not forget that $\mathbf{R}^p = \Pi(\mathbf{R}^n)$), then from Theorem 1.22 it follows that

$$\bar{\partial}\Psi(\vec{v}) = \Lambda^T\bar{\partial}\hat{\Psi}(\Lambda\vec{v}) \quad \forall \vec{v} \in \mathbf{R}^n. \tag{3.101}$$

Let $(\vec{u}, \vec{\Xi}) \in \mathbf{R}^n \times \mathbf{R}^m$ be a solution to $(\vec{P})$. Since $\Xi_i \in c_i\hat{b}(x^i, (\Lambda\vec{u})_i)$ $\forall i = 1, ..., m$ we see from (3.99), (3.100) that $\vec{\Xi} \in \bar{\partial}\hat{\Psi}(\Lambda\vec{u})$ and $\Lambda^T\vec{\Xi} \equiv \vec{\Xi}'$ belongs to $\partial\Psi(\vec{u})$ as follows from (3.101). The rest of the proof is obvious. $\square$

Now we shall describe how from the knowledge of a substationary point $\vec{u}$ one can obtain the corresponding element $\vec{\Xi}$ of the subgradient of $\mathcal{L}$ at $\vec{u}$.

Let $\vec{u} \in \mathbf{R}^n$ be a substationary point of $\mathcal{L}$, i.e., there exists $\vec{\Xi}' \in \bar{\partial}\Psi(\vec{u})$ such that

$$(A\vec{u}, \vec{v})_{\mathbf{R}^n} + (\vec{\Xi}', \vec{v})_{\mathbf{R}^n} = (\vec{f}, \vec{v})_{\mathbf{R}^n} \quad \forall \vec{v} \in \mathbf{R}^n. \tag{3.102}$$

Since $\vec{\Xi}' = \Lambda^T \vec{\Xi}$, where $\vec{\Xi} \in \bar{\partial}\hat{\Psi}(\Lambda \vec{u})$, we obtain from (3.102) that

$$(A\vec{u}, \vec{v})_{\mathbf{R}^n} + (\vec{\Xi}, P(\Pi\vec{v}))_{\mathbf{R}^m} = (\vec{f}, \vec{v})_{\mathbf{R}^n} \quad \forall \vec{v} \in \mathbf{R}^n \qquad (3.103)$$

or, equivalently,

$$A\vec{u} + \Lambda^T \vec{\Xi} = \vec{f} \qquad (3.104)$$

making use of the definition of Λ. Having $\vec{u}$ at our disposal, we can compute $\vec{\Xi}$ from (3.104) as the solution of

$$\Lambda^T \vec{\Xi} = \vec{f} - A\vec{u}. \qquad (3.105)$$

A special attention will be paid to the case when (3.105) has a unique solution, i.e., when $\text{Ker}\,\Lambda^T = \{0\}$. From the classical results of the linear algebra we know that $(\text{Ker}\,\Lambda^T)^\perp = \text{Im}\,\Lambda$. Thus $\text{Ker}\,\Lambda^T = \{0\}$ iff $\text{Im}\,\Lambda = \mathbf{R}^m$, i.e., when $\Lambda = P\Pi$ maps $\mathbf{R}^n$ onto $\mathbf{R}^m$. Since at the same time $\Pi\mathbf{R}^n = \mathbf{R}^p$ this is equivalent to say that P maps $\mathbf{R}^p$ onto $\mathbf{R}^m$. Going back to the functional analytic setting this means that P maps W onto Y (see also Theorem 3.2). Summarizing, we proved:

Theorem 3.7 *Let P be the mapping of W onto Y and let $\vec{u}$ be a substationary point of $\mathcal{L}$. Then the linear system*

$$\Lambda^T \vec{\Xi} = \vec{f} - A\vec{u}$$

has a unique solution $\vec{\Xi}$ and the couple $(\vec{u}, \vec{\Xi})$ solves $(\vec{\mathrm{P}})$.

Let us give an explicit form of the mapping Λ, for the problem presented in Example 3.2. Let the nonmonotone friction law be given by (3.78) (we assume the same geometry of Γ_c as in Example 3.2). Then the mapping $\Pi : V \to W$ is represented by the matrix Π which with any nodal displacement vector $\vec{v} \in \mathbf{R}^n$ associates a subvector $\vec{v}\,' \in \mathbf{R}^p$, containing the x_1-component of $\vec{v}$ at the contact nodes, i.e., the nodes of $\mathcal{D}_h$, lying on $\overline{\Gamma}_c \setminus \overline{\Gamma}_u$. The form of Λ depends on the type of elements used for the construction of Y and on the mapping P. If Type IV element is used, P is given by (3.67) and $\overline{\Gamma}_u \cap \overline{\Gamma}_c = \emptyset$ then

$$(\Lambda\vec{v})_i = \begin{cases} 3/4v_1' + 1/4v_2' & \text{if } i = 1 \\ 1/4\dfrac{l_{i-1}}{l_{i-1}+l_i}v_{i-1}' + 3/4v_i' + 1/4\dfrac{l_i}{l_{i-1}+l_i}v_{i+1}' & \text{if } 2 \le i \le p-1 \\ 1/4v_{p-1}' + 3/4v_p' & \text{if } i = p, \end{cases}$$

where the symbol l_i has the same meaning as in Remark 3.14, $\vec{v}\,' = (v_1', ..., v_p')$ and p is the number of the contact nodes (an appropriate modification of Λ is necessary when $\overline{\Gamma}_u \cap \overline{\Gamma}_c \neq \emptyset$). If P is given by (3.76) then

$$(\Lambda\vec{v})_i = v_i' \quad 1 \le i \le p.$$

3.6 CONSTRAINED HEMIVARIATIONAL INEQUALITIES

At the beginning of this chapter an abstract setting of a class of scalar hemi-variational inequalities formulated and solved in the product of two *spaces V* and *Y* was introduced. Now, we shall analyse another type of hemivariational inequalities by imposing additional constraints on the first component u: the space V will be now replaced by a nonempty, closed and convex subset K of V. The new problem reads as follows:

$$\begin{cases} \text{Find } (u,\Xi) \in K \times Y \text{ such that} \\ a(u, v - u) + \langle \Xi, \Pi v - \Pi u \rangle_{Y \times Z} \geq \langle f, v - u \rangle \quad \forall v \in K \\ \Xi(x) \in \hat{b}(x, \Pi u(x)) \quad \text{for a.a. } x \in \omega. \end{cases} \tag{P}$$

The meaning of symbols remains the same as in the previous sections.

Remark 3.15 *The new formulation* (P) *results from the problem of finding substationary points of* $\mathcal{L}$ *with respect to* K, *where*

$$\mathcal{L}(v) = \frac{1}{2} a(v, v) - \langle f, v \rangle + \Phi(v) \tag{3.106}$$

with

$$\Phi(v) = \int_{\omega} \int_{0}^{\Pi v(x)} b(x, t) dt d\mu.$$

Indeed, from Definition 1.7 it follows that $u \in K$ *is a substationary point of* $\mathcal{L}$ *with respect to* K *iff*

$$0 \in \bar{\partial} \mathcal{L}(u) + N_K(u),$$

where $N_K(u)$ *is the normal cone of* K *at* u. *If from* $w \in \bar{\partial}\Phi(u)$ *it follows that* $w(x) \in \hat{b}(x, \Pi u(x))$ *for a.a.* $x \in \omega$ *then the previous inclusion and* (3.106) *lead to* (P).

Now we shall study the approximation of constrained hemivariational inequalities. As we shall see, the approach will be similar to this one we have already used in the unconstrained case. For this reason, we shall not repeat those parts of proofs, which are identical.

First, we introduce a system $\{K_h\}$, $h \to 0+$ of *closed, convex subsets* K_h of V_h, $\dim V_h = n(h) < \infty$. As before we define spaces Y_h by (3.22) and appropriate discretizations a_h, f_h of a, f, respectively.

The *discrete constrained hemivariational inequality* is defined as a problem of finding a couple $(u_h, \Xi_h) \in K_h \times Y_h$ such that

$$\begin{cases} a_h(u_h, v_h - u_h) + \int_{\omega} \Xi_h P_h(\Pi v_h - \Pi u_h) d\mu \\ \geq \langle f_h, v_h - u_h \rangle_h \quad \forall v_h \in K_h \\ \Xi_h(x) \in \hat{b}(\sum_{i=1}^{m} \mathcal{X}_{\text{int}_{\omega} K_i}(x) x_h^i, P_h(\Pi u_h)(x)) \quad \text{for a.a. } x \in \omega \end{cases} \tag{P$_h$}$$

again with the same meaning of symbols as in the previous sections. Next we shall prove the existence of solutions to (P)$_h$ and analyse the relation between (P) and (P)$_h$, when $h \to 0+$.

Before we start, let us summarize the assumptions on data, which will be used in the sequel: we suppose that

(i) the function b satisfies (3.2), (3.3), (3.5), (3.16) and the stronger growth condition (3.47);

(ii) the system of approximated bilinear forms $\{a_h\}$ satisfies (3.24), (3.25), (3.52);

(iii) the system of approximated linear forms $\{f_h\}$ satisfies (3.26) and (3.53);

(iv) the mapping Π satisfies (3.48);

(v) the mapping P_h satisfies (3.49);

(vi) the system $\{K_h\}$ is such that $0 \in K_h$ for any $h > 0$ and (1.172) and (1.173) are satisfied.

To prove that (P)$_h$ has solutions for any $h > 0$ we introduce an auxiliary problem based on the *penalization* of the constraint $v \in K_h$ and the *regularization* of the nonsmooth function b. To this end we introduce a *penalty functional* $\beta_h \in V_h^*$ such that

- β_h is convex and continuously Fréchet differentiable in V_h;   (3.107)
- the Fréchet derivative $\beta_h'(v) = 0$ iff $v \in K_h$ for any $h > 0$.   (3.108)

The auxiliary *regularized and penalized problem* reads as follows:

$$
\begin{cases}
\text{Find } u_h^{\kappa\varepsilon} \in V_h \text{ such that} \\
a_h(u_h^{\kappa\varepsilon}, v_h) + \int_\omega b^\kappa \big(\sum_{i=1}^m \mathcal{X}_{\mathrm{int}_\omega K_i}(x) x_h^i, P_h(\Pi u_h^{\kappa\varepsilon})(x)\big) \\
P_h(\Pi v_h)(x) d\mu + \frac{1}{\varepsilon}\langle \beta_h'(u_h^{\kappa\varepsilon}), v_h\rangle_h = \langle f_h, v_h\rangle_h \quad \forall v_h \in V_h,
\end{cases}
\qquad (\mathrm{P})_h^{\kappa\varepsilon}
$$

where b^κ is defined by (3.11) and ε, κ are positive parameters. Suppose that $h > 0$ is *fixed*. Then we have:

Lemma 3.8 *Let (i)-(iii) from above be satisfied and let $0 \in K_h$ for any $h > 0$. Then* (P)$_h^{\kappa\varepsilon}$ *has at least one solution* $u_h^{\kappa\varepsilon}$ *for any* h, κ *and* $\varepsilon > 0$ *which is uniformly bounded with respect to* $h, \kappa, \varepsilon > 0$:

$$\exists c > 0 : \|u_h^{\kappa\varepsilon}\| \le c \quad \forall h, \kappa, \varepsilon > 0.$$

Proof: The proof is parallel to this one of Lemma 3.4. Define a mapping $T_h^{\kappa\varepsilon} : V_h \to V_h^*$ by

$$
\langle T_h^{\kappa\varepsilon} y_h, z_h\rangle_h \equiv a_h(y_h, z_h) + \int_\omega b^\kappa \big(\sum_{i=1}^m \mathcal{X}_{\mathrm{int}_\omega K_i}(x) x_h^i, P_h(\Pi y_h)(x)\big)
$$

$$
P_h(\Pi z_h)(x) d\mu + \frac{1}{\varepsilon}\langle \beta_h'(y_h), z_h\rangle_h - \langle f_h, z_h\rangle_h.
$$

Then $T_h^{\kappa\varepsilon}$ is continuous and there exists a constant $c > 0$, which does not depend on h, κ and ε such that

$$\langle T_h^{\kappa\varepsilon} z_h, z_h \rangle_h > 0 \quad \forall \|z_h\| = c.$$

Indeed, the only difference between $T_h^{\kappa\varepsilon}$ and T_h^κ from Lemma 3.4 is the presence of the penalty term. But

$$\frac{1}{\varepsilon} \langle \beta_h'(z_h), z_h \rangle_h = \frac{1}{\varepsilon} \langle \beta_h'(z_h) - \beta_h'(0), z_h - 0 \rangle_h \geq 0$$

making use that $0 \in K_h$, (3.108) and the monotonicity of the mapping $z_h \mapsto \beta_h'(z_h)$, $z_h \in V_h$ valid for smooth convex functions. Thus the penalty term can be neglected and we finally find that

$$\langle T_h^{\kappa\varepsilon} z_h, z_h \rangle_h \geq \tilde{\alpha} \|z_h\|^2 - \tilde{\beta} \|z_h\| - c,$$

where $\tilde{\alpha}, \tilde{\beta}$ are constants from (3.25) and (3.26), respectively, and $c > 0$ is another constant. The rest of the proof follows from Theorem 1.26. $\qquad\square$

Next we shall study the limit passage in $(P)_h^{\kappa\varepsilon}$ for $\kappa \to 0+$ keeping h, ε fixed.

Let $\{u_h^{\kappa\varepsilon}\}$ be a sequence of bounded solutions to $(P)_h^{\kappa\varepsilon}$ the existence of which follows from Lemma 3.8. Then also $\{b^\kappa(x_h^i, P_h(\Pi u_h^{\kappa\varepsilon})(x_h^i))\}$ is bounded and one can find subsequences and functions $u_h^\varepsilon \in V_h$ and $\Xi_h^\varepsilon \in Y_h$ such that

$$u_h^{\kappa\varepsilon} \to u_h^\varepsilon, \quad b^\kappa(x_h^i, P_h(\Pi u_h^{\kappa\varepsilon})(x_h^i)) \to \Xi_h^\varepsilon(x_h^i) \quad \forall i = 1, ..., m$$

as $\kappa \to 0+$. Then it is readily seen that the couple $(u_h^\varepsilon, \Xi_h^\varepsilon) \in V_h \times Y_h$ satisfies the equation

$$a_h(u_h^\varepsilon, v_h) + \int_\omega \Xi_h^\varepsilon P_h(\Pi v_h) d\mu + \frac{1}{\varepsilon} \langle \beta_h'(u_h^\varepsilon), v_h \rangle_h \qquad (3.109)$$
$$= \langle f_h, v_h \rangle_h \quad \forall v_h \in V_h.$$

The inclusions

$$\Xi_h^\varepsilon(x_h^i) \in \hat{b}(x_h^i, P_h(\Pi u_h^\varepsilon)(x_h^i)), \qquad (3.110)$$

$i = 1, ..., m$ can be proven in the same way as in Theorem 3.1. The equation (3.109) together with (3.110) present the *penalized form* of $(P)_h$. Now we are ready to prove

Theorem 3.8 *Let all the assumptions of Lemma 3.8 be satisfied. Then $(P)_h$ has at least one solution for any $h > 0$.*

Proof: Let $(u_h^\varepsilon, \Xi_h^\varepsilon) \in V_h \times Y_h$ be a solution of (3.109) and (3.110). We may assume that $\{u_h^\varepsilon\}$ is bounded in V, independently of h and $\varepsilon > 0$. Therefore for a given fixed $h > 0$, the sequence $\{\Xi_h^\varepsilon\}$ is bounded with respect to ε, as well. Thus we can pass to appropriate subsequences such that

$$\begin{cases} u_h^\varepsilon \to u_h, \\ \Xi_h^\varepsilon \to \Xi_h, \quad \varepsilon \to 0+, \end{cases} \qquad (3.111)$$

where $(u_h, \Xi_h) \in V_h \times Y_h$. We prove that (u_h, Ξ_h) is a solution to (P)$_h$. From (3.109) it follows that

$$|\langle \beta'_h(u_h^\varepsilon), v_h \rangle_h| \le C\varepsilon \quad \forall \varepsilon > 0, \ \forall \|v_h\| \le 1,$$

where $C > 0$ does not depend on ε. Letting $\varepsilon \to 0+$ we see that

$$\langle \beta'_h(u_h), v_h \rangle_h = \lim_{\varepsilon \to 0+} \langle \beta'_h(u_h^\varepsilon), v_h \rangle_h = 0.$$

Thus $u_h \in K_h$.

Let $v_h \in K_h$ be an arbitrary element. Inserting $v_h := v_h - u_h^\varepsilon$ into (3.109) we obtain:

$$a_h(u_h^\varepsilon, v_h - u_h^\varepsilon) + \int_\omega \Xi_h^\varepsilon P_h(\Pi v_h - \Pi u_h^\varepsilon) d\mu$$

$$+ \frac{1}{\varepsilon}\langle \beta'_h(u_h^\varepsilon) - \beta'_h(v_h), v_h - u_h^\varepsilon \rangle_h = \langle f_h, v_h - u_h \rangle_h.$$

Due to the monotonicity of the mapping $z_h \mapsto \beta'_h(z_h)$, $z_h \in V_h$ the penalty term is nonpositive. Therefore

$$a_h(u_h^\varepsilon, v_h - u_h^\varepsilon) + \int_\omega \Xi_h^\varepsilon P_h(\Pi v_h - \Pi u_h^\varepsilon) d\mu \qquad (3.112)$$

$$\ge \langle f_h, v_h - u_h^\varepsilon \rangle_h$$

holds for any $v_h \in K_h$ and any $\varepsilon > 0$. Letting $\varepsilon \to 0+$ in (3.112) and using (3.111) we finally arrive at

$$a_h(u_h, v_h - u_h) + \int_\omega \Xi_h P_h(\Pi v_h - \Pi u_h) d\mu$$

$$\ge \langle f_h, v_h - u_h \rangle_h.$$

Since the sequences $\{\Xi_h^\varepsilon(x_h^i)\}$ and $\{P_h(\Pi u_h^\varepsilon)(x_h^i)\}$ converge to $\Xi_h(x_h^i)$ and $P_h(\Pi u_h)(x_h^i)$, respectively, as $\varepsilon \to 0+$ and $\Xi_h^\varepsilon(x_h^i) \in \hat{b}(x_h^i, P_h(\Pi u_h^\varepsilon)(x_h^i)) \ \forall i = 1, ..., m$, we see that

$$\Xi_h(x_h^i) \in \hat{b}(x_h^i, P_h(\Pi u_h)(x_h^i)) \quad \forall i = 1, ..., m$$

making use of Proposition 3.1. $\qquad \square$

The relation between (P)$_h$ and (P), when $h \to 0+$ is established in

Theorem 3.9 *Let (i)-(vi) be satisfied. Let $\{(u_h, \Xi_h)\}$ be a sequence of solutions to (P)$_h$ with $\{u_h\}$ being bounded in V independently of h. Then there exist: a subsequence of $\{(u_h, \Xi_h)\}$, the number $q' > 1$ from (3.47) and a pair $(u, \Xi) \in K \times L^{q'}(\omega)$ such that*

$$\begin{cases} u_h \to u, & \text{in } V; \\ \Xi_h \rightharpoonup \Xi, & \text{in } L^{q'}(\omega), \ h \to 0+. \end{cases} \qquad (3.113)$$

Moreover (u, Ξ) solves (P). *Furthermore any cluster point of $\{(u_h, \Xi_h)\}$ in the sense of (3.113) solves* (P).

Proof: We may assume that $u_h \rightharpoonup u$ in V. From (3.47)-(3.49) it follows that $\{\Xi_h\}$ is bounded in $L^{q'}(\omega)$ so that one can find a subsequence of $\{\Xi_h\}$ and $\Xi \in L^{q'}(\omega)$ such that $(3.113)_2$ holds. From (1.173) it follows that $u \in K$. Let $\bar{v} \in K$ be given. Then accordingly to (1.172) there exists a sequence $\{\bar{v}_h\}$, $\bar{v}_h \in K_h$ such that $\bar{v}_h \to \bar{v}$. Due to (3.49) we have that

$$\begin{cases} P_h(\Pi u_h) \to \Pi u \\ P_h(\Pi \bar{v}_h) \to \Pi \bar{v} \end{cases} \quad \text{in } L^s(\omega), \ s \geq q. \tag{3.114}$$

Let us suppose for the moment that we have already shown that $\{u_h\}$ tends to u in the norm of V. Then passing to the limit with $h \to 0+$ in $(P)_h$, using (3.52), (3.53) and (3.114) we arrive at

$$a(u, \bar{v} - u) + \int_\omega \Xi(\Pi \bar{v} - \Pi u) d\mu \geq \langle f, \bar{v} - u \rangle.$$

The inclusion $\Xi(x) \in \hat{b}(x, \Pi u(x))$ for a.a. $x \in \omega$ follows from the proof of Theorem 3.3.

It remains to show that $u_h \to u$ (strongly) in V. Let $\{\bar{v}_h\}$, $\bar{v}_h \in K_h$, be a sequence such that $\bar{v}_h \to u$ in V. Then from (3.25) it follows that

$$\tilde{\alpha} \|u_h - \bar{v}_h\|^2 \leq a_h(u_h - \bar{v}_h, u_h - \bar{v}_h)$$
$$= a_h(u_h, u_h - \bar{v}_h) - a_h(\bar{v}_h, u_h - \bar{v}_h) \to 0, \quad \text{as } h \to 0+.$$

Indeed, the term $a_h(\bar{v}_h, u_h - \bar{v}_h) \to 0$ as $h \to 0+$ because of (3.52). From the definition of $(P)_h$ we see that

$$a_h(u_h, u_h - \bar{v}_h)$$
$$\leq - \int_\omega \Xi_h(P_h(\Pi u_h - \Pi \bar{v}_h)) d\mu + \langle f_h, u_h - \bar{v}_h \rangle_h$$
$$\leq \|\Xi_h\|_{L^{q'}(\omega)} \|P_h(\Pi u_h - \Pi \bar{v}_h)\|_{L^q(\omega)} + \langle f_h, u_h - \bar{v}_h \rangle_h$$

Thus

$$\limsup_{h \to 0+} a_h(u_h, u_h - \bar{v}_h)$$
$$\leq \lim_{h \to 0+} \left\{ \|\Xi_h\|_{L^{q'}(\omega)} \|P_h(\Pi u_h - \Pi \bar{v}_h)\|_{L^q(\omega)} + \langle f_h, u_h - \bar{v}_h \rangle_h \right\} = 0$$

making use of (3.53) and (3.114) with $\bar{v} := u$. Since

$$0 \leq \liminf_{h \to 0+} \|u_h - \bar{v}_h\|^2 \leq \limsup_{h \to 0+} \|u_h - \bar{v}_h\|^2 \leq 0$$

we have that

$$\|u_h - \bar{v}_h\| \to 0 \quad \text{as } h \to 0+.$$

The strong convergence of $\{u_h\}$ to u in V now follows from the triangle inequality:

$$\|u_h - u\| \leq \|u_h - \bar{v}_h\| + \|\bar{v}_h - u\| \to 0 + .$$

$\square$

Remark 3.16 *The number $q' > 1$, appearing in the assertion of Theorem 3.9 comes from the growth condition (3.47). The pairing $\langle \cdot, \cdot \rangle_{Y \times Z}$ is realized by the integral over ω of the product $\Xi \in L^{q'}(\omega)$ with $\Pi v \in L^q(\omega)$, $1/q' + 1/q = 1$.*

We shall apply the previous abstract results to the approximation of a non-monotone unilateral boundary value problem.

Example 3.3 Let us consider a plane elastic body, represented by a polygonal domain Ω, whose boundary is decomposed into Γ_u, Γ_P and Γ_c. Along the contact part Γ_c the body is supported by a *deformable* foundation represented by a halfplane $\mathbf{R}^2_-$ and Γ_c itself is placed in the x_1-axis. On Γ_c the following *unilateral conditions* will be prescribed:

$$\begin{cases} u_2(x) \geq \varphi(x), \quad -T_2(x) \leq \Xi(x) \\ (u_2(x) - \varphi(x))(T_2(x) + \Xi(x)) = 0 \qquad \text{for a.a. } x \in \Gamma_c, \\ \Xi(x) \in \hat{b}(x, u_2(x)) \end{cases} \tag{3.115}$$

where φ is a given *nonpositive* function on Γ_c. The first kinematical constraint in (3.115) says that the deformable foundation restricts the deformation of Ω in a such a way that the vertical displacement u_2 is limited from below by a function φ. The reaction-displacement law, relating the normal component T_2 of the stress vector T to the normal (=vertical) component u_2 of the displacement field u is represented by a (generally) nonsmooth and nonmonotone multifunction $\hat{b}$ as follows:

- if $u_2(x) > \varphi(x)$ then $-T_2(x) \in \hat{b}(x, u_2(x))$

- if $u_2(x) = \varphi(x)$ then $-T_2(x)$ can take any

value from (γ, ∞) (see Fig.3.10).

To complete the conditions on Γ_c we suppose that the tangential component of the stress vector is prescribed:

$$T_1(x) = S_1(x) \quad \text{for a.a. } x \in \Gamma_c.$$

On the remaining parts Γ_u and Γ_P of $\partial\Omega$ the same conditions as in Example 3.2 will be given. Now, we are ready to present the variational formulation of this problem. We set

$$K = \{v = (v_1, v_2) \in H^1(\Omega; \mathbf{R}^2) \mid v = 0 \text{ on } \Gamma_u, \ v_2 \geq \varphi \text{ a.e. in } \Gamma_c\},$$

$$a(u, v) \equiv \int_\Omega c_{ijkl} \varepsilon_{ij}(u) \varepsilon_{kl}(v) dx,$$

$$\langle f, v \rangle \equiv \int_\Omega F_i v_i dx + \int_{\Gamma_P} P_i v_i ds + \int_{\Gamma_c} S_1 v_1 dx_1,$$

$$\Pi v = v_2|_{\Gamma_c}, \quad \omega = \Gamma_c, \quad Z = L^2(\omega), \quad Y = L^{q'}(\omega),$$

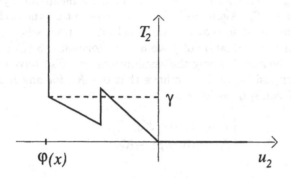

Figure 3.10.

where q' comes from the growth condition (3.47). Next we shall suppose that

$$\begin{cases} - F \in L^2(\Omega; \mathbf{R}^2), \ P \in L^2(\Gamma_P; \mathbf{R}^2), \ S_1 \in L^2(\Gamma_c); \\ - \text{the function } b \text{ defining } \hat{b}, \text{ satisfies } (i) \text{ from} \\ \quad \text{the beginning of this section, especially} \\ \quad \text{the growth condition (3.47);} \\ - \varphi \text{ is the trace on } \Gamma_c \text{ of a function } \bar{\varphi} \in H^2(\Omega) \cap V. \end{cases} \qquad (3.116)$$

The constrained hemivariational inequality, describing the equilibrium state of Ω reads as follows:

$$\begin{cases} \text{Find } (u, \Xi) \in K \times L^{q'}(\Gamma_c) \text{ such that} \\ a(u, v - u) + \int_{\Gamma_c} \Xi(v_2 - u_2) dx_1 \geq \langle f, v - u \rangle \quad \forall v \in K \\ \Xi(x) \in \hat{b}(x, u_2(x)) \quad \text{for a.a. } x \in \Gamma_c. \end{cases} \qquad (3.117)$$

It is an easy exercise to show (by using the Green's formula (1.16)) that if u is sufficiently smooth, then from (3.117) the equilibrium equation in Ω and the boundary conditions on Γ_u, Γ_P and Γ_c, follow.

Now, let us pass to the discretization of (3.117). Let $\{\mathcal{D}_h\}$ be a regular system of triangulations of $\overline{\Omega}$, which is consistent with the decomposition of $\partial\Omega$ into Γ_u, Γ_P and Γ_c. With any $\mathcal{D}_h$ the finite dimensional space V_h and its closed convex subset K_h will be associated:

$$V_h = \{v_h \in C(\overline{\Omega}; \mathbf{R}^2) \mid v_h|_T \in (P_1(T))^2 \ \forall T \in \mathcal{D}_h, \ v_h = 0 \text{ on } \Gamma_u\},$$

$$K_h = \{v_h = (v_{h1}, v_{h2}) \in V_h \mid v_{h2}(A) \geq \varphi(A) \ \forall A \in \mathcal{N}_h\},$$

where $\mathcal{N}_h$ stands for the set of all the contact nodes of $\mathcal{D}_h$, i.e., the nodes in $\overline{\Gamma}_c \setminus \overline{\Gamma}_u$. Let us notice that K_h *is not* a subset of K, in general.

Remark 3.17 *Since φ is the trace of a function $\bar{\varphi} \in H^2(\Omega) \cap V \hookrightarrow C(\overline{\Omega})$, the definition of K_h makes sense.*

The space Y_h given by (3.22) is constructed by means of Type III or IV elements defined on Γ_c. Again, as before, we suppose that the body is made of a homogenous material so that no numerical integration, when computing a, is necessary. For the evaluation of f we use the formulas (3.80) and (3.81). In order to apply Theorem 3.9 only the assumptions on $\{K_h\}$ have to be verified.

Since φ is nonpositive on Γ_c, we have that $0 \in K_h$ for any $h > 0$. Further both K_h and K can split as follows:

$$\begin{cases} K = (0, \bar\varphi) + K_0 \\ K_h = (0, r_h\bar\varphi) + K_{0h}, \end{cases} \tag{3.118}$$

where

$$K_0 = \{v = (v_1, v_2) \in H^1(\Omega; \mathbf{R}^2) \mid v = 0 \text{ on } \Gamma_u, \ v_2 \geq 0 \text{ on } \Gamma_c\}$$
$$K_{0h} = \{v_h = (v_{h1}, v_{h2}) \in V_h \mid v_{h2}(A) \geq 0 \ \forall A \in \mathcal{N}_h\}$$

and $r_h\bar\varphi$ stands for the piecewise linear lagrange interpolate of $\bar\varphi$. From Lemma 1.6 we know that the family $\{K_h\}$ is dense in K in the sense of (1.172). Let us verify (1.173), i.e.,

$$v_h \rightharpoonup v, \ v_h \in K_h \implies v \in K.$$

From $(3.118)_2$ it follows that

$$v_h = (0, r_h\bar\varphi) + w_h, \ w_h \in K_{0h}.$$

Since $\bar\varphi \in H^2(\Omega) \cap V$ we know that $r_h\bar\varphi \to \bar\varphi$ in $H^1(\Omega)$. On the other hand $w_h \in K_{0h} \subset K_0$ for any $h > 0$. Therefore the weak limit w of $\{w_h\}$ belongs to K_0, as well. Consequently, $v = (0, \bar\varphi) + w \in K$.

Since Π and P_h satisfy (3.48) and (3.49) with $s = q = 2$, respectively, we see that, if b satisfies (3.47) with $q' = q = 2$, then $\Xi \in L^2(\Gamma_c)$ and

$$\begin{cases} u_h \to u \text{ in } V, \\ \Xi_h \rightharpoonup \Xi \text{ in } L^2(\Gamma_c), \ h \to 0+, \end{cases}$$

as follows from Theorem 3.9.

Let us describe very briefly the derivation of the algebraic form of $(P)_h$. We proceed exactly in the same way as in the case of unconstrained hemivariational inequalities. Also the definition of all the symbols used in what follows is the same as in Section 3.5. Let us recall only that the spaces V_h, W_h and Y_h are isomorphic with $\mathbf{R}^n$, $\mathbf{R}^p$ and $\mathbf{R}^m$, respectively. The convex set K_h, defining the constraints imposed on u_h can be identified with a closed, convex subset $\mathcal{K} \subset \mathbf{R}^n$. The algebraic form of the constrained hemivariational inequality reads as follows:

$$\begin{cases} \text{Find } (\vec{u}, \vec\Xi) \in \mathcal{K} \times \mathbf{R}^m \text{ such that} \\ (\mathbf{A}\vec{u}, \vec{v} - \vec{u})_{\mathbf{R}^n} + (\vec\Xi, \Lambda(\vec{v} - \vec{u}))_{\mathbf{R}^m} \geq (\vec{f}, \vec{v} - \vec{u})_{\mathbf{R}^n} \ \ \forall \vec{v} \in \mathcal{K} \\ \Xi_i \in c_i \hat{b}(x^i, (\Lambda\vec{u})_i) \ \ \forall i = 1, ..., m. \end{cases} \tag{$\vec{P}$}$$

Let the function $\mathcal{L} : \mathbf{R}^n \rightarrow \mathbf{R}$ be given by (3.90), where $\Psi(\vec{v}) \equiv \hat{\Psi}(\Lambda\vec{v})$ with $\hat{\Psi}$ defined by (3.91). As before we can prove the following relation between substationary points of $\mathcal{L}$ on $\mathcal{K}$ and solutions to $(\vec{P})$:

Theorem 3.10 *It holds:*

(i) *for every substationary point $\vec{u}$ of $\mathcal{L}$ on $\mathcal{K}$ there exists $\vec{\Xi} \in \mathbf{R}^m$ such that $\Lambda^T\vec{\Xi} \in \bar{\partial}\Psi(\vec{u})$ and $(\vec{u}, \vec{\Xi})$ solves $(\vec{P})$;*

(ii) *let for any $i = 1, ..., m$ there exist one-sided limits $b(x^i, \xi\pm)$ for any $\xi \in \mathbf{R}$ and let P map $\mathbf{R}^p$ onto $\mathbf{R}^m$. If $(\vec{u}, \vec{\Xi}) \in \mathcal{K} \times \mathbf{R}^m$ is a solution to $(\vec{P})$ then $\vec{u}$ is a substationary point of $\mathcal{L}$ on $\mathcal{K}$ and $\Lambda^T\vec{\Xi} \in \bar{\partial}\Psi(\vec{u})$.*

Proof proceeds exactly in the same way as in Theorems 3.5 and 3.6.

In the unconstrained case we described, how from the knowledge of a substationary point $\vec{u}$ of $\mathcal{L}$ one can get the information on the second component $\vec{\Xi} \in \mathbf{R}^m$. Now we shall discuss the same question in the case of constrained hemivariational inequalities.

Let $\vec{u} \in \mathcal{K}$ be a substationary point of $\mathcal{L}$ in $\mathcal{K}$. Then from the first part of Theorem 3.10 we know that there exists $\vec{\Xi} \in \mathbf{R}^m$ such that

$$(A\vec{u}, \vec{v} - \vec{u})_{\mathbf{R}^n} + (\vec{\Xi}, \Lambda(\vec{v} - \vec{u}))_{\mathbf{R}^m} \geq (\vec{f}, \vec{v} - \vec{u})_{\mathbf{R}^n} \qquad (3.119)$$

holds for any $\vec{v} \in \mathcal{K}$. First, let us consider the case, when $\vec{u} \in \operatorname{int}\mathcal{K}$. Then from (3.119) it follows that

$$(A\vec{u}, \vec{z})_{\mathbf{R}^n} + (\vec{\Xi}, \Lambda\vec{z})_{\mathbf{R}^m} = (\vec{f}, \vec{z})_{\mathbf{R}^n} \quad \forall\vec{z} \in \mathbf{R}^n,$$

i.e., $\vec{\Xi}$ satisfies the same system of equations as in the unconstrained case and we can use Theorem 3.7.

The assumption on $\vec{u}$ to be an inner point of $\mathcal{K}$ is *very restrictive* and seldom satisfied. Let us consider a more general case.

Let us recall that the mapping $\Lambda : \mathbf{R}^n \rightarrow \mathbf{R}^m$ is the product of two matrices Π and P, representing the linear mappings $\Pi : V_h \rightarrow W_h$ and $P_h : W_h \rightarrow Y_h$, respectively. Assume that the set $\mathcal{K}$ defining the constraints on $\vec{u}$ is given by

$$\mathcal{K} = \{\vec{x} \in \mathbf{R}^n \mid \Pi\vec{x} \in C \iff (\Pi x)_i \in C_i \; \forall i = 1, ..., p\},$$

where a closed convex subset $C \subset \mathbf{R}^p$ is the cartesian product of p closed convex subsets $C_i \subset \mathbf{R}$, $i = 1, ..., p$:

$$C = C_1 \times C_2 \times \cdots \times C_p.$$

Suppose that the substationary point $\vec{u}$ is such that $(\Pi\vec{u})_i \in \operatorname{int}C_i$ for some $i \in \{1, ..., p\}$. We shall show that in a special case one can compute the corresponding i-th component Ξ_i of $\vec{\Xi}$. Indeed, denote by

$$X_i = \{\vec{z} \in \mathbf{R}^n \mid (\Pi\vec{z})_j = 0 \; \forall j \neq i\}.$$

Then all the vectors of the form $\vec{v} = \vec{u} + t\vec{z}$, $\vec{z} \in X_i$, belong to $\mathcal{K}$ provided that $|t| \leq \delta$ and $\delta > 0$ is small enough. Inserting such $\vec{v}$ into (3.119) we easily get that

$$(\mathbf{A}\vec{u}, \vec{z})_{\mathbf{R}^n} + (\mathbf{P}^T\vec{\Xi})_i (\mathbf{\Pi}\vec{z})_i (\text{no sum}) = (\vec{f}, \vec{z})_{\mathbf{R}^n} \qquad (3.120)$$

holds for any $\vec{z} \in X_i$. Now suppose that $\dim W_h = \dim Y_h$, i.e., $p = m$ and P_h is defined by (3.76) so that $\mathbf{P}$ is the $(m \times m)$ identity matrix. Then $(\mathbf{P}^T\vec{\Xi})_i = \Xi_i$ and from (3.120) the i-th component Ξ_i can be computed.

To illustrate these results, let us consider the nonmonotone unilateral boundary value problem. This time we shall consider a plane body, represented by a polygonal domain, which is unilaterally supported by a deformable foundation with a general smooth boundary. The space V_h will be the same as in Example 3.3. Define the closed convex subset K_h of V_h by

$$K_h = \{v_h \in V_h \mid (v_h.\nu)(A) \geq \varphi(A) \ \forall A \in \mathcal{N}_h\},$$

where φ and $\mathcal{N}_h$ are the same as in Example 3.3 and ν is the unit outward normal vector to Γ_c. The mapping $\Pi : V \to L^2(\Gamma_c)$ is defined by

$$\Pi v \equiv v.\nu \quad \text{on } \Gamma_c.$$

As before we shall suppose that Γ_c is a straight segment. Let W_h be the image of V_h with respect to Π. It is readily seen that

$$W_h = \{w_h \in C(\overline{\Gamma}_c) \mid w_h|_{T'} \in P_1(T') \ \forall T' \in \mathcal{D}_h|_{\Gamma_c}\} \cap \Pi(V),$$

where the symbol $\mathcal{D}_h|_{\Gamma_c}$ denotes the partition of Γ_c induced by $\mathcal{D}_h$. Then $\dim W_h = p$, where p is the number of all the contact nodes $A \in \overline{\Gamma}_c \setminus \overline{\Gamma}_u$. As far as the space Y_h is concerned, we use Type IV elements. To get that $\dim Y_h = \dim W_h$ we have to distinguish two cases: (i) $\overline{\Gamma}_u \cap \overline{\Gamma}_c = \emptyset$, (ii) $\overline{\Gamma}_u \cap \overline{\Gamma}_c \neq \emptyset$. If (i) holds then the partition $\mathcal{T}_h$ used when defining Y_h is made of segments K_i, which join the midpoints of two edges $T' \subset \partial\Omega$ of the triangles $T \in \mathcal{D}_h$ sharing the same contact node $A \in \mathcal{N}_h$ completed by two half segments "at the beginning and the end" of Γ_c (see Fig. 3.11). If $\overline{\Gamma}_u \cap \overline{\Gamma}_c \neq \emptyset$, then at least one of these two halfsegments, namely this one which contains a point from the intersection $\overline{\Gamma}_u \cap \overline{\Gamma}_c$ has to be appended to its inner neighbour to get a new element (see Fig.3.12). With any $A \in \mathcal{N}_h$, the index set $\mathcal{I}(A) \subset \{1,...,n\}$ containing just two elements (i_1, i_2) will be associated:

$$(i_1, i_2) \in \mathcal{I}(A) \iff \text{the } i_1, i_2\text{-th component of } \vec{v} \in \mathbf{R}^n \text{ corresponds}$$
$$\text{to the value of } v_h = (v_{h1}, v_{h2}) \in V_h \text{ at } A :$$
$$v_{i_1} = v_{h1}(A), \quad v_{i_2} = v_{h2}(A).$$

Observe that $\mathcal{I}(A) \cap \mathcal{I}(B) = \emptyset$ when $A \neq B$, $A, B \in \mathcal{N}_h$. Then the set K_h can be identified with

$$\mathcal{K} = \{\vec{v} \in \mathbf{R}^n \mid v_{i_1}\nu_1 + v_{i_2}\nu_2 \geq \varphi(A), \ (i_1, i_2) \in \mathcal{I}(A), \ A \in \mathcal{N}_h\},$$

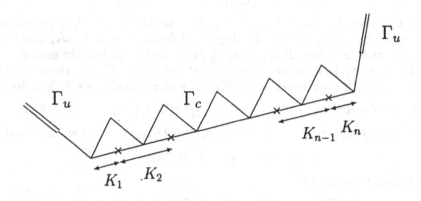

Figure 3.11.

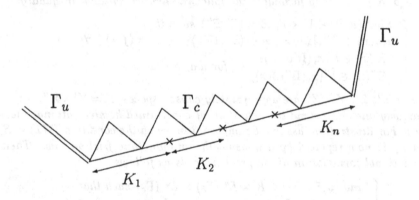

Figure 3.12.

where $\nu = (\nu_1, \nu_2)$.

Taking into account the definition of Π and K we see that the $(p \times n)$ matrix Π, representing Π is given by

$$\begin{pmatrix} 0 & \nu_1 & 0 & 0 & \dots & \nu_2 & 0 \\ 0 & 0 & \nu_1 & 0 & \dots & 0 & \nu_2 \\ \cdot & \cdot & \cdot & \cdot & \cdot & \cdot & \cdot & \cdot \end{pmatrix},$$

where each row of Π contains at most two nonzero elements at the position determined by $\mathcal{I}(A)$, $A \in \mathcal{N}_h$, while each column contains at most one nonzero element. The convex set $C \subset \mathbf{R}^p$ is the cartesian product:

$$C = C_1 \times C_2 \times \cdots \times C_p,$$

where $\mathcal{C}_i = [\varphi(A), \infty)$ for $A \in \mathcal{N}_h$. We immediately see that

$$\vec{v} \in \mathcal{K} \quad \text{iff} \quad (\mathbf{\Pi}\vec{v})_i \in \mathcal{C}_i \quad \forall i = 1, ..., p.$$

The mapping $P_h : W_h \to Y_h$ will be defined by (3.76), i.e., P_h is the lagrange interpolation operator with the degrees of freedom at $A \in \mathcal{N}_h$, $\dim W_h = \dim Y_h$, and the matrix $\mathbf{P}$ representing P_h is the $(p \times p)$ identity matrix.

Let $A_i \in \mathcal{N}_h$ be a contact node at which $(\mathbf{\Pi}\vec{u})_i \in \operatorname{int}\mathcal{C}_i$, i.e., the body is not in contact with the support at A_i. Then the corresponding set X_i is defined by

$$X_i = \{\vec{z} \in \mathbf{R}^n \mid z_{j_1}\nu_1 + z_{j_2}\nu_2 = 0 \ (j_1, j_2) \in \mathcal{I}(A_j), \ \forall j \neq i, \ A_j \in \mathcal{N}_h\}.$$

Taking $\vec{z}_0 \in X_i$ such that $z_{0i_1}\nu_1 + z_{0i_2}\nu_2 = 1$, $(i_1, i_2) \in \mathcal{I}(A_i)$ we have that

$$\Xi_i = (\vec{f}, \vec{z}_0)_{\mathbf{R}^n} - (\mathbf{A}\vec{u}, \vec{z}_0)_{\mathbf{R}^n},$$

as follows from (3.120).

Remark 3.18 *The abstract formulation of unconstrained and constrained hemivariational inequalities presented in the previous sections, in fact, involves more general situations. Consider, for example, the case when the mapping Π has "two components" Π' and Π'' and the same holds for the multifunction $\hat{b} = (\hat{b}', \hat{b}'')$. We may formulate the following hemivariational inequality:*

$$\left\{ \begin{array}{l} Find \ (u, \Xi) \in V \times Y, \ \Xi = (\Xi', \Xi'') \ such \ that \\ a(u, v) + \langle \Xi', \Pi'v \rangle_{Y' \times Z'} + \langle \Xi'', \Pi''v \rangle_{Y'' \times Z''} = \langle f, v \rangle \quad \forall v \in V \\ \Xi'(x) \in \hat{b}'(x, (\Pi'u)(x)) \\ \Xi''(x) \in \hat{b}''(x, (\Pi''u)(x)) \end{array} \right. \quad for \ a.a. \ x \in \omega, \qquad (P)$$

where Y', Z', Y'', Z'' are appropriately chosen spaces, $Y = Y' \times Y''$. In a similar way one can extend the definition of constrained hemivariational inequalities. For illustration, assume Example 3.3 in which the condition $T_1(x) = S_1(x)$ on Γ_c is now replaced by a nonsmooth nonmonotone friction law. Then the variational formulation of the problem reads as follows:

$$\left\{ \begin{array}{l} Find \ (u, \Xi_1, \Xi_2) \in K \times L^{q_1'}(\Gamma_c) \times L^{q_2'}(\Gamma_c) \ such \ that \\ a(u, v - u) + \int_{\Gamma_c} \Xi_1(v_1 - u_1)dx_1 + \int_{\Gamma_c} \Xi_2(v_2 - u_2)dx_1 \\ \geq \langle f, v - u \rangle \quad \forall v \in K \\ \Xi_1(x) \in \hat{b}_1(x, u_1(x)) \\ \Xi_2(x) \in \hat{b}_2(x, u_2(x)) \end{array} \right. \quad for \ a.a. \ x \in \Gamma_c, \qquad (3.121)$$

where $\hat{b}_1, \hat{b}_2$ are multifunctions describing the respective constitutive laws and q_1', q_2' are the exponents from the growth condition (3.47) satisfied by $\hat{b}_1$ and $\hat{b}_2$.

Everything what has been said for the approximation of constrained and unconstrained hemivariational inequalities can be easily extended to this more general situation, except Theorem 3.6 and 3.10 (ii), showing the equivalence between two possible formulations of the problem: hemivariational inequality versus substationarity of superpotentials. However, it is easy to see, by checking proofs, that if Π' and Π'', being the matrix representatives of $\Pi'|_{V_h}$ and $\Pi''|_{V_h}$, respectively, operate only with their own variables, the above mentioned equivalence holds true, again.

3.7   APPROXIMATION OF VECTOR-VALUED HEMIVARIATIONAL INEQUALITIES

In practice we often meet problems in which *vector-valued* mechanical quantities are related by a nonmonotone, possibly multivalued constitutive law. To be able to treat such a type of problems we have to adapt the approach presented in the previous sections. We introduce a new class of hemivariational inequalities, termed vector-valued permitting us to introduce such laws into the mathematical model. As before we start with an abstract setting of a class of *unconstrained* vector-valued hemivariational inequalities. We shall describe how to discretize them and we establish the convergence results similar to those from the previous sections. We shall follow very closely Sections 3.2-3.6 except the existence proof for discrete problems: this time we use tools of the set-valued analysis. We end up this section by extending these results to constrained hemivariational inequalities.

Let $\Omega \subset \mathbf{R}^n$ be a bounded domain with the Lipschitz boundary $\partial\Omega$, $\mathbf{V} \subset H^1(\Omega; \mathbf{R}^d)$ $(d > 1)$ be a space of vector-valued functions[†] defined in Ω. Let $a : \mathbf{V} \times \mathbf{V} \to \mathbf{R}$ be a bounded, $\mathbf{V}$-elliptic bilinear form and $f \in \mathbf{V}^*$ be given. As before, the symbol ω stands for a set, where a nonmonotone law is prescribed. We shall consider two cases: *(i)* and *(ii)* from the beginning of this chapter. Finally, let $\mathbf{Y} = L^2(\omega; \mathbf{R}^d)$.

A nonmonotone law will be characterized by a *locally Lipschitz* function $j : \mathbf{R}^d \to \mathbf{R}$, satisfying the following *generalized sign condition* expressed by means of the generalized directional derivative of j (see Definition 1.4):

$$j^\circ(\xi; -\xi) \le C_1 + C_2|\xi|^q \quad \forall \xi \in \mathbf{R}^d, \tag{3.122}$$

where C_1, C_2 are positive constants independent of ξ and $q \in [1, 2)$. Further, we suppose that j fulfills the following *growth condition* in terms of the generalized gradient of j:

$$\forall \eta \in \bar{\partial} j(\xi) \implies |\eta| \le C_3(1 + |\xi|) \quad \forall \xi \in \mathbf{R}^d, \tag{3.123}$$

where C_3 is a positive constant which is independent of ξ and η.

Remark 3.19 *Here we present one example of j, satisfying (3.122) and (3.123). Let $\varphi_1, \varphi_2 \in C^1(\mathbf{R}^d)$ be two functions which are quadratic "far away" from the origin: $\exists R > 0$ such that*

$$\varphi_i(\xi) = \frac{1}{2}(\mathbf{A}_i \xi, \xi)_{\mathbf{R}^d} - (y_i, \xi)_{\mathbf{R}^d} \quad \forall |\xi| \ge R, \ \xi \in \mathbf{R}^d,$$

where $y_i \in \mathbf{R}^d$ and $\mathbf{A}_i \in \mathcal{L}(\mathbf{R}^d, \mathbf{R}^d)$, $i = 1, 2$, are symmetric and positive semidefinite matrices. Then $j(\xi) \equiv \min\{\varphi_1(\xi), \varphi_2(\xi)\}$ satisfies (3.122)

[†]Here and in what follows, bold characters will be used for denoting vector-valued function spaces.

and (3.123). Indeed, from Proposition 1.11 (note that $\min_i\{\varphi_i(\xi)\} = -\max_i\{-\varphi_i(\xi)\}$) we know that

$$\bar{\partial}j(\xi) = \nabla\varphi_i(\xi) \quad if \ \xi \in I_i \setminus I_{i+1}, \ (I_3 \equiv I_1) \tag{3.124}$$

where

$$I_i = \{\xi \in \mathbf{R}^d \mid \varphi_i(\xi) = j(\xi)\}, \quad i = 1, 2$$

and

$$\bar{\partial}j(\xi) \subset \{\eta \in \mathbf{R}^d \mid \eta = \lambda\nabla\varphi_1(\xi) + (1-\lambda)\nabla\varphi_2(\xi) \tag{3.125}$$
$$\forall \lambda \in [0,1]\} \quad if \ \xi \in I_1 \cap I_2.$$

Since the gradients of φ_1, φ_2 are continuous in $\mathbf{R}^d$ there exists a constant $C > 0$ such that

$$|\nabla\varphi_i(\xi)| \le C \quad \forall |\xi| \le R, \quad i = 1, 2 \tag{3.126}$$

while

$$\nabla\varphi_i(\xi) = \mathbf{A}_i\xi - y_i, \quad i = 1, 2$$

if $|\xi| \ge R$. From this and (3.124)-(3.126), the generalized growth condition (3.123) follows. Using (1.28) we can write:

$$j^\circ(\xi; -\xi) = \max\{(\eta, -\xi)_{\mathbf{R}^d} \mid \eta \in \bar{\partial}j(\xi)\}. \tag{3.127}$$

Again, if $|\xi| \le R$ then $j^\circ(\xi; -\xi) \le \tilde{C}$ for some $\tilde{C} > 0$. Since $\mathbf{A}_i, \ i = 1, 2$, are positive semidefinite matrices then

$$(\nabla\varphi_i(\xi), -\xi)_{\mathbf{R}^d} = -\frac{1}{2}(\mathbf{A}_i\xi, \xi)_{\mathbf{R}^d} + (y_i, \xi)_{\mathbf{R}^d} \le (y_i, \xi)_{\mathbf{R}^d}$$

holds for any $|\xi| > R$. This and (3.127) yield the generalized sign condition with $q = 1$.

Now we are ready to give

Definition 3.2 *A pair of functions $(u, \Xi) \in \mathbf{V} \times \mathbf{Y}$ is said to be a solution of the vector-valued hemivariational inequality iff*

$$\begin{cases} a(u, v) + \displaystyle\int_\omega \Xi.\Pi v d\mu = \langle f, v \rangle \quad \forall v \in \mathbf{V}, \\ \Xi(x) \in \bar{\partial}j(\Pi u(x)) \quad for \ a.a. \ x \in \omega, \end{cases} \tag{P}$$

where Π is a given linear and continuous mapping from $\mathbf{V}$ into $\mathbf{Y}$.[‡]

[‡] *Recall that $\int_\omega \Xi.\Pi v d\mu = \sum_{i=1}^d \int_\omega \Xi_i(\Pi v)_i d\mu$.*

Remark 3.20 *As we shall see later the fact that* $\Xi \in L^2(\omega; \mathbf{R}^d)$ *is a consequence of the generalized growth condition (3.123).*

It is possible to prove directly (see Naniewicz and Panagiotopoulos, 1995) that under the above assumptions, problem (**P**) has a solution. The same result will follow from the convergence analysis presented below.

Now we shall introduce an approximation of (**P**). The finite element spaces $\mathbf{V}_h$ and $\mathbf{Y}_h$ will be constructed exactly in the same way as described in Section 3.2, i.e., $\mathbf{V}_h$ is defined by (3.21) and $\mathbf{Y}_h = (Y_h)^d$ with Y_h given by (3.22). Also $\{a_h\},\{f_h\}$ keep the same meaning as in Section 3.2: the symbols a_h, f_h will denote approximations of a and f, respectively. Similarly to the scalar case we introduce linear continuous mappings $P_h : \mathbf{W}_h \to \mathbf{Y}_h$, where $\mathbf{W}_h = \Pi(\mathbf{V}_h)$ and satisfying the following assumptions:

$$y_h \to y \text{ in } \mathbf{V}, \ y_h \in \mathbf{V}_h \implies \|P_h(\Pi y_h) - \Pi y\|_{L^2(\omega; \mathbf{R}^d)} \to 0, \ h \to 0+, \ (3.128)$$

$$\exists C > 0 : \|P_h w_h\|_{L^2(\omega; \mathbf{R}^d)} \leq C \|w_h\|_{L^2(\omega; \mathbf{R}^d)}, \quad \forall w_h \in \mathbf{W}_h, \ \forall h > 0. \ (3.129)$$

The approximation of the vector-valued hemivariational inequality (**P**) *reads as follows:*

$$
\begin{cases}
\text{Find } (u_h, \Xi_h) \in \mathbf{V}_h \times \mathbf{Y}_h \text{ such that} \\
a_h(u_h, v_h) + \displaystyle\int_\omega \Xi_h . P_h(\Pi v_h) d\mu = \langle f_h, v_h \rangle_h \quad \forall v_h \in \mathbf{V}_h, \qquad (\mathbf{P})_h \\
\Xi_h(x) \in \bar{\partial}j(P_h(\Pi u_h)(x)) \quad \text{for a.a. } x \in \omega,
\end{cases}
$$

where $\langle \cdot, \cdot \rangle_h$ denotes the duality pairing between $\mathbf{V}_h$ and $\mathbf{V}_h^*$.

Remark 3.21 *Any* $\eta \in \bar{\partial}j(P_h(\Pi v_h))$, $v_h \in \mathbf{V}_h$ *can be viewed to be an element of* $\mathbf{Y}_h$. *The last inclusion in the definition of* $(\mathbf{P})_h$ *is again equivalent to* m *inclusions at each point* x_h^i *(see Remark 3.5):*

$$\Xi_h(x_h^i) \in \bar{\partial}(P_h(\Pi u_h)(x_h^i)) \quad i = 1, ..., m.$$

Next we prove the existence of solutions to $(\mathbf{P})_h$ by using Theorem 1.27. Before doing that we give an equivalent operator form of $(\mathbf{P})_h$ to which this theorem will be applied. To this end we introduce the mapping $A_h \in \mathcal{L}(\mathbf{V}_h, \mathbf{V}_h^*)$ by means of

$$\langle A_h u_h, v_h \rangle_h = a_h(u_h, v_h) \quad \forall u_h, v_h \in \mathbf{V}_h$$

and the set-valued mapping $T_h : \mathbf{V}_h \to \mathbf{V}_h^*$ as follows:

$$T_h v_h \equiv \Lambda_h^T(\bar{\partial}j(\Lambda_h v_h)) = \{w_h \in \mathbf{V}_h^* \mid \exists z_h \in \mathbf{Y}_h : \qquad (3.130)$$
$$w_h = \Lambda_h^T z_h, \ z_h \in \bar{\partial}j(\Lambda_h v_h)\}, \ v_h \in \mathbf{V}_h,$$

where $\Lambda_h \equiv P_h \Pi \in \mathcal{L}(\mathbf{V}_h, \mathbf{Y}_h)$ and $\Lambda_h^T \in \mathcal{L}(\mathbf{Y}_h, \mathbf{V}_h^*)$ is the transpose of Λ_h (we identify $\mathbf{Y}_h$ with $\mathbf{Y}_h^*$). Using the definition of T_h one can easily prove:

Lemma 3.9 *Let* $(u_h, \Xi_h) \in \mathbf{V}_h \times \mathbf{Y}_h$ *solve* $(\mathbf{P})_h$. *Then*

$$0 \in A_h u_h - f_h + T_h u_h \qquad (3.131)$$

and $\Lambda_h^T \Xi_h \in T_h u_h$. *On the contrary, if* u_h *is a solution to (3.131) then there exists* $\Xi_h \in \mathbf{Y}_h$ *such that* $\Lambda_h^T \Xi_h \in T_h u_h$ *and the couple* (u_h, Ξ_h) *solves* $(\mathbf{P})_h$.

Therefore it is sufficient to show that (3.131) has a solution.

Theorem 3.11 *Let (3.24)-(3.26), (3.122), (3.123) and (3.129) be satisfied. Then there exists at least one solution* (u_h, Ξ_h) *to* $(\mathbf{P})_h$ *for any* $h > 0$. *Moreover, any solution of* $(\mathbf{P})_h$ *is bounded in* $\mathbf{V} \times \mathbf{Y}$ *uniformly with respect to* $h > 0$.

Proof: We shall apply Theorem 1.27 to the set-valued mapping $\Psi_h : \mathbf{V}_h \to \mathbf{V}_h^*$ defined by

$$\Psi_h v_h \equiv A_h v_h - f_h + T_h v_h, \quad v_h \in \mathbf{V}_h. \qquad (3.132)$$

The mapping T_h is upper semicontinuous in $\mathbf{V}_h$ and $T_h v_h$ is a nonempty, closed and convex subset of $\mathbf{V}_h^*$ for any $v_h \in \mathbf{V}_h$ as follows from (3.130), Proposition 1.9 and the fact that Λ_h^T is linear and continuous. Clearly, all these properties are satisfied by Ψ_h, as well. Only what it remains to verify is the coerciveness of Ψ_h in the sense of Theorem 1.27.

Let $v_h \in \mathbf{V}_h$ and $z_h \in \bar{\partial}j(\Lambda_h v_h)$. Then

$$\int_\omega z_h . P_h(\Pi v_h) d\mu = - \int_\omega z_h . (-P_h(\Pi v_h)) d\mu \qquad (3.133)$$

$$\geq - \int_\omega j^\circ(P_h(\Pi v_h); -P_h(\Pi v_h)) d\mu$$

as follows from (1.28). The generalized sign condition (3.122) makes possible to estimate the last integral in (3.133) from below:

$$- \int_\omega j^\circ(P_h(\Pi v_h); -P_h(\Pi v_h)) d\mu \geq - \int_\omega (C_1 + C_2 |P_h(\Pi v_h)|^q) d\mu$$

$$\geq -\tilde{C}_1 - \tilde{C}_2 \|v_h\|^q \quad \forall v_h \in \mathbf{V}_h,$$

where $q \in [1, 2)$ using (3.129) and the fact that $\Pi \in \mathcal{L}(\mathbf{V}, \mathbf{Y})$. This and (3.133) yield:

$$\int_\omega z_h . P_h(\Pi v_h) d\mu \geq -\tilde{C}_1 - \tilde{C}_2 \|v_h\|^q \quad \forall v_h \in \mathbf{V}_h, \qquad (3.134)$$

$$\forall z_h \in \bar{\partial}j(P_h(\Pi v_h)).$$

From this the coerciveness of Ψ_h easily follows. Indeed, let $v_h \in \mathbf{V}_h$ and $w_h \in \Psi_h(v_h)$. Then $w_h = A_h v_h - f_h + \Lambda_h^T z_h$ for some $z_h \in \bar{\partial}j(P_h(\Pi v_h))$ and

$$\langle w_h, v_h \rangle_h = \langle A_h v_h - f_h + \Lambda_h^T z_h, v_h \rangle_h$$

$$= a_h(v_h, v_h) - \langle f_h, v_h \rangle_h + \int_\omega z_h . P_h(\Pi v_h) d\mu$$

$$\geq \tilde{\alpha} \|v_h\|^2 - \tilde{\beta} \|v_h\| - \tilde{C}_1 - \tilde{C}_2 \|v_h\|^q, \quad q \in [1, 2)$$

making use of (3.25),(3.26) and (3.134). From this we conclude that Ψ_h is coercive in $\mathbf{V}_h$. Consequently, (3.131) has a solution u_h and there exists $\Xi_h \in \mathbf{Y}_h$, $\Xi_h \in \bar{\partial}j(P_h(\Pi u_h))$ such that (u_h, Ξ_h) solves $(\mathbf{P})_h$.

Next we prove that any solution (u_h, Ξ_h) of $(\mathbf{P})_h$ is bounded in $\mathbf{V} \times \mathbf{Y}$ uniformly with respect to $h > 0$. Substituting $v_h := u_h$ into $(\mathbf{P})_h$, using (3.25),(3.26) and (3.134) we arrive at

$$\bar{\alpha}\|u_h\|^2 \le a_h(u_h, u_h) = -\int_\omega \Xi_h . P_h(\Pi u_h)d\mu + \langle f_h, u_h \rangle_h$$

$$\le \tilde{C}_1 + \tilde{C}_2\|u_h\|^q + \tilde{\beta}\|u_h\|, \quad q \in [1, 2)$$

with $\tilde{C}_1, \tilde{C}_2$ and $\tilde{\beta}$ independent of h, implying the boundedness of $\{u_h\}$ in $\mathbf{V}$. This and the growth condition (3.123) yield

$$\int_\omega |\Xi_h|^2 d\mu \le C \quad \forall h > 0.$$

The proof is complete. □

The convergence analysis will follow the same guidelines of Section 3.3 with minor modifications only. Let the family $\{\mathbf{V}_h\}$ of spaces $\mathbf{V}_h$ introduced above satisfy the standard density assumption:

$$\forall v \in \mathbf{V} \,\exists\{v_h\}, \; v_h \in \mathbf{V}_h : v_h \to v \quad \text{in } \mathbf{V}, \quad h \to 0+. \tag{3.135}$$

We prove

Theorem 3.12 *Let all the assumptions of Theorem 3.11 be satisfied. Let moreover $\{\mathbf{V}_h\}$ be dense in the sense of (3.135) and the approximated forms $\{a_h\},\{f_h\}$ satisfy (3.52),(3.53), respectively. Then from any sequence $\{(u_h, \Xi_h)\}$ of solutions to $(\mathbf{P})_h$ one can find its subsequence and elements $u \in \mathbf{V}$, $\Xi \in \mathbf{Y}$ such that*

$$\begin{cases} u_h \to u & \text{in } \mathbf{V} \\ \Xi_h \rightharpoonup \Xi & \text{in } \mathbf{Y}, \text{ as } h \to 0+. \end{cases} \tag{3.136}$$

The couple (u, Ξ) is a solution to $(\mathbf{P})$ and any cluster point of $\{(u_h, \Xi_h)\}$ in the sense of (3.136) solves $(\mathbf{P})$.

Proof: The existence of a subsequence of $\{(u_h, \Xi_h)\}$ such that

$$\begin{cases} u_h \rightharpoonup u & \text{in } \mathbf{V} \\ \Xi_h \rightharpoonup \Xi & \text{in } \mathbf{Y}, \; h \to 0+ \end{cases} \tag{3.137}$$

for some $(u, \Xi) \in \mathbf{V} \times \mathbf{Y}$ is a consequence of the boundedness of $\{(u_h, \Xi_h)\}$ in $\mathbf{V} \times \mathbf{Y}$. Using the assumptions of the theorem one can prove exactly in the same way as in Section 3.3 that

$$a(u, v) + \int_\omega \Xi . \Pi v d\mu = \langle f, v \rangle \quad \forall v \in \mathbf{V}.$$

To conclude that (u, Ξ) is a solution of (P) it remains to show that

$$\Xi(x) \in \bar{\partial}j(\Pi u(x)) \quad \text{for a.a. } x \in \omega. \tag{3.138}$$

First let us observe that due to (3.128) and (3.137)$_1$ we may assume (passing to a new subsequence if necessary) that

$$P_h(\Pi u_h(x)) \to \Pi u(x) \quad \text{for a.a. } x \in \omega. \tag{3.139}$$

To prove (3.138) one has to show that

$$(\Xi(x), z)_{\mathbf{R}^d} \le j^\circ(\Pi u(x); z) \quad \forall z \in \mathbf{R}^d \text{ and a.a. } x \in \omega$$

as follows from Definition 1.5 or equivalently: for each $\delta > 0$ there exists $\omega_0 \subset \omega$, meas $\omega_0 < \delta$, such that

$$\int_{\omega \backslash \omega_0} \Xi.\phi d\mu \le \int_{\omega \backslash \omega_0} j^\circ(\Pi u; \phi)d\mu \quad \forall \phi \in L^\infty(\omega; \mathbf{R}^d).$$

Let $\delta > 0$ and $\phi \in L^\infty(\omega; \mathbf{R}^d)$ be given. Then

$$\int_{\omega \backslash \omega_0} \Xi.\phi d\mu \overset{(3.137)_2}{=} \lim_{h \to 0+} \int_{\omega \backslash \omega_0} \Xi_h.\phi d\mu$$

$$(\text{Def. } 1.5 \implies) \quad \le \limsup_{h \to 0+} \int_{\omega \backslash \omega_0} j^\circ(P_h(\Pi u_h); \phi)d\mu$$

$$(\text{Th. } 1.6 \implies) \quad \le \int_{\omega \backslash \omega_0} \limsup_{h \to 0+} j^\circ(P_h(\Pi u_h); \phi)d\mu$$

$$(\text{Prop. } 1.9 \ (ii) + (3.139) \implies) \quad \le \int_{\omega \backslash \omega_0} j^\circ(\Pi u; \phi)d\mu.$$

For the last but one inequality we use the same arguments as in the proof of Theorem 3.3: due to (3.139) and Egoroff's theorem there exists $\omega_0 \subset \omega$ such that $\{P_h(\Pi u_h)\}$ converges to Πu uniformly in $\omega \backslash \omega_0$. Further, from (3.123) and (1.28) it follows that $j^\circ(P_h(\Pi u_h(x)); \phi(x)) \le C(1 + |P_h(\Pi u_h(x)|)|\phi(x)|$ for a.a. $x \in \omega \backslash \omega_0$. Then we use Theorem 1.6.

It remains to prove that the weak convergence in (3.137)$_1$ can be replaced by the strong one. From (3.135), the existence of $\{\bar{u}_h\}$, $\bar{u}_h \in \mathbf{V}_h$ such that $\bar{u}_h \to u$ in $\mathbf{V}$ follows. Using (3.25) and the definition of (P)$_h$ we see that

$$\tilde{\alpha}\|u_h - \bar{u}_h\|^2 \le a_h(u_h - \bar{u}_h, u_h - \bar{u}_h) = \langle f_h, u_h - \bar{u}_h \rangle_h$$

$$- \int_\omega \Xi_h.P_h(\Pi(u_h - \bar{u}_h))d\mu - a_h(\bar{u}_h, u_h - \bar{u}_h).$$

For $h \to 0+$ the right hand side of this inequality tends to zero as follows (3.52),(3.53) and (3.128). From the triangle inequality we get the strong convergence of $\{u_h\}$ to u in $\mathbf{V}$. $\qquad\square$

Now we pass to the case of *constrained hemivariational inequalities*. Let $\mathbf{K} \subset \mathbf{V}$ be *a nonempty, closed* and *convex subset* of $\mathbf{V}$. The meaning of other symbols remains.

A pair of functions $(u, \Xi) \in \mathbf{K} \times \mathbf{Y}$ is declared to be a solution of the constrained hemivariational inequality iff

$$\begin{cases} a(u, v - u) + \int_{\omega} \Xi.\Pi(v - u)d\mu \geq \langle f, v - u \rangle \quad \forall v \in \mathbf{K}, \\ \Xi(x) \in \bar{\partial}j(\Pi u(x)) \quad \text{for a.a. } x \in \omega. \end{cases} \qquad (\mathbf{P})^c$$

To define convergent approximations of $(\mathbf{P})^c$ we introduce a family $\{\mathbf{K}_h\}$ of nonempty, closed convex subsets $\mathbf{K}_h$ of $\mathbf{V}_h$, satisfying (1.172) and (1.173). The approximation of $(\mathbf{P})^c$ is now defined as follows:

$$\begin{cases} \text{Find } (u_h, \Xi_h) \in \mathbf{K}_h \times \mathbf{Y}_h \text{ such that} \\ a_h(u_h, v_h - u_h) + \int_{\omega} \Xi_h.P_h(\Pi v_h - \Pi u_h)d\mu \\ \geq \langle f_h, v_h - u_h \rangle_h \quad \forall v_h \in \mathbf{K}_h, \\ \Xi_h(x) \in \bar{\partial}j(P_h(\Pi u_h)(x)) \quad \text{for a.a. } x \in \omega. \end{cases} \qquad (\mathbf{P})_h^c$$

The existence of solutions to $(\mathbf{P})_h^c$ will be now established by using Corallary 1.2. For the same reason as in the unconstrained case we first rewrite $(\mathbf{P})_h^c$ into an equivalent operator form.

Let $\Psi_h : \mathbf{V}_h \to \mathbf{V}_h^*$ be the set-valued mapping defined by (3.132). It is readily seen that if $(u_h, \Xi_h) \in \mathbf{K}_h \times \mathbf{Y}_h$ solves $(\mathbf{P})_h^c$ then $y_h \equiv A_h u_h - f_h + \Lambda_h^T \Xi_h \in \Psi_h(u_h)$ and $\langle y_h, v_h - u_h \rangle_h \geq 0 \ \forall v_h \in \mathbf{K}_h$. On the contrary, let $u_h \in \mathbf{K}_h$ be such that there exists $y_h \in \Psi_h(u_h)$ and $\langle y_h, v_h - u_h \rangle_h \geq 0 \ \forall v_h \in \mathbf{K}_h$. Then $y_h = A_h u_h - f_h + \Lambda_h^T \Xi_h$ for some $\Lambda_h^T \Xi_h \in T_h u_h$ and the couple (u_h, Ξ_h) solve $(\mathbf{P})_h^c$.

Now we prove

Theorem 3.13 *Let (1.172),(1.173) and all the assumptions of Theorem 3.11 be satisfied. Then the constrained hemivariational inequality $(\mathbf{P})_h^c$ has at least one solution (u_h, Ξ_h) for any $h > 0$. Moreover, any solution of $(\mathbf{P})_h^c$ is bounded in $\mathbf{V} \times \mathbf{Y}$ uniformly with respect to $h > 0$.*

Proof: All we have to do is to verify the coerciveness of Ψ_h on $\mathbf{K}_h$ with respect to an element from $\mathbf{K}_h$. Recall that Ψ_h is defined by (3.132).

Let $u_0 \in \mathbf{K}$ be given. Accordingly to (1.172) there exists a sequence $\{u_{0h}\}$, $u_{0h} \in \mathbf{K}_h$ such that

$$u_{0h} \to u_0 \quad \text{in } \mathbf{V}, \ h \to 0+. \qquad (3.140)$$

Let $v_h \in \mathbf{V}_h$ and $z_h \in \bar{\partial} j(\Lambda_h v_h)$ be given. Then

$$\int_\omega z_h . P_h(\Pi v_h - \Pi u_{0h}) d\mu \tag{3.141}$$

$$= -\int_\omega z_h .(-P_h(\Pi v_h)) d\mu - \int_\omega z_h . P_h(\Pi u_{0h}) d\mu$$

$$\geq -\int_\omega j^\circ (P_h(\Pi v_h); -P_h(\Pi v_h)) d\mu - \int_\omega C_3(1 + |P_h(\Pi v_h)|)|P_h(\Pi u_{0h})| d\mu$$

$$\geq -\int_\omega (C_1 + C_2 |P_h(\Pi v_h)|^q) d\mu$$

$$- \left(\int_\omega (C_3(1 + P_h(\Pi v_h)))^2 d\mu \right)^{\frac{1}{2}} \left(\int_\omega |P_h(\Pi u_{0h})|^2 d\mu \right)^{\frac{1}{2}}$$

$$\geq -\tilde{C}_1 - \tilde{C}_2 \|v_h\|^q - \tilde{C}_3(1 + \|v_h\|) \|u_{0h}\|$$

with $q \in [1, 2)$, making use of (3.122), (3.123), the boundedness of $\Pi : \mathbf{V} \to \mathbf{Y}$ and (3.129). From this the coerciveness of Ψ_h with respect to $u_{0h} \in \mathbf{K}_h$ easily follows.

Indeed, let $v_h \in \mathbf{V}_h$ and $w_h \in \Psi_h(v_h)$. Then $w_h = A_h v_h - f_h + \Lambda_h^T z_h$ for some $z_h \in \bar{\partial} j(P_h(\Pi v_h))$ and

$$\langle w_h, v_h - u_{0h} \rangle_h = \langle A_h v_h - f_h + \Lambda_h^T z_h, v_h - u_{0h} \rangle_h$$

$$= a_h(v_h, v_h - u_{0h}) - \langle f_h, v_h - u_{0h} \rangle_h + \int_\omega z_h . P_h(\Pi v_h - \Pi u_{0h}) d\mu$$

$$\geq \tilde{\alpha} \|v_h\|^2 - \tilde{M} \|v_h\| \|u_{0h}\| - \tilde{\beta}(\|v_h\| + \|u_{0h}\|)$$

$$- \tilde{C}_1 - \tilde{C}_2 \|v_h\|^q - \tilde{C}_3(1 + \|v_h\|) \|u_{0h}\|, \quad q \in [1, 2)$$

as follows from (3.24)-(3.26) and (3.141), implying the desired property.

The remaining assumptions of Corollary 1.2 are trivially satisfied (see also the proof of Theorem 3.11). Thus $(\mathbf{P})_h^c$ has at least one solution (u_h, Ξ_h) for any $h > 0$.

Let us show that $\{(u_h, \Xi_h)\}$, where (u_h, Ξ_h) is a solution to $(\mathbf{P})_h^c$, is bounded in $\mathbf{V} \times \mathbf{Y}$. Inserting $v_h := u_{0h}$ into the first inequality in $(\mathbf{P})_h^c$ we obtain:

$$\tilde{\alpha} \|u_h\|^2 \leq a_h(u_h, u_{0h}) + \int_\omega \Xi_h . P_h(\Pi u_{0h} - \Pi u_h) d\mu \tag{3.142}$$

$$+ \langle f_h, u_{0h} - u_h \rangle_h.$$

The second term on the right hand side of (3.142) can be estimated as follows:

$$\int_\omega \Xi_h . P_h(\Pi u_{0h} - \Pi u_h) d\mu \tag{3.143}$$

$$\leq \tilde{C}_3(1 + \|u_h\|) \|u_{0h}\| + \int_\omega j^\circ (P_h(\Pi u_h); -P_h(\Pi u_h)) d\mu$$

$$\leq \tilde{C}_3(1 + \|u_h\|) \|u_{0h}\| + \tilde{C}_1 + \tilde{C}_2 \|u_h\|^q, \quad q \in [1, 2).$$

Since the constants $\tilde{C}_i$, $i = 1, 2, 3$, are independent of h, the boundedness of $\{u_h\}$ in the norm of $\mathbf{V}$ easily follows from (3.142) and (3.143). The boundedness of $\{\Xi_h\}$ in $\mathbf{Y}$ is the direct consequence of the growth condition (3.123). $\square$

Now we are ready to establish the following convergence result:

Theorem 3.14 *Let the family $\{\mathbf{K}_h\}$ satisfy (1.172),(1.173) and let all the assumptions of Theorem 3.12 concerning of $\{a_h\},\{f_h\}$ and j be satisfied. Then from any sequence $\{(u_h, \Xi_h)\}$ of solutions to $(\mathbf{P})^c_h$ one can find its subsequence and elements $u \in \mathbf{K}$, $\Xi \in \mathbf{Y}$ such that*

$$\begin{cases} u_h \to u & \text{in } \mathbf{V}, \\ \Xi_h \rightharpoonup \Xi & \text{in } \mathbf{Y}, \ h \to 0 + . \end{cases} \tag{3.144}$$

The couple (u, Ξ) is a solution to $(\mathbf{P})^c_h$ and any cluster point of $\{(u_h, \Xi_h)\}$ in the sense of (3.144) solves $(\mathbf{P})^c_h$.

Proof: The boundedness of $\{(u_h, \Xi_h)\}$ in $\mathbf{V} \times \mathbf{Y}$ guaranteed by Theorem 3.13 yields the existence of a subsequence tending weakly to a couple $(u, \Xi) \in \mathbf{V} \times \mathbf{Y}$. From (1.173) it follows that $u \in \mathbf{K}$. Using exactly the same approach as in the proof of Theorem 3.9 (with $q' = q = 2$) it is possible to show that a subsequence of $\{u_h\}$ tends strongly to u in V, provided that (3.52) and (3.53) hold true. From this and (1.172) we easily obtain that (u, Ξ) satisfies the inequality in $(\mathbf{P})^c_h$. The inclusion $\Xi(x) \in \bar{\partial}j(\Pi(u(x))$ for a.a. $x \in \omega$ has been already proven in Theorem 3.12. $\square$

Now we shortly present the algebraic form of $(\mathbf{P})_h$ and $(\mathbf{P})^c_h$. The derivation is exactly the same as in the scalar case, discussed in Section 3.5.

Let $h > 0$ be fixed. Then $\mathbf{V}_h$, $\mathbf{K}_h$, $\mathbf{Y}_h = (Y_h)^d$ can be identified with $\mathbf{R}^n$, $\mathcal{K}$ and $(\mathbf{R}^d)^m$, respectively, where $\dim \mathbf{V}_h = n$, $\mathcal{K}$ is a closed convex subset of $\mathbf{R}^n$, $\dim \mathbf{Y}_h = md$ with $m = \dim Y_h$. The matrix form of $(\mathbf{P})_h$ reads as follows:

$$\begin{cases} \text{Find } (\vec{u}, \vec{\Xi}) \equiv (\vec{u}, \vec{\Xi}_1, ..., \vec{\Xi}_m) \in \mathbf{R}^n \times (\mathbf{R}^d)^m \text{ such that} \\ (\mathbf{A}\vec{u}, \vec{v})_{\mathbf{R}^n} + (\vec{\Xi}, \Lambda\vec{v})_{(\mathbf{R}^d)^m} = (\vec{f}, \vec{v})_{\mathbf{R}^n} \quad \forall \vec{v} \in \mathbf{R}^n \\ \vec{\Xi}_i \in c_i \bar{\partial}j((\Lambda\vec{u})_i) \quad \forall i = 1, ..., m \end{cases} \qquad (\vec{\mathbf{P}})$$

or in the case of $(\mathbf{P})^c_h$

$$\begin{cases} \text{Find } (\vec{u}, \vec{\Xi}) \equiv (\vec{u}, \vec{\Xi}_1, ..., \vec{\Xi}_m) \in \mathcal{K} \times (\mathbf{R}^d)^m \text{ such that} \\ (\mathbf{A}\vec{u}, \vec{v} - \vec{u})_{\mathbf{R}^n} + (\vec{\Xi}, \Lambda(\vec{v} - \vec{u}))_{(\mathbf{R}^d)^m} \geq (\vec{f}, \vec{v} - \vec{u})_{\mathbf{R}^n} \quad \forall \vec{v} \in \mathcal{K} \\ \vec{\Xi}_i \in c_i \bar{\partial}j((\Lambda\vec{u})_i) \quad \forall i = 1, ..., m. \end{cases} \qquad (\vec{\mathbf{P}})^c$$

where $\mathbf{A}$ is the $n \times n$ stiffness matrix, $\vec{f}$ is the load vector, Λ is a $dm \times n$ matrix, representing the linear mapping $\Lambda_h \equiv P_h\Pi : \mathbf{V}_h \to \mathbf{Y}_h$ and $c_i = \text{meas } K_i$ $\forall i = 1, ..., m$ (see also Section 3.5).

Instead of solving $(\vec{\mathbf{P}})$, $(\vec{\mathbf{P}})^c$ directly, we use the same approach as in the scalar case: we look for substationary points $\vec{u} \in \mathbf{R}^n$ of the corresponding superpotential $\mathcal{L}$ given by

$$\mathcal{L}(\vec{v}) = \frac{1}{2}(\vec{v}, \mathbf{A}\vec{v})_{\mathbf{R}^n} - (\vec{f}, \vec{v})_{\mathbf{R}^n} + \Psi(\vec{v})$$

with

$$\Psi(\vec{v}) = \sum_{i=1}^{m} c_i j((\Lambda \vec{v})_i),$$

i.e.,

$$\vec{0} \in \bar{\partial}\mathcal{L}(\vec{u})$$

in the unconstrained case and

$$\vec{0} \in \bar{\partial}\mathcal{L}(\vec{u}) + N_{\mathcal{K}}(\vec{u})$$

for the constrained hemivariational inequality, where $N_{\mathcal{K}}(\vec{u})$ stands for the normal cone of $\mathcal{K}$ at $\vec{u}$.

The previous results will be now applied to a nonmonotone skin friction problem in plane elasticity (see Naniewicz and Panagiotopoulos, 1995).

Let us consider a plane elastic body represented by a *polygonal* domain Ω. In order to describe skin effects we split the body forces F into two parts: $F = \overline{F} + \overline{\overline{F}}$, where $\overline{\overline{F}} \in L^2(\Omega; \mathbf{R}^2)$ is given *a priori* and $\overline{F}$ is induced by skin effects on a subdomain $\Omega_0 \subset\subset \Omega$. We consider the multivalued reaction-displacement law in the form

$$-\overline{F}(x) \in \bar{\partial}j(u(x)) \quad \text{for a.a. } x \in \Omega_0,$$

where $j : \mathbf{R}^2 \to \mathbf{R}$ is a locally Lipschitz function satisfying (3.122) and (3.123). For simplicity we suppose that the zero displacements are prescribed on $\partial\Omega$ and Ω_0 is *polygonal*. The corresponding hemivariational inequality reads as follows:

$$\begin{cases} \text{Find } (u, \Xi) \in \mathbf{V} \times \mathbf{Y} \text{ such that} \\ a(u, v) + \int_{\Omega_0} \Xi.v dx = (\overline{\overline{F}}, v)_{0,\Omega} \quad \forall v \in \mathbf{V}, \\ \Xi(x) \in \bar{\partial}j(u(x)) \quad \text{for a.a. } x \in \Omega_0, \end{cases} \qquad (3.145)$$

where $\mathbf{V} = H_0^1(\Omega; \mathbf{R}^2)$, $\mathbf{Y} = L^2(\Omega_0; \mathbf{R}^2)$ and the bilinear form a is defined by (1.79). Let us note that Π is the identity mapping in this case so that $\Pi \in \mathcal{L}(\mathbf{V}; \mathbf{Y})$ and the second component Ξ is equal to $-\overline{F}$.

Let $\{\mathcal{D}_h\}$, $h \to 0+$, be a *strongly* regular family of triangulations of $\overline{\Omega}$ such that the restriction $\mathcal{D}_h|_{\overline{\Omega}_0}$ defines the triangulation of $\overline{\Omega}_0$ for any $h > 0$ (do not forget that Ω_0 is supposed to be polygonal). For a given $h > 0$ the space $\mathbf{V}_h$ consists of all continuous piecewise linear functions over $\mathcal{D}_h$ vanishing on $\partial\Omega$. It is readily seen that (3.135) is fulfilled. The space $\mathbf{Y}_h = (Y_h)^2$ with Y_h

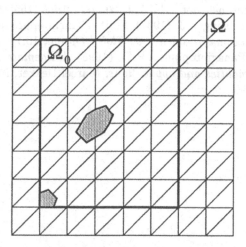

Figure 3.13.

given by (3.22) will be constructed by means of Type II elements defined over the partition $\mathcal{T}_h$ of $\overline{\Omega}_0$ induced by $\mathcal{D}_h|_{\overline{\Omega}_0}$ (see Section 3.4 and Fig. 3.13). The mapping $P_h : \mathbf{W}_h \to \mathbf{Y}_h$, where $\mathbf{W}_h = \Pi(\mathbf{V}_h) = \mathbf{V}_h|_{\Omega_0}$ is defined by (3.59) applied to each component of $w_h \in \mathbf{W}_h$. From Lemma 3.5 and Consequence 3.1 it follows that (3.128) is satisfied. Let us verify (3.129):

$$\|P_h w_h\|_{L^2(\Omega_0;\mathbf{R}^2)} \leq \|P_h w_h - w_h\|_{L^2(\Omega_0;\mathbf{R}^2)} + \|w_h\|_{L^2(\Omega_0;\mathbf{R}^2)}$$
$$\leq Ch\|w_h\|_{H^1(\Omega_0;\mathbf{R}^2)} + \|w_h\|_{L^2(\Omega_0;\mathbf{R}^2)}$$
$$\leq C\|w_h\|_{L^2(\Omega_0;\mathbf{R}^2)} \quad \forall w_h \in \mathbf{W}_h, \ \forall h > 0,$$

where $c > 0$ does not depend on h. Here Lemma 3.5 and the inverse inequality (1.258) were used.

The approximation of (3.145) with $\mathbf{V}_h$, $\mathbf{Y}_h$ and P_h defined above, satisfies all the assumptions of Theorem 3.12. Thus (3.145) has a solution and the approximate solutions are close on subsequences in the sense of Theorem 3.12.

References

Chang, K. C. (1981). Variational methods for non-differentiable functionals and their applications to partial differential equations. *J. Math. Anal. Appl.*, 80:102–129.

Ciarlet, P. G. (1978). *The Finite Element Method for Elliptic Problems*. North Holland, Amsterdam, New York, Oxford.

Glowinski, R. (1984). *Numerical Methods for Nonlinear Variational Problems*. Springer-Verlag, New York.

Glowinski, R., Lions, J. L., and Trémoliéres, R. (1981). *Numerical analysis of variational inequalities*, volume 8 of *Studies in Mathematics and its Applications*. North Holland, Amsterdam, New York.

Miettinen, M. and Haslinger, J. (1995). Approximation of nonmonotone multi-valued differential inclusions. *IMA J. Numer. Anal.*, 15:475–503.

Miettinen, M. and Haslinger, J. (1997). Finite element approximation of vector-valued hemivariational problems. *J. Global Optim.*, 10:17–35.

Naniewicz, Z. and Panagiotopoulos, P. D. (1995). *Mathematical theory of hemivariational inequalities and applications.* Marcel Dekker, New York.

4 TIME DEPENDENT CASE

This chapter is devoted to finite element approximations of scalar time dependent hemivariational inequalities. We start with the parabolic case following closely Miettinen and Haslinger, 1998. At the end of this chapter we discuss, how the results can be extended to constrained problems. Our presentation will follow the structure used for the static case in Chapter 3. First, we introduce an abstract formulation of a class of parabolic hemivariational inequalities (see Miettinen, 1996, Miettinen and Panagiotopoulos, 1999).

Let $\Omega \subset \mathbf{R}^n$ be a bounded domain with the Lipschitz boundary $\partial\Omega$. Let V be a Hilbert space such that the imbedding $V \subset H^1(\Omega; \mathbf{R}^d)$ is dense and continuous. Then $V \subseteq H \equiv L^2(\Omega; \mathbf{R}^d) \subseteq V^*$ form an evolution triple. We denote by $\|\cdot\|$, $\|\cdot\|_*$ and $|\cdot|$ the norms in V, V^* and H, respectively. The duality pairing between V and V^* is denoted by $\langle\cdot,\cdot\rangle$ and the inner product in $L^2(\Omega; \mathbf{R}^d)$ by $(\cdot,\cdot)$. Finally, let $W(V) \equiv \{v \in L^2(0,T;V) : v' \in L^2(0,T;V^*)\}$ and $\|v\|_{W(V)} = \|v\|_{L^2(0,T;V)} + \|v'\|_{L^2(0;T;V^*)}$, $T > 0$.

Let $a : V \times V \to \mathbf{R}$ be a bounded, V-elliptic bilinear form. Further, we assume that the initial state u_0 is an element of H and the right hand side f belongs to $L^2(0,T;V^*)$.

The function $b : \omega \times \mathbf{R} \to \mathbf{R}$ defining the multivalued nonmonotone relation will satisfy the conditions (3.2), (3.3), (3.16) and the growth condition

$$\exists \text{ constants } c_1 > 0 \text{ and } c_2 > 0 \text{ such that} \tag{4.1}$$
$$|b(x,\xi)| \le c_1 + c_2|\xi| \qquad \text{for all } x \in \omega, \text{ a.a. } \xi \in \mathbf{R},$$

163

i.e., the condition (3.47) with $q = q' = 2$. Therefore, the spaces Y and Z, introduced in Chapter 3 are now both equal to $L^2(\omega)$ and the corresponding duality is represented by the $L^2(\omega)$-scalar product. The multifunction $\hat{b} : \omega \times \mathbf{R} \rightarrow 2^{\mathbf{R}}$ is defined by (3.8). The mapping $\Pi \in \mathcal{L}(V, L^2(\omega))$ has the same meaning as in Chapter 3.

We shall consider only the case when ω is a subdomain of Ω with the Lipschitz boundary $\partial\omega$. Further, we set $\omega_T = \omega \times (0, T)$.

Remark 4.1 *Here it is important to note that because of (4.1) the assumption (3.5) will not be needed (which says that the graph of b is essentially growing, but does not involve any growth condition). The reason for that will be evident from the proof of Lemma 4.1. Throughout this chapter we also consider the case when the bilinear form a does not depend on t.*

Now we are ready to give the definition of a parabolic hemivariational inequality:

Definition 4.1 *A pair of functions $(u, \Xi) \in W(V) \times L^2(\omega_T)$ is said to be a solution of a parabolic hemivariational inequality (P) iff*

$$
\begin{cases}
\displaystyle \int_0^T \langle u'(t), v(t) \rangle dt + \int_0^T a(u(t), v(t)) dt + \int_0^T \int_\omega \Xi(t) \Pi v(t) dx dt \\
\displaystyle = \int_0^T \langle f(t), v(t) \rangle dt \qquad \forall v \in L^2(0, T; V), \\
\Xi(x, t) \in \hat{b}(x, \Pi u(x, t)) \qquad \text{for a.a. } (x, t) \in \omega_T \\
\text{and the function } u \text{ satisfies the initial condition } u(0) = u_0.
\end{cases}
\tag{P}
$$

Remark 4.2 *(An equivalent version of Definition 4.1). It is readily seen that the first relation in Definition 4.1 can be rewritten equivalently as follows::*

$$
\langle u'(t), v \rangle + a(u(t), v) + \int_\omega \Xi(t) \Pi v dx \tag{4.2}
$$
$$
= \langle f(t), v \rangle \qquad \forall v \in V \text{ and for a.a. } t \in (0, T).
$$

In order to obtain an operator form of (4.2), we introduce the mapping $A : V \rightarrow V^*$ defined by

$$
a(u, v) = \langle Au, v \rangle \qquad \forall u, v \in V \tag{4.3}
$$

and the function $\tilde{\Xi} \in L^2(0, T; V^*)$ (recall that Π is a continuous linear mapping from V to $L^2(\omega)$) defined by

$$
\langle \tilde{\Xi}(t), v \rangle = \int_\omega \Xi(t) \Pi v dx \quad \forall v \in V \text{ and for a.a. } t \in (0, T). \tag{4.4}
$$

Then the operator form of (4.2) is given by

$$
u'(t) + Au(t) + \tilde{\Xi}(t) = f(t) \quad \text{in } V^*, \text{ for a.a. } t \in (0, T). \tag{4.5}
$$

Example 4.1 As a model example let us consider a heat conduction problem with a nonmonotone relation (a temperature control problem without assuming any monotonicity for the control device (see Section 1.3), e.g.). Let Ω be a bounded domain in $\mathbf{R}^2$, representing a body, in which the temperature distribution is governed by the time dependent heat equation

$$u'(t) - \Delta u(t) = h(t) \quad \text{in } \Omega, \text{ for a.a. } t \in (0, T),$$

with h decomposed as follows:

$$\begin{cases} h = g - \Xi, \\ g \text{ is given and } \Xi(x, t) \in \hat{b}(x, u(x, t)) \text{ for a.a. } (x, t) \in \omega_T, \ \omega \subset \Omega. \end{cases}$$

On the boundary $\partial\Omega$ the temperature u satisfies the homogenous Dirichlet boundary condition

$$u(t) = 0 \quad \text{on } \partial\Omega \text{ for a.a. } t \in (0, T).$$

Moreover, at $t = 0$ the temperature is given:

$$u(x, 0) = u_0(x).$$

Then

$$V = H_0^1(\Omega), \quad H = L^2(\Omega), \quad \Pi v = v|_\omega,$$

$$a(u, v) = \int_\Omega \nabla u \cdot \nabla v dx,$$

$$\langle f(t), v \rangle = \int_\Omega g(t) v dx, \quad g \in L^2(0, T; L^2(\Omega))$$

and the corresponding parabolic hemivariational inequality reads as follows:

$$\begin{cases} \text{Find } (u, \Xi) \in W(V) \times L^2(\omega_T) \text{ such that} \\ \int_0^T \langle u'(t), v(t) \rangle dt + \int_0^T \int_\Omega \nabla u(t) \cdot \nabla v(t) dx dt + \int_0^T \int_\omega \Xi(t) v(t) dx dt \\ = \int_0^T \int_\Omega g(t) v(t) dx dt \quad \forall v \in L^2(0, T; V), \\ \Xi(x, t) \in \hat{b}(x, u(x, t)) \quad \text{for a.a. } (x, t) \in \omega_T \\ \text{and } u(0) = u_0. \end{cases}$$

Remark 4.3 *Some extensions of problem (P) studied in Miettinen and Panagiotopoulos, 1999 are possible. If instead of (4.1) we assume (3.4),(3.5) (as in Chapter 3), then one can show that there exists a solution $(u, \Xi) \in L^2(0, T; V) \cap$*

$C([0,T];H_w) \times L^1(\omega_T)$ *such that*

$$(u(T), v(T)) - (u(0), v(0)) - \int_0^T (u(t), v'(t)) \, dt \qquad (4.6)$$

$$+ \int_0^T a(u(t), v(t)) dt + \int_0^T \int_\omega \Xi(t) \Pi v(t) dx dt$$

$$= \int_0^T \langle f(t), v(t) \rangle dt \qquad \forall v \in C^1(0, T; \tilde{V})$$

and the last two conditions of Definition 4.1 are satisfied. By $C([0,T];H_w)$ we denoted the set of functions $u : [0,T] \to H$ which are continuous with respect to the weak topology of H, i.e., $(u(s), v)$ converges to $(u(t), v)$ as $s \to t$ for all $v \in H$, and by $\tilde{V} = V \cap L^\infty(\Omega)$. If, moreover, a is symmetric, $f \in L^2(0, T; H)$ and $u_0 \in V \cap L^\infty(\Omega)$ we can improve (4.6) as follows: The solution $(u, \Xi) \in L^\infty(0, T; V) \cap H^1(0, T; H) \times L^1(\omega_T) \cap L^2(0, T; V^*)$ and satisfies

$$\int_0^T (u'(t), v(t)) dt + \int_0^T a(u(t), v(t)) dt + \int_0^T \langle \tilde{\Xi}(t), v(t) \rangle dt \qquad (4.7)$$

$$= \int_0^T (f(t), v(t)) dt \qquad \forall v \in L^2(0, T; V),$$

where $\tilde{\Xi}$ is defined by (4.4).

If ω is a subset of $\partial\Omega$, then it is also possible to prove some existence results. If, for example, b satisfies (3.2),(3.3),(3.16),(4.1), then a solution in the sense of Definition 4.1 exists, and, if b fulfills the conditions (3.4),(3.5) instead of (4.1), a is symmetric, $f \in L^2(0, T; H)$ and $u_0 \in V \cap C(\overline{\Omega})$, then there exists a solution in the sense of (4.7) (for details and proofs see Miettinen and Panagiotopoulos, 1999).

The reason why we restricted ourselves to condition (4.1) and to ω as a subdomain of Ω throughout this chapter is that the derivation of the a priori estimates and the convergence analysis for the fully discrete approximation remain unsolved in the case of more general growth conditions and $\omega \subseteq \partial\Omega$ (see Remark 4.10).

4.1   DISCRETIZATION

This section deals with an approximation of (P) by applying the *standard Galerkin method* (see Glowinski et al., 1981, Thomée, 1984) In contrast to the static case we approximate both, the space and the time variable. Therefore, two different discretizations are needed: the space variable will be approximated by finite elements while a finite difference method will be used for the time discretization.

Let $h > 0$ and $k > 0$ be the *discretization parameters* of the space and time variables, respectively. Let Δ_k be an equidistant partition of $[0,T]$ into r subintervals of length $k = T/r$. The time derivative $u'(t)$ will be approximated by the following finite difference quotients:

$$u'(t) \approx \frac{u(t+k) - u(t)}{k} \quad \text{or} \quad u'(t) \approx \frac{u(t) - u(t-k)}{k}. \tag{4.8}$$

By λ^{i+1} we denote the characteristic functions of the intervals $]ik, (i+1)k]$, $i = 0, ..., r-1$.

For the approximation of the space variable finite elements will be used. As before, the discretization parameter $h > 0$ is related to the norm of partitions $\mathcal{D}_h$, $\mathcal{T}_h$ of $\overline{\Omega}$ and $\overline{\omega}$, respectively. Let $V_h \subset V$ and $Y_h \subset Y \equiv L^2(\omega)$ be finite element spaces associated with $\mathcal{D}_h$, $\mathcal{T}_h$ and defined by (3.21), (3.22), respectively. Further, let W_h be the image of V_h by Π and P_h a linear mapping from W_h into Y_h.

We start with the approximations of the spaces $L^2(0, T; V)$, $L^2(0, T; L^2(\omega))$, which will be denoted by $L^2(\Delta_k; V_h)$, $L^2(\Delta_k; Y_h)$, respectively:

$$L^2(\Delta_k; V_h) = \{v_h \in L^\infty(0, T; V_h) \mid v_h = \sum_{i=1}^{r} v^i \lambda^i, \ v^i \in V_h\}$$

$$L^2(\Delta_k; Y_h) = \{v_h \in L^\infty(0, T; Y_h) \mid v_h = \sum_{i=1}^{r} v^i \lambda^i, \ v^i \in Y_h\},$$

i.e., both consist of functions, which are piecewise constant in time on the partition Δ_k and take their values in V_h, Y_h, respectively.

If $\{u^i\}_{i=0}^{r}$ is the set of values of a sufficiently smooth function u at the time levels $t_i = ik$:

$$u^i \equiv u(ik) \qquad i = 0, 1, ..., r,$$

then the symbol $u^{i+\theta}$, $\theta \in [0, 1]$, stands for the convex combination of the values at two successive time steps i and $(i+1)$:

$$u^{i+\theta} \equiv (1 - \theta)u^i + \theta u^{i+1}, \qquad i = 0, ..., r-1.$$

Similarly to the static case, we shall study the *full* approximation of the problem, including an approximation of the bilinear form a-and of the linear functional f. To this end, let $\{a_h\}$, $a_h : V_h \times V_h \to \mathbf{R}$, be a family of approximations of the bilinear form a satisfying the assumptions (3.24) and (3.25).

Further we assume that for each pair (h, k) there exist functions $\{f_{h,k}^i\}_{i=0}^{r}$, $f_{h,k}^i \in V_h^*$, which are uniformly bounded with respect to h and k in the following sense:

$$\exists C > 0 : \left| \langle f_{h,k}^i, v_h \rangle_h \right| \leq C \|v_h\| \quad \forall v_h \in V_h, \ i = 0, ..., r \text{ and } \forall h, k > 0, \tag{4.9}$$

where $\langle \cdot, \cdot \rangle_h$ denotes the duality pairing between V_h and V_h^*. Finally, we shall approximate the initial state u_0 by a sequence $\{u_{0h}\}$, $u_{0h} \in V_h$.

A pair of functions $(u_{h,k}^\theta, \Xi_{h,k}^\theta) \in L^2(\Delta_k; V_h) \times L^2(\Delta_k; Y_h)$ of the form

$$u_{h,k}^\theta = \sum_{i=0}^{r-1} u_{h,k}^{i+\theta} \lambda^{i+1}, \quad \Xi_{h,k}^\theta = \sum_{i=0}^{r-1} \Xi_{h,k}^{i+\theta} \lambda^{i+1},$$

is called to be a solution of the *approximation scheme* $(P)_{h,k}^{\theta}$, if its time level values $(u_{h,k}^{i+\theta}, \Xi_{h,k}^{i+\theta})$ for all $i = 0, ..., r - 1$ solve the problem:

$$
\begin{cases}
\text{Find } (u_{h,k}^{i+\theta}, \Xi_{h,k}^{i+\theta}) \in V_h \times Y_h \text{ such that} \\
\left(\dfrac{u_{h,k}^{i+1} - u_{h,k}^{i}}{k}, v_h \right) + a_h(u_{h,k}^{i+\theta}, v_h) + \displaystyle\int_\omega \Xi_{h,k}^{i+\theta} P_h(\Pi v_h) dx \\
= \left\langle f_{h,k}^{i+\theta}, v_h \right\rangle_h \quad \forall v_h \in V_h, \\
\Xi_{h,k}^{i+\theta}(x) \in \hat{b}(\displaystyle\sum_{j=1}^{m} \mathcal{X}_{\text{int}_\omega K_j}(x) x_h^j, P_h(\Pi u_{h,k}^{i+\theta})(x)) \text{ for a.a. } x \in \omega,
\end{cases} \quad (P)_{h,k}^{\theta}
$$

and $u_{h,k}^0 = u_{0h}$. The approximate schemes corresponding to $\theta = 1, \frac{1}{2}, 0$ are termed: *implicit, Crank-Nicholson* and *explicit*, respectively.

Remark 4.4 *Instead of the duality pairing* $\left\langle \frac{u_{h,k}^{i+1} - u_{h,k}^{i}}{k}, v_h \right\rangle$, *the inner product* $\left(\frac{u_{h,k}^{i+1} - u_{h,k}^{i}}{k}, v_h \right)$ *is used in* $(P)_{h,k}^{\theta}$. *This is due to the fact that* $\langle v, w \rangle$ *is realized by* (v, w) *if* v *belongs to* H *(* $\frac{u_{h,k}^{i+1} - u_{h,k}^{i}}{k} \in V_h \subset H$ *). Note also that* $\left(\frac{u_{h,k}^{i+1} - u_{h,k}^{i}}{k}, v_h \right)$ *is evaluated exactly.*

Remark 4.5 *One can also use other difference quotients than (4.8). For example, in Glowinski, 1984 the two step implicit scheme using the approximation*

$$
u'(t) \approx \frac{\frac{3}{2}u(t) - 2u(t - k) + \frac{1}{2}u(t - 2k)}{k}
$$

has been proposed. The other terms are treated as in the standard implicit method.

At the end of this section we prove the existence of solutions to $(P)_{h,k}^{\theta}$.

Lemma 4.1 *Let all the assumptions concerning of* a_h, $\{f_{h,k}^i\}$, u_{0h}, P_h *and* b *be satisfied. Then* $(P)_{h,k}^{\theta}$ *has at least one solution* $(u_{h,k}^{\theta}, \Xi_{h,k}^{\theta}) \in L^2(\Delta_k; V_h) \times L^2(\Delta_k; Y_h)$ *for any* $\theta \in [0, 1]$, $h > 0$ *and* $k > 0$ *(k sufficiently small).*

Proof: The idea is to transform $(P)_{h,k}^{\theta}$ to the discrete elliptic problem for each time step and to use the results of Section 3.2. Let $0 < \theta \leq 1$. Then we can rewrite $(P)_{h,k}^{\theta}$ as follows:

$$
\begin{cases}
\text{Find } (u_{h,k}^{i+\theta}, \Xi_{h,k}^{i+\theta}) \in V_h \times Y_h \text{ for all } i = 0, ..., r - 1 \text{ s.t.} \\
ka_h(u_{h,k}^{i+\theta}, v_h) + \dfrac{1}{\theta}(u_{h,k}^{i+\theta}, v_h) + k \displaystyle\int_\omega \Xi_{h,k}^{i+\theta} P_h(\Pi v_h) dx \\
= k\langle f_{h,k}^{i+\theta}, v_h \rangle_h + \dfrac{1}{\theta}(u_{h,k}^i, v_h) \quad \forall v_h \in V_h, \\
\Xi_{h,k}^{i+\theta}(x) \in \hat{b}(\displaystyle\sum_{j=1}^{m} \mathcal{X}_{\text{int}_\omega K_j}(x) x_h^j, P_h(\Pi u_{h,k}^{i+\theta})(x)) \text{ for a.e. } x \in \omega.
\end{cases} \quad (P)_{h,k}^{\theta}
$$

Let us assume that $(P)_{h,k}^{\theta}$ has been already solved for the time steps $i = 0, ..., n-1$. Hence, the functions $u_{h,k}^1, ..., u_{h,k}^n \in V_h$ and $\Xi_{h,k}^{\theta}, ..., \Xi_{h,k}^{n-1+\theta} \in Y_h$ are known. Then we define

$$\bar{a}(v_h, w_h) \equiv ka_h(v_h, w_h) + \frac{1}{\theta}(v_h, w_h)$$

$$-\frac{1}{\theta}k\gamma \int_{\omega} P_h(\Pi v_h) P_h(\Pi w_h) dx \quad \forall v_h, w_h \in V_h, \tag{4.10}$$

$$b_0(x, \xi) \equiv kb(x, \xi) + \frac{1}{\theta}k\gamma\xi,$$

$$\langle \bar{f}, v_h \rangle_h \equiv k\langle f_{h,k}^{n+\theta}, v_h \rangle_h + \frac{1}{\theta}(u_{h,k}^n, v_h) \quad \forall v_h \in V_h,$$

where γ is an appropriate positive constant. Using these notations, the problem $(P)_{h,k}^{\theta}$ for $i = n$ can be expressed as follows:

$$\begin{cases} \text{Find } (u_{h,k}^{n+\theta}, \Xi_{h,k}^{n+\theta}) \in V_h \times Y_h \text{ such that} \\ \bar{a}(u_{h,k}^{n+\theta}, v_h) + \int_{\omega} \Xi_{h,k}^{n+\theta} P_h(\Pi v_h) dx = \langle \bar{f}, v_h \rangle_h \quad \forall v_h \in V_h, \\ \Xi_{h,k}^{n+\theta}(x) \in \hat{b}_0(\sum_{j=1}^{m} \mathcal{X}_{\text{int}_{\omega} K_j}(x) x_h^j, P_h(\Pi u_{h,k}^{n+\theta})(x)) \quad \text{for a.a. } x \in \omega. \end{cases} \tag{4.11}$$

Now all the assumptions of Theorem 3.1 except (3.5) and (3.25) are obviously satisfied. Choosing the constant γ strictly greater that $c_2\theta$ we easily verify that if

$$\bar{\xi} = \frac{c_1}{\frac{\gamma}{\theta} - c_2}$$

then

$$\frac{b_0(x, \xi)}{k} = b(x, \xi) + \frac{1}{\theta}\gamma\xi \geq (-c_1 - c_2\xi) + \frac{1}{\theta}\gamma\xi$$

$$= (\frac{\gamma}{\theta} - c_2)\xi - c_1 \geq 0 \quad \text{for all } x \in \omega, \text{ a.a. } \xi \geq \bar{\xi}$$

and, similarly,

$$\frac{b_0(x, \xi)}{k} \leq 0 \quad \text{for all } x \in \omega, \text{ a.a. } \xi \leq -\bar{\xi}.$$

Therefore, we have

$$\operatorname*{ess\,sup}_{\xi \in (-\infty, -\bar{\xi})} \sup_{x \in \omega} b_0(x, \xi) \leq 0 \leq \operatorname*{ess\,inf}_{\xi \in (\bar{\xi}, \infty)} \inf_{x \in \omega} b_0(x, \xi),$$

i.e., (3.5) holds.

Using the linearity of the mappings P_h, Π and the fact that all norms are equivalent in finite dimensional spaces, we have

$$|P_h(\Pi v_h)|_{0,\omega} \leq c_0|v_h| \quad \forall v_h \in V_h$$

for some positive constant c_0. Then (3.25) follows from the estimate:

$$\bar{a}(v_h, v_h) = ka_h(v_h, v_h) + \frac{1}{\theta}((v_h, v_h))$$

$$-k\gamma \int_\omega P_h(\Pi v_h) P_h(\Pi v_h) dx$$

$$\geq k\tilde{a}\|v_h\|^2 + \frac{1}{\theta}\left(|v_h|^2 - k\gamma|P_h(\Pi v_h)|^2_{0,\omega}\right)$$

$$\geq k\tilde{a}\|v_h\|^2 + \frac{1}{\theta}\left(1 - k\gamma c_0^2\right)|v_h|^2 \geq k\tilde{a}\|v_h\|^2,$$

if the time increment k satisfies the condition:

$$k \leq \frac{1}{\gamma c_0^2} < \frac{1}{c_2 \theta c_0^2}. \tag{4.12}$$

Thus, $(P)_{h,k}^\theta$ is solvable under the condition (4.12).

If $\theta = 0$ problem $(P)_{h,k}^0$ is equivalent to the following one:

$$\begin{cases} \text{Find } u_{h,k}^{i+1} \in V_h \text{ and } \Xi_{h,k}^i \in Y_h \text{ for all } i = 1, ..., r-1 \text{ such that} \\ (u_{h,k}^{i+1}, v_h) = -ka_h(u_{h,k}^i, v_h) + (u_{h,k}^i, v_h) \\ -k \int_\omega \Xi_{h,k}^i P_h(\Pi v_h) dx + k\langle f_{h,k}^i, v_h \rangle_h \quad \forall v_h \in V_h, \\ \Xi_{h,k}^i(x) \in \hat{b}\left(\sum_{j=1}^m \mathcal{X}_{\text{int}_\omega K_j}(x) x_h^j, P_h(\Pi u_{h,k}^i)(x)\right) \quad \text{for a.a. } x \in \omega, \end{cases}$$

which is trivially solvable. □

Remark 4.6 *Note that, if b satisfies also (3.5), we do not need to impose any condition on k in the previous lemma.*

4.2   CONVERGENCE ANALYSIS

The aim of this section is to show that solutions of $(P)_{h,k}^\theta$ and (P) are close on subsequences in the weak topology of $L^2(0,T;V) \times L^2(\omega_T)$. We shall consider the cases $\theta \in [1/2, 1]$ and $\theta \in [0, 1/2)$ separately. We prove that in the first case the θ-scheme is *unconditionally stable*, while the second case leads to the *conditionally stable* scheme. A scheme is said to be conditionally stable if the mesh sizes h, k are in a certain relation.

To this end we suppose that:

$$\text{the family } \{V_h\} \text{ is dense in } V: \tag{4.13}$$
$$\forall v \in V \; \exists\{v_h\}, v_h \in V_h : v_h \to v \text{ in } V \text{ as } h \to 0+;$$

$$\text{the inverse inequality holds between the } V\text{- and } H\text{-norms:} \tag{4.14}$$
$$\exists \; s(h) = const. > 0 \; such \; that$$
$$\|v\| \leq s(h)|v| \qquad \forall v \in V_h.$$

Remark 4.7 *Note that (4.13) is a weaker condition than (3.35). This is due to the fact that the function b satisfies the growth condition (4.1).*

Let $\rightsquigarrow$ be a rule associating with any V_h the unique Y_h (cf. Section 3.2). Similarly to the elliptic case, we shall suppose that the mappings Π and P_h satisfy the following assumptions:

$$v_h \rightharpoonup v \text{ in } L^2(0,T;V) \quad \text{and} \quad v_h \to v \text{ in } L^2(0,T;H), \qquad (4.15)$$
$$v_h \in L^2(0,T;V_h), \text{ as } h \to 0+$$
$$\implies z_h \to \Pi v \text{ in } L^2(\omega_T), \text{ as } h \to 0+,$$

where $z_h(\cdot) \equiv P_h(\Pi(v_h(\cdot))) \in L^2(0,T;Y_h)$ and $\Pi v(t) \equiv \Pi(v(t))$ (note that, if $v \in L^2(0,T;V)$, then Πv defined in this way is really an element of $L^2(\omega_T)$). Further, the family $\{P_h\}$, $h \to 0+$, is assumed to be *uniformly bounded* with respect to h in the following sense:

$$|P_h(\Pi v_h)|_{0,\omega} \equiv \|P_h(\Pi v_h)\|_{L^2(\omega)} \leq c(h)\|v_h\| + C|v_h| \quad \forall v_h \in V_h, \ \forall h > 0, (4.16)$$

where $C, c(h)$ are positive constants, C independent of h and $c(h) \to 0$ as $h \to 0+$.

Finally we suppose that the constants $c(h)$ and $s(h)$ satisfy the condition

$$c(h)s(h) \leq C \qquad \forall h > 0, \tag{4.17}$$

where C is a positive constant independent of h.

As far as the remaining approximating data $\{a_h\}$, $\{u_{0h}\}$ and $\{f_{h,k}^\theta\}$, where $f_{h,k}^\theta = \sum_{i=0}^{r-1} \left((1-\theta)f_{h,k}^i + \theta f_{h,k}^{i+1}\right)\lambda^{i+1}$, are concerned, the following properties will be required: $\{a_h\}$ satisfies (3.38),

$$u_{0h} \to u_0 \text{ in } H \text{ as } h \to 0+ \tag{4.18}$$

and

$$v_h \to v \text{ in } L^2(0,T;V), v_h \in L^2(0,T;V_h), \text{ as } h \to 0+ \implies \qquad (4.19)$$
$$\int_0^T \langle f_{h,k}^\theta(t), v_h(t)\rangle_h dt \to \int_0^T \langle f(t), v(t)\rangle dt \text{ as } h,k \to 0+.$$

Remark 4.8 *Taking into account the density of polynomials $a_0 + a_1t + ... + a_nt^n$, $a_i \in V_h$, $n \in \mathbf{N}$, in $L^2(0,T;V)$ (see Proposition 1.1) it is easy to see that the condition (3.38) implies*

$$u_h \rightharpoonup u, \ v_h \to v \text{ in } L^2(0,T;V), \tag{4.20}$$
$$u_h, v_h \in L^2(0,T;V_h), \text{ as } h \to 0+$$
$$\implies \int_0^T a_h(u_h(t), v_h(t))dt \to \int_0^T a(u(t), v(t))dt.$$

For convenience of the reader, we collect all the assumptions which will be used in what follows:

(i) the function b satisfies (3.2), (3.3), (3.16) and (4.1);

(ii) the system of approximated bilinear forms $\{a_h\}$ satisfies (3.24), (3.25) and (3.38);

(iii) the system of approximated linear forms $\{f^\theta_{h,k}\}$ satisfies (4.9) and (4.19);

(iv) the sequence of approximated initial conditions $\{u_{0h}\}$, $u_{0h} \in V_h$, satisfies (4.18);

(v) Π is a continuous linear mapping from V to $L^2(\omega)$;

(vi) the mapping P_h satisfies (4.15)-(4.17);

(vii) the system $\{V_h\}$ satisfies (4.13) and (4.14).

Then it holds:

Theorem 4.1 *Let all the above assumptions (i)-(vii) be satisfied. Let $\{(u^\theta_{h,k}, \Xi^\theta_{h,k})\}$ be a sequence of solutions to $(P)^\theta_{h,k}$. Then:*

a) $\theta \in [1/2, 1]$. There exist: a subsequence of $\{(u^\theta_{h,k}, \Xi^\theta_{h,k})\}$ and an element $(u, \Xi) \in W(V) \times L^2(\omega_T)$ such that

$$u^\theta_{h,k} \rightharpoonup u \quad in \ L^2(0,T;V), \tag{4.21}$$

$$u^\theta_{h,k} \to u \quad in \ L^2(0,T;H), \tag{4.22}$$

$$\Xi^\theta_{h,k} \rightharpoonup \Xi \quad in \ L^2(\omega_T), \tag{4.23}$$

as $h, k \to 0+$. Moreover, (u, Ξ) is a solution of (P).

b) $\theta \in [0, 1/2)$. If, moreover, the pairs (h, k) satisfy the stability condition

$$1 - 2(1 - \theta)ks(h)^2 \frac{\tilde{M}^2}{\tilde{\alpha}} \geq c > 0, \tag{4.24}$$

where c is a positive constant, then the conclusions of the case a) remain true. Constants $\tilde{M}, \tilde{\alpha}, s(h)$ are the same as in (3.24),(3.25),(4.14), respectively.

Furthermore any cluster point of $\{(u^\theta_{h,k}, \Xi^\theta_{h,k})\}$ in the sense of (4.21)-(4.23) is a solution of (P).

In the proof we shall need the discrete analogue of Gronwall's inequality (see, e.g., Fairweather, 1978):

Lemma 4.2 *(discrete Gronwall's inequality) Let $f(t)$, $g(t)$ and $h(t)$ be non-negative functions defined on $\{t \in [0,T] : t = ik, \ i = 0, 1, ..., r, \ rk = T\}$ and let $g(t)$ be non-decreasing. If*

$$f(nk) + h(nk) \leq g(nk) + Ck \sum_{i=0}^{n-1} f(ik), \quad n = 1, 2, ..., r,$$

where C is a positive constant, then

$$f(nk) + h(nk) \leq g(nk)e^{Cnk}, \quad n = 1, 2, ..., r.$$

Proof of Theorem 4.1: We shall treat the cases $0 \leq \theta < \frac{1}{2}$ and $\frac{1}{2} \leq \theta \leq 1$ separately. The proofs consist of several steps: the derivation of *a priori* estimates, permitting us to pass to convergent subsequences and a limit procedure in discrete problems.

The case $\frac{1}{2} \leq \theta \leq 1$

Step I: A priori estimates. First, we recall two classical relations, which will be frequently used in what follows:

$$(a - b)a = \frac{1}{2}a^2 - \frac{1}{2}b^2 + \frac{1}{2}(a - b)^2, \qquad \forall a, b \in \mathbf{R}, \tag{4.25}$$

$$ab \leq \varepsilon a^2 + \frac{1}{4\varepsilon}b^2, \qquad \forall a, b \in \mathbf{R}, \quad \forall \varepsilon > 0. \tag{4.26}$$

We prove that the sequence $\{u^\theta_{h,k}\}$ remains within bounded subsets of $L^2(0, T; V)$ and $L^\infty(0, T; H)$. Indeed, inserting $v = u^{i+\theta}_{h,k}$ in $(\mathrm{P})^\theta_{h,k}$ we have that

$$\left(\frac{u^{i+1}_{h,k} - u^i_{h,k}}{k}, u^{i+\theta}_{h,k}\right) + a_h\left(u^{i+\theta}_{h,k}, u^{i+\theta}_{h,k}\right) \tag{4.27}$$

$$+ \int_\omega \Xi^{i+\theta}_{h,k} P_h(\Pi u^{i+\theta}_{h,k})dx = \langle f^{i+\theta}_{h,k}, u^{i+\theta}_{h,k}\rangle_h.$$

Taking into account the definition of $u^{i+\theta}_{h,k}$ and (4.25) we have:

$$\frac{1}{k}\left(u^{i+1}_{h,k} - u^i_{h,k}, u^{i+\theta}_{h,k}\right) \tag{4.28}$$

$$= \frac{1}{2k}\left(|u^{i+1}_{h,k}|^2 - |u^i_{h,k}|^2\right) + \frac{(2\theta - 1)}{2k}|u^{i+1}_{h,k} - u^i_{h,k}|^2.$$

From (3.25) it follows that

$$a_h(u^{i+\theta}_{h,k}, u^{i+\theta}_{h,k}) \geq \tilde{\alpha}\|u^{i+\theta}_{h,k}\|^2 \tag{4.29}$$

and, (4.1),(4.16),(4.26) give an upper estimate for*

$$\left|\int_\omega \Xi^{i+\theta}_{h,k} P_h(\Pi u^{i+\theta}_{h,k})dx\right| \tag{4.30}$$

$$\leq \int_\omega C\left(1 + |P_h(\Pi u^{i+\theta}_{h,k})|\right)|P_h(\Pi u^{i+\theta}_{h,k})|dx$$

$$\leq C + \tilde{c}(h)\|u^{i+\theta}_{h,k}\|^2 + C|u^{i+\theta}_{h,k}|^2,$$

---

*Here and in what follows, the symbol C stands for a positive constant, which is independent of h, k and θ. If C depends on a particular parameter ξ, this dependence (if necessary) is pointed out by writing $C(\xi)$.

where $\tilde{c}(h)$ is a positive constant satisfying $\tilde{c}(h) \to 0$ as $h \to 0+$. From (4.9) we get an upper bound for the right hand sides of (4.27):

$$\left| \langle f_{h,k}^{i+\theta}, u_{h,k}^{i+\theta} \rangle_h \right| \leq C \| u_{h,k}^{i+\theta} \| \leq \varepsilon \| u_{h,k}^{i+\theta} \|^2 + C(\varepsilon) \tag{4.31}$$

Summing up (4.27) for $i = 0$ to $i = n - 1$, $n \leq r$, multiplying it by $2k$ and taking into account (4.28)–(4.31) we arrive at

$$\sum_{i=0}^{n-1} \left\{ \left| u_{h,k}^{i+1} \right|^2 - \left| u_{h,k}^i \right|^2 \right\} + \sum_{i=0}^{n-1} (2\theta - 1) \left| u_{h,k}^{i+1} - u_{h,k}^i \right|^2 \tag{4.32}$$

$$+ 2 \sum_{i=0}^{n-1} k \left(\tilde{\alpha} - (\varepsilon + \tilde{c}(h)) \right) \left\| u_{h,k}^{i+\theta} \right\|^2 \leq C(\varepsilon) + C \sum_{i=0}^{n-1} k \left| u_{h,k}^{i+\theta} \right|^2.$$

From the definition of $u_{h,k}^{i+\theta}$ we get:

$$(1 - Ck) \left| u_{h,k}^n \right|^2 + \sum_{i=0}^{n-1} (2\theta - 1) \left| u_{h,k}^{i+1} - u_{h,k}^i \right|^2 \tag{4.33}$$

$$+ 2 \sum_{i=0}^{n-1} k \left(\tilde{\alpha} - (\varepsilon + \tilde{c}(h)) \right) \left\| u_{h,k}^{i+\theta} \right\|^2 \leq C(\varepsilon) + C \sum_{i=0}^{n-1} k \left| u_{h,k}^i \right|^2$$

for all $n = 1, ..., r$. Next, we take h, k and ε small enough so that both constants $(1-Ck)$ and $(\tilde{\alpha}-(\varepsilon+\tilde{c}(h)))$ are positive. This fixes the constant $C(\varepsilon)$ appearing on the right hand side of (4.33). Then using the discrete Gronwall's inequality we deduce from (4.33) that

$$\max_{1 \leq i \leq r} \left| u_{h,k}^i \right|, \left| u_{h,k}^{i-1+\theta} \right| \leq C, \tag{4.34}$$

$$\| u_{h,k}^\theta \|_{L^2(0,T;V)}^2 = \sum_{i=0}^{r-1} k \left\| u_{h,k}^{i+\theta} \right\|^2 \leq C, \tag{4.35}$$

$$(2\theta - 1) \sum_{i=0}^{r-1} \left| u_{h,k}^{i+1} - u_{h,k}^i \right|^2 \leq C \tag{4.36}$$

provided that h, k are small enough. Thus, $\{ u_{h,k}^\theta \}$ is bounded in the $L^\infty(0, T; H)$- and $L^2(0, T; V)$-norms.

Now we show that $\{ \Xi_{h,k}^\theta \}$ is bounded in $L^2(\omega_T)$. Indeed, from (4.1), (4.16) and the fact that $V \hookrightarrow H$ we see that

$$\| \Xi_{h,k}^\theta \|_{L^2(\omega_T)}^2 = \sum_{i=0}^{r-1} k \int_\omega \left| \Xi_{h,k}^{i+\theta} \right|^2 dx \tag{4.37}$$

$$\leq \sum_{i=0}^{r-1} k \int_\omega \left(C(1 + |P_h(\Pi u_{h,k}^{i+\theta})|) \right)^2 dx$$

$$\leq C \left(1 + \sum_{i=0}^{r-1} k \| P_h(\Pi u_{h,k}^{i+\theta}) \|_{L^2(\omega)}^2 \right) \leq C \left(1 + \sum_{i=0}^{r-1} k \| u_{h,k}^{i+\theta} \|^2 \right),$$

which together with (4.35) completes the proof of step I.

Step II: Convergences of subsequences. From the *a priori* estimates (4.35) and (4.37) we see that the sequences $\{u_{h,k}^\theta\}$ and $\{\Xi_{h,k}^\theta\}$ are uniformly bounded in $L^2(0,T;V)$ and $L^2(\omega_T)$, respectively. Therefore, one can extract subsequences and find elements $u \in L^2(0,T;V)$ and $\Xi \in L^2(\omega_T)$ such that

$$u_{h,k}^\theta \rightharpoonup u \qquad \text{in } L^2(0,T;V), \tag{4.38}$$

$$\Xi_{h,k}^\theta \rightharpoonup \Xi \qquad \text{in } L^2(\omega_T), \tag{4.39}$$

as $h, k \to 0+$, i.e., (4.21) and (4.23) hold.

Next, we prove (4.22):

$$u_{h,k}^\theta \to u \qquad \text{in } L^2(0,T;H). \tag{4.40}$$

To this end we use Proposition 1.7. Because of (4.34),(4.35) it remains to prove the weak convergence of $\{u_{h,k}^\theta(t)\}$ to $u(t)$ in $L^1(\Omega;\mathbf{R}^d)$ a.e. in $[0,T]$. Instead of this we show that the respective subsequence tends to u weakly in H which is a stronger property.

Fix $t \in (0,T]$. We prove that $\{u_{h,k}^\theta(t)\}$ is a Cauchy sequence in the weak topology of H. Due to the density of V in H this means to show that

$$\left|(u_{h,k}^\theta(t) - u_{h',k'}^\theta(t), v)\right| \to 0 \quad \text{as } h, h', k, k' \to 0+ \tag{4.41}$$

for all $v \in V$.

Let $v \in V$ be given. By virtue of (4.13) there exists a sequence $\{v_h\}$, $v_h \in V_h$, converging to v strongly in V. Then by adding and subtracting appropriate terms we get:

$$\left|(u_{h,k}^\theta(t) - u_{h',k'}^\theta(t), v)\right| \le \left|(u_{h,k}^\theta(t), v_h) - (u_{h',k'}^\theta(t), v_{h'})\right| \tag{4.42}$$
$$+ \left|(u_{h,k}^\theta(t), v - v_h)\right| + \left|(u_{h',k'}^\theta(t), v - v_{h'})\right|.$$

The last two terms in (4.42) are easily estimated as follows:

$$\left|(u_{h,k}^\theta(t), v - v_h)\right| \le |u_{h,k}^\theta(t)||v - v_h| \to 0 \tag{4.43}$$

$$\left|(u_{h',k'}^\theta(t), v - v_{h'})\right| \le |u_{h',k'}^\theta(t)||v - v_h'| \to 0 \tag{4.44}$$

as $h, h', k, k' \to 0+$, because $\{u_{h,k}^{\theta}(t)\}$ is bounded in H for all $t \in [0, T]$. The chosen t belongs to $](n-1)k, nk]$ for some $n \le r$. Then it holds

$$\left(u_{h,k}^{\theta}(t), v_h\right) = \left(\theta u_{h,k}^{n} + (1-\theta)u_{h,k}^{n-1}, v_h\right) \tag{4.45}$$

$$= \theta k \sum_{i=0}^{n-1} \frac{1}{k}(u_{h,k}^{i+1} - u_{h,k}^{i}, v_h) + \theta(u_{h,k}^{0}, v_h)$$

$$+ (1-\theta)k \sum_{i=0}^{n-2} \frac{1}{k}(u_{h,k}^{i+1} - u_{h,k}^{i}, v_h) + (1-\theta)(u_{h,k}^{0}, v_h)$$

$$= \theta k \sum_{i=0}^{n-1} \left(-a_h(u_{h,k}^{i+\theta}, v_h) - (\Xi_{h,k}^{i+\theta}, P_h(\Pi v_h))_{0,\omega} + \langle f_{h,k}^{i+\theta}, v_h \rangle_h \right)$$

$$+ (1-\theta)k \sum_{i=0}^{n-2} \left(-a_h(u_{h,k}^{i+\theta}, v_h) - (\Xi_{h,k}^{i+\theta}, P_h(\Pi v_h))_{0,\omega} + \langle f_{h,k}^{i+\theta}, v_h \rangle_h \right)$$

$$+ \left(u_{h,k}^{0}, v_h\right)$$

making use of the definition of $(P)_{h,k}^{\theta}$.

In order to prove the convergence of the terms appearing after the last sign of the equality in (4.45) we define two auxiliary functions from $L^2(0, T; V_h)$:

$$v_{h,k}^{1}(s) = \begin{cases} v_h, & s \in [0, t] \\ 0, & s \in (t, T], \end{cases} \tag{4.46}$$

$$v_{h,k}^{2}(s) = \begin{cases} v_h, & s \in (t, nk] \\ 0, & \text{otherwise.} \end{cases} \tag{4.47}$$

It is readily seen that

$$v_{h,k}^{1} \to v^1 \equiv v^1(s) = \begin{cases} v, & s \in [0, t] \\ 0, & s \in (t, T], \end{cases} \tag{4.48}$$

$$v_{h,k}^{2} \to 0 \tag{4.49}$$

in $L^2(0, T; V)$ as $h, k \to 0+$. From this, Remark 4.8 and (4.38) we obtain that

$$k \sum_{i=0}^{n-1} a_h(u_{h,k}^{i+\theta}, v_h) \tag{4.50}$$

$$= \int_0^t a_h(u_{h,k}^{\theta}(s), v_h)ds + \int_t^{nk} a_h(u_{h,k}^{\theta}(s), v_h)ds$$

$$= \int_0^T a_h(u_{h,k}^{\theta}(s), v_{h,k}^{1}(s))ds + \int_0^T a_h(u_{h,k}^{\theta}(s), v_{h,k}^{2}(s))ds$$

$$\to \int_0^T a(u(s), v^1(s))ds = \int_0^t a(u(s), v)ds.$$

Because of (4.48),(4.49) we can use (4.15) giving

$$P_h(\Pi v_{h,k}^{1}) \to \Pi v^1, \qquad P_h(\Pi v_{h,k}^{2}) \to 0 \tag{4.51}$$

in $L^2(\omega_T)$ as $h, k \to 0+$.

From (4.39) and (4.51) it follows that

$$k \sum_{i=0}^{n-1} \left(\Xi_{h,k}^{i+\theta}, P_h(\Pi v_h) \right)_{0,\omega} \tag{4.52}$$

$$= \int_0^T \left(\Xi_{h,k}^\theta(s), P_h(\Pi v_{h,k}^1)(s) \right)_{0,\omega} ds$$

$$+ \int_0^T \left(\Xi_{h,k}^\theta(s), P_h(\Pi v_{h,k}^2)(s) \right)_{0,\omega} ds$$

$$\to \int_0^T \left(\Xi(s), \Pi v^1(s) \right)_{0,\omega} ds = \int_0^t \left(\Xi(s), \Pi v \right)_{0,\omega} ds.$$

In a similar way, using (4.19),(4.48),(4.49) we get that

$$k \sum_{i=0}^{n-1} \langle f_{h,k}^{i+\theta}, v_h \rangle_h \to \int_0^t \langle f(s), v \rangle ds. \tag{4.53}$$

Thus from (4.50),(4.52) and (4.53) we finally obtain that

$$\theta k \sum_{i=0}^{n-1} \left(-a_h(u_{h,k}^{i+\theta}, v_h) - (\Xi_{h,k}^{i+\theta}, P_h(\Pi v_h))_{0,\omega} + \langle f_{h,k}^{i+\theta}, v_h \rangle_h \right) \tag{4.54}$$

$$\to \theta \int_0^t \left(-a(u(s), v) - (\Xi(s), \Pi v)_{0,\omega} + \langle f(s), v \rangle \right) ds.$$

Using exactly the same approach we get:

$$(1-\theta) k \sum_{i=0}^{n-2} \left(-a_h(u_{h,k}^{i+\theta}, v_h) \right. \tag{4.55}$$

$$\left. -(\Xi_{h,k}^{i+\theta}, P_h(\Pi v_h))_{0,\omega} + \langle f_{h,k}^{i+\theta}, v_h \rangle_h \right)$$

$$\to (1-\theta) \int_0^t \left(-a(u(s), v) - (\Xi(s), \Pi v)_{0,\omega} + \langle f(s), v \rangle \right) ds.$$

The only difference in the proof is that the auxiliary functions $v_{h,k}^1, v_{h,k}^2$ are now defined as follows:

$$v_{h,k}^1(s) = \begin{cases} v_h, & s \in [0, t] \\ 0, & s \in (t, T], \end{cases} \tag{4.56}$$

$$v_{h,k}^2(s) = \begin{cases} -v_h, & s \in ((n-1)k, t] \\ 0, & \text{otherwise.} \end{cases} \tag{4.57}$$

In addition, by virtue of (4.18) it holds that

$$(u_{h,k}^0, v_h) = (u_{0h}, v_h) \to (u_0, v) \tag{4.58}$$

Hence, from (4.45),(4.54),(4.55) and (4.58) we conclude that

$$\left(u^\theta_{h,k}(t), v_h\right) \to \int_0^t \left(-a(u(s),v)\right.$$

$$\left. -(\Xi(s), \Pi v)_{0,\omega} + \langle f(s), v\rangle\right) ds + (u_0, v) \tag{4.59}$$

as h, k tend to 0+. Clearly, the same result holds true for the limit of $\left(u^\theta_{h',k'}(t),\right.$ $\left.v_{h'}\right)$ implying

$$\left|\left(u^\theta_{h,k}(t), v_h\right) - \left(u^\theta_{h',k'}(t), v_{h'}\right)\right| \to 0 \tag{4.60}$$

as h, k, h', k' tend to 0+. This completes the proof that $\{u^\theta_{h,k}(t)\}$ is a Cauchy sequence in the weak topology of H for all $t \in [0,T]$.

Now, since H is weakly complete, there exists a weak limit $\hat u(t) \in H$ of the sequence $\{u^\theta_{h,k}(t)\}$ for all $t \in [0,T]$. It remains to show that $\hat u(t) = u(t)$ a.e. in [0,T]. Recalling that $\{u^\theta_{h,k}\}$ tends weakly to u in $L^2(0,T;V)$ we also have that $u^\theta_{h,k} \rightharpoonup u$ in $L^2(0,T;H)$. Thus, the Mazur's lemma implies the existence of convex combinations $\{u_l\} = \sum_{h,k\le l} \eta^l_{h,k} u^\theta_{h,k}$ of $\{u^\theta_{h,k}\}$, which converge strongly to u in $L^2(0,T;H)$ as $l \to 0+$ and, consequently, $u_l(t) \to u(t)$ in H for a.a. $t \in [0,T]$. Therefore,

$$(u(t),v) = \lim_{l\to 0+} \sum_{h,k\le l} \eta^l_{h,k}\left(u^\theta_{h,k}(t),v\right) = (\hat u(t),v) \tag{4.61}$$

holds for all $v \in V$ and a.a. $t \in [0,T]$. In the last equality we used that $\sum_{h,k\le l} \eta^l_{h,k} = 1 \ \forall l > 0$ and

$$\left(u^\theta_{h,k}(t),v\right) \to \left(\hat u(t),v\right) \qquad \forall t \in [0,T] \text{ and } v \in H. \tag{4.62}$$

Consequently, $\hat u(t) = u(t)$ for a.a. $t \in [0,T]$. This and Proposition 1.7 lead to (4.40).

Step III: Limit procedure. We begin by proving that

$$\begin{cases} u \in W(V) \quad \text{and} \\ u'(t) + Au(t) + \tilde\Xi(t) = f(t) \quad \text{in } V^*, \text{ for a.a. } t \in (0,T) \end{cases} \tag{4.63}$$

(see Remark 4.2).

Let $\phi \in C_0^\infty((0,T))$ and $v \in V$ be given. Then from (4.13) it follows that there exists a sequence $\{v_h\}$, $v_h \in V_h$, converging to v strongly in V. Let us denote by $\Phi(x,t) = v(x)\phi(t)$ and $\Phi_h(x,t) = v_h(x)\phi(t)$. We define a piecewise linear interpolate $\bar u_{h,k} \in W(V)$ made of $\{u^i_{h,k}\}$ as follows:

$$\bar u_{h,k}(t) = \left(\frac{u^{i+1}_{h,k} - u^i_{h,k}}{k}\right)(t - ik) + u^i_{h,k}, \tag{4.64}$$

$$\forall t \in (ik, (i+1)k], \ i = 0, ..., r - 1.$$

From this it is readily seen that

$$\bar{u}'_{h,k}(t) = \frac{u_{h,k}^{i+1} - u_{h,k}^i}{k} \qquad \forall t \in]ik, (i+1)k], \ i = 0, ..., r-1, \quad (4.65)$$

and, therefore, $\bar{u}_{h,k}$ is an element of $H^1(0,T;V_h) = \{v \in L^2(0,T;V_h) : v' \in L^2(0,T;V_h)\}$. Applying the integration by parts, we obtain:

$$\int_0^T \left(\bar{u}'_{h,k}(t), \Phi_h(t)\right) dt = -\int_0^T \left(\bar{u}_{h,k}(t), \Phi'_h(t)\right) dt. \quad (4.66)$$

Next, we shall show that also the sequence of $\bar{u}_{h,k}$ converges to u strongly in $L^2(0,T;H)$ (actually the weak convergence is sufficient, as we shall see). First let $\theta > \frac{1}{2}$. Then from (4.36) we obtain that

$$\|\bar{u}_{h,k} - u_{h,k}^\theta\|_{L^2(0,T;H)}^2 = \int_0^T |\bar{u}_{h,k}(t) - u_{h,k}^\theta(t)|^2 dt \quad (4.67)$$

$$= \sum_{i=0}^{r-1} \int_{ik}^{(i+1)k} \left| \left(\frac{u_{h,k}^{i+1} - u_{h,k}^i}{k}\right)(t - ik) + u_{h,k}^i - u_{h,k}^{i+\theta} \right|^2 dt$$

$$= \sum_{i=0}^{r-1} \int_{ik}^{(i+1)k} \left| (u_{h,k}^{i+1} - u_{h,k}^i)\left(\frac{t - ik - \theta k}{k}\right) \right|^2 dt$$

$$= \sum_{i=0}^{r-1} |u_{h,k}^{i+1} - u_{h,k}^i|^2 \int_{ik}^{(i+1)k} \left(\frac{t - ik - \theta k}{k}\right)^2 dt$$

$$\leq \sum_{i=0}^{r-1} k|u_{h,k}^{i+1} - u_{h,k}^i|^2$$

$$\leq Ck \to 0 \qquad \text{as } h, k \to 0+.$$

From (4.40) the strong convergence of $\{\bar{u}_{h,k}\}$ to u in $L^2(0,T;H)$ follows.

On the other hand, if $\theta = \frac{1}{2}$ then only the corresponding weak convergence can be proven. From (4.34) it is readily seen that $\{\bar{u}_{h,k}\}$ is bounded in $L^2(0,T;H)$. Therefore, it is enough to prove that[†]

$$\int_0^T \left(u_{h,k}^\theta(t) - \bar{u}_{h,k}(t), \Psi(t)\right) dt \to 0 \quad \text{as } h, k \to 0+ \quad (4.68)$$

for all Ψ belonging to a dense subset of $L^2(0,T;H)$. Recall that the set of all polynomials of the form $a_0 + a_1 t + ... + a_n t^n$, $a_i \in H$ and $n \in \mathbf{N}$, is dense in $L^2(0,T;H)$. If we replace the condition $a_i \in H$ by $a_i \in V$, the density is, of course, still preserved. Therefore, it is enough to verify (4.68) when $\Psi = vt^l$, $v \in V$ and $l \in \mathbf{N}$.

[†]Here we prove the weak convergence for any $\theta \in [\frac{1}{2}, 1]$.

Let $\{v_h\}$, $v_h \in V_h$ be a sequence such that $v_h \to v$ in V. Then it holds that

$$\left| \int_0^T (u_{h,k}^\theta(t) - \bar{u}_{h,k}(t), (v - v_h)t') dt \right| \to 0 \qquad \text{as } h, k \to 0+ \quad (4.69)$$

and

$$\left| \int_0^T (u_{h,k}^\theta(t) - \bar{u}_{h,k}(t), v_h t') dt \right| \qquad (4.70)$$

$$= \left| \sum_{i=0}^{r-1} \int_{ik}^{(i+1)k} \left(\left(\frac{u_{h,k}^{i+1} - u_{h,k}^i}{k} \right)(t - ik - \theta k), v_h \right) t' dt \right|$$

$$= \left| \sum_{i=0}^{r-1} \left(\frac{u_{h,k}^{i+1} - u_{h,k}^i}{k}, v_h \right) \int_{ik}^{(i+1)k} (t - ik - \theta k) t' dt \right|$$

$$\leq Ck^2 \left| \sum_{i=0}^{r-1} \left(\frac{u_{h,k}^{i+1} - u_{h,k}^i}{k}, v_h \right) \right|$$

$$= Ck^2 \left| \sum_{i=0}^{r-1} \left(-a_h(u_{h,k}^{i+\theta}, v_h) - (\Xi_{h,k}^{i+\theta}, P_h(\Pi v_h))_{0,\omega} + \langle f_{h,k}^{i+\theta}, v_h \rangle_h \right) \right|$$

$$= Ck \left| \int_0^T \left(-a_h(u_{h,k}^\theta(t), v_h) - (\Xi_{h,k}^\theta(t), P_h(\Pi v_h))_{0,\omega} + \langle f_{h,k}^\theta(t), v_h \rangle_h \right) \right|$$

$$\leq Ck \to 0 \qquad \text{as } h, k \to 0+ .$$

From (4.69) and (4.70) the desired convergence result (4.68) easily follows.

It remains to show that the pair (u, Ξ) being a limit of $\{u_{h,k}^\theta\}$, $\{\Xi_{h,k}^\theta\}$, respectively, is a solution to (P).

Let Φ, Φ_h be the same as above. Inserting (4.65) into $(P)_{h,k}^\theta$, integrating over the time interval $(0, T)$ and using (4.66) we obtain:

$$-\int_0^T (\bar{u}_{h,k}(t), \Phi_h'(t)) dt + \int_0^T a_h(u_{h,k}^\theta(t), \Phi_h(t)) dt \qquad (4.71)$$

$$+ \int_0^T (\Xi_{h,k}^\theta(t), P_h(\Pi \Phi_h)(t))_{0,\omega} dt = \int_0^T \langle f_{h,k}^\theta(t), \Phi_h(t) \rangle_h dt$$

Since the sequences $\{\Phi_h\}$, $\{\Phi_h'\}$ converge strongly to Φ, Φ' in $L^2(0, T; V)$, $L^2(0, T; H)$, respectively, then using (4.15),(4.19),(4.20),(4.38),(4.39) and letting $h, k \to 0+$ in (4.71) we obtain that

$$-\int_0^T (u(t), \Phi'(t)) dt + \int_0^T a(u(t), \Phi(t)) dt \qquad (4.72)$$

$$+ \int_0^T (\Xi(t), \Pi \Phi(t))_{0,\omega} dt = \int_0^T \langle f(t), \Phi(t) \rangle dt.$$

Rewriting (4.72) in the form

$$\int_0^T \big(u(t), v\big)\phi'(t)dt \tag{4.73}$$

$$= -\int \langle -Au(t) - \bar{\Xi}(t) + f(t), v\rangle\phi(t)dt \quad \forall v \in V \text{ and } \phi \in C_0^\infty((0,T)),$$

and using Proposition 1.3 we conclude that there exists the generalized derivative $u' \in L^2(0,T;V^*)$ of u and $u' = -Au - \bar{\Xi} + f$ in $L^2(0,T;V^*)$.

The initial condition $u(0) = u_0$ follows the same guidelines. The main difference is that now we choose ϕ from $C^\infty([0,T])$ such that $\phi(0) = 1$, $\phi(T) = 0$, and replace (4.66) by

$$\int_0^T \big(\bar{u}'_{h,k}(t), \Phi_h(t)\big)dt \tag{4.74}$$

$$= -\int_0^T \big(\bar{u}_{h,k}(t), \Phi'_h(t)\big)dt + \big(\bar{u}_{h,k}(0), \Phi_h(0)\big).$$

Inserting (4.65) into $(P)_{h,k}^\theta$, integrating over the interval $(0,T)$, using (4.74) and passing to the limit we get that

$$-\int_0^T \big(u(t), \Phi'(t)\big)dt + \big(u_0, \Phi(0)\big) + \int_0^T a(u(t), \Phi(t))dt \tag{4.75}$$

$$+ \int_0^T \big(\Xi(t), \Pi\Phi(t)\big)_{0,\omega}dt = \int_0^T \langle f(t), \Phi(t)\rangle dt.$$

Here we used the fact that $\Phi_h \to \Phi$ in $C([0,T];H)$. Integration by parts gives us

$$\int_0^T \langle u'(t), \Phi(t)\rangle dt = -\int_0^T \big(u(t), \Phi'(t)\big)dt - \big(u(0), \Phi(0)\big). \tag{4.76}$$

Combining (4.76) with (4.75), taking into account that $u' = -Au - \bar{\Xi} + f$ and the definition of Φ we arrive at

$$(u_0 - u(0), v) = 0 \quad \forall v \in V. \tag{4.77}$$

Then due to the density of V in H we obtain $u(0) = u_0$.

It remains to verify that

$$\Xi(x,t) \in \hat{b}(x, \Pi u(x,t)) \quad \text{a.e. in } \omega_T. \tag{4.78}$$

From (4.15) and (4.38)–(4.40) we have that

$$P_h(\Pi u_{h,k}^\theta) \to \Pi u \quad \text{in } L^2(\omega_T), \tag{4.79}$$

$$\Xi_{h,k}^\theta \rightharpoonup \Xi \quad \text{in } L^2(\omega_T). \tag{4.80}$$

From (4.79) it follows (by passing to a subsequence if necessary) that $P_h(\Pi u_{h,k}^\theta)$ tends to Πu a.e. in ω_T. Then (4.78) can be proven exactly in the same way as in the static case.

The case $0 \le \theta < \frac{1}{2}$

Step I: A priori estimates. We proceed as in the previous case. First, we substitute $v = u_{h,k}^{i+1}$ into $(P)_{h,k}^{\theta}$ yielding

$$
\left(\frac{u_{h,k}^{i+1} - u_{h,k}^i}{k}, u_{h,k}^{i+1} \right) + a_h(u_{h,k}^{i+\theta}, u_{h,k}^{i+1}) \tag{4.81}
$$

$$
+ \int_{\omega} \Xi_{h,k}^{i+1} P_h(\Pi u_{h,k}^{i+1}) dx = \left\langle f_{h,k}^{i+\theta}, u_{h,k}^{i+1} \right\rangle_h.
$$

Then, using (4.25), the first term in (4.81) is equivalent to

$$
\frac{1}{k} \left(u_{h,k}^{i+1} - u_{h,k}^i, u_{h,k}^{i+1} \right) = \frac{1}{2k} \left(|u_{h,k}^{i+1}|^2 - |u_{h,k}^i|^2 + |u_{h,k}^{i+1} - u_{h,k}^i|^2 \right). \tag{4.82}
$$

On the other hand, the second term in (4.81) can be estimated from below by

$$
a_h(u_{h,k}^{i+\theta}, u_{h,k}^{i+1}) = \theta a_h(u_{h,k}^{i+1}, u_{h,k}^{i+1}) \tag{4.83}
$$

$$
+ (1-\theta) a_h(u_{h,k}^i, u_{h,k}^i) + (1-\theta) a_h(u_{h,k}^i, u_{h,k}^{i+1} - u_{h,k}^i)
$$

$$
\ge \theta \tilde{\alpha} \|u_{h,k}^{i+1}\|^2 + (1-\theta) \tilde{\alpha} \|u_{h,k}^i\|^2
$$

$$
- \frac{1}{4} \tilde{\alpha}(1-\theta) \|u_{h,k}^i\|^2 - (1-\theta) \frac{\tilde{M}^2}{\tilde{\alpha}} s(h)^2 |u_{h,k}^{i+1} - u_{h,k}^i|^2,
$$

making use of (3.25) and the estimate:

$$
|a_h(u_{h,k}^i, u_{h,k}^{i+1} - u_{h,k}^i)| \le \tilde{M} \|u_{h,k}^i\| \|u_{h,k}^{i+1} - u_{h,k}^i\| \tag{4.84}
$$

$$
\le \frac{1}{4} \tilde{\alpha} \|u_{h,k}^i\|^2 + \frac{\tilde{M}^2}{\tilde{\alpha}} \|u_{h,k}^{i+1} - u_{h,k}^i\|^2
$$

$$
\le \frac{1}{4} \tilde{\alpha} \|u_{h,k}^i\|^2 + \frac{\tilde{M}^2}{\tilde{\alpha}} s(h)^2 |u_{h,k}^{i+1} - u_{h,k}^i|^2,
$$

which follows from (3.24),(4.14) and (4.26). From (4.9),(4.14) and (4.26) we obtain that

$$
\langle f_{h,k}^{i+\theta}, u_{h,k}^{i+1} \rangle_h = \theta \langle f_{h,k}^{i+\theta}, u_{h,k}^{i+1} \rangle_h \tag{4.85}
$$

$$
+ (1-\theta) \langle f_{h,k}^{i+\theta}, u_{h,k}^i \rangle_h + (1-\theta) \langle f_{h,k}^{i+\theta}, u_{h,k}^{i+1} - u_{h,k}^i \rangle_h
$$

$$
\le \theta C \|u_{h,k}^{i+1}\| + (1-\theta) C \|u_{h,k}^i\| + (1-\theta) C \|u_{h,k}^{i+1} - u_{h,k}^i\|
$$

$$
\le C(\tilde{\alpha}, \varepsilon) + \frac{1}{4} \theta \tilde{\alpha} \|u_{h,k}^{i+1}\|^2 + \frac{1}{4} (1-\theta) \tilde{\alpha} \|u_{h,k}^i\|^2
$$

$$
+ (1-\theta) \varepsilon s(h)^2 |u_{h,k}^{i+1} - u_{h,k}^i|^2
$$

holds for any $\varepsilon > 0$. Above we also used the estimate

$$
(1-\theta) C \|u_{h,k}^{i+1} - u_{h,k}^i\| \le C(\varepsilon) + (1-\theta) \varepsilon \|u_{h,k}^{i+1} - u_{h,k}^i\|^2
$$

$$
\le C(\varepsilon) + (1-\theta) \varepsilon s(h)^2 |u_{h,k}^{i+1} - u_{h,k}^i|^2.
$$

Finally, by using (4.1) we estimate the third term in (4.81):

$$\left| \int_\omega \Xi_{h,k}^{i+\theta} P_h(\Pi u_{h,k}^{i+1}) dx \right| \tag{4.86}$$

$$\leq C + C|P_h(\Pi u_{h,k}^{i+1})|_{0,\omega}^2 + C|P_h(\Pi u_{h,k}^i)|_{0,\omega}^2.$$

From (4.14),(4.16),(4.17) we have that

$$|P_h(\Pi v_h)|_{0,\omega} \leq c(h)\|v_h\| + C|v_h| \tag{4.87}$$

$$\leq c(h)s(h)|v_h| + C|v_h| \leq C|v_h| \quad \forall v_h \in V_h.$$

From this, (4.86) can be estimated from above by

$$\left| \int_\omega \Xi_{h,k}^{i+\theta} P_h(\Pi u_{h,k}^{i+1}) dx \right| \leq C + C|u_{h,k}^{i+1}|^2 + C|u_{h,k}^i|^2. \tag{4.88}$$

Summing up (4.81) from $i = 0$ to $i = n - 1$, $n \leq r$, multiplying by $2k$ and taking into account (4.18) and (4.82)–(4.88) we get that

$$(1 - Ck)|u_{h,k}^n|^2 + \theta\tilde{\alpha} \sum_{i=0}^{n-1} k\|u_{h,k}^{i+1}\|^2 + (1 - \theta)\tilde{\alpha} \sum_{i=0}^{n-1} k\|u_{h,k}^i\|^2 \tag{4.89}$$

$$+ \left(1 - 2(1 - \theta)ks(h)^2(\frac{\tilde{M}^2}{\tilde{\alpha}} + \varepsilon)\right) \sum_{i=0}^{n-1} |u_{h,k}^{i+1} - u_{h,k}^i|^2$$

$$\leq C(\varepsilon) + C \sum_{i=0}^{n-1} k|u_{h,k}^i|^2.$$

Next we restrict ourselves to such h,k that $1 - Ck > 0$ and the stability condition (4.24) is satisfied. Then the discrete Gronwall's inequality implies the following *a priori* estimates:

$$\max_{1 \leq i \leq r} |u_{h,k}^i|, |u_{h,k}^{i-1+\theta}| \leq C, \tag{4.90}$$

$$\|u_{h,k}^\theta\|_{L^2(0,T;V)}^2 = \sum_{i=0}^{r-1} k\|u_{h,k}^{i+\theta}\|^2 \leq C, \tag{4.91}$$

$$\sum_{i=0}^{r-1} |u_{h,k}^{i+1} - u_{h,k}^i|^2 \leq C. \tag{4.92}$$

Similarly as in (4.37) we can conclude that

$$\|\Xi_{h,k}^\theta\|_{L^2(\omega_T)} \leq C. \tag{4.93}$$

The rest of the proof proceeds exactly in the same way as we did in the case of $\frac{1}{2} \leq \theta \leq 1$.   □

Remark 4.9 *From the proof of Theorem 4.1 we see that for any $\theta \in [0, \frac{1}{2}) \cup (\frac{1}{2}, 1]$ it holds:*

$$\bar{u}_{h,k} \to u \quad in \ L^2(0, T; H)$$

and

$$\bar{u}_{h,k} \rightharpoonup u \quad in \ L^2(0, T; H)$$

if $\theta = \frac{1}{2}$, where $\bar{u}_{h,k}$ is the piecewise linear function defined by (4.64).

Next, we shall study problems (P) and $(P)^\theta_{h,k}$ provided that the bilinear form a and its approximations $\{a_h\}$ are *symmetric*. This enables us to improve the convergence results of Theorem 4.1 and to increase the regularity of the solution (u, Ξ).

First, we strengthen the assumptions concerning of f and its approximations $\{f^i_{h,k}\}$:

$$f \in L^2(0, T; H), \tag{4.94}$$

and the approximate functions $\{f^i_{h,k}\}^r_{i=0}$, $f^i_{h,k} \in H$, are uniformly bounded with respect to h and k in the following sense:

$$|(f^i_{h,k}, v_h)_h| \le C|v_h| \quad \forall v_h \in V_h \text{ and } i = 0, ..., r, \tag{4.95}$$

and the sequence $\{f^\theta_{h,k}\}$ satisfies:

$$v_h \to v \text{ in } L^2(0, T; H), \ v_h \in L^2(0, T; V_h) \tag{4.96}$$

$$\Longrightarrow \int_0^T (f^\theta_{h,k}(t), v_h(t))_h \, dt \to \int_0^T (f(t), v(t)) \, dt, \text{ as } h, k \to 0+,$$

where $(\cdot, \cdot)_h$ denotes the approximation of the inner product $(\cdot, \cdot)$.

Finally, in addition to (4.18) we assume that

$$\text{the sequence } \{u_{0h}\} \text{ is bounded in } V. \tag{4.97}$$

Theorem 4.2 *Let all the assumptions of Theorem 4.1 and (4.94)-(4.97) be satisfied, and a, $\{a_h\}$ be symmetric. Let $\{(u^\theta_{h,k}, \Xi^\theta_{h,k})\}$ be a sequence of solutions to $(P)^\theta_{h,k}$. Then:*

a) $\theta \in [1/2, 1]$: There exist: a subsequence of $\{(u^\theta_{h,k}, \Xi^\theta_{h,k})\}$ and an element $(u, \Xi) \in L^\infty(0, T; V) \cap H^1(0, T; H) \times L^\infty(0, T; L^2(\omega))$ such that

$$u^\theta_{h,k}, \bar{u}_{h,k} \stackrel{*}{\rightharpoonup} u \quad in \ L^\infty(0, T; V), \tag{4.98}$$

$$u^\theta_{h,k} \to u \quad in \ L^2(0, T; V), \tag{4.99}$$

$$\bar{u}_{h,k} \to u \quad in \ L^2(0, T; V), \quad (\theta \in (1/2, 1]) \tag{4.100}$$

$$\bar{u}_{h,k} \rightharpoonup u \quad in \ H^1(0, T; H), \tag{4.101}$$

$$\bar{u}_{h,k} \to u \quad in \ C([0, T]; H), \tag{4.102}$$

$$\Xi^\theta_{h,k} \stackrel{*}{\rightharpoonup} \Xi \quad in \ L^\infty(0, T; L^2(\omega)), \ h, k \to 0+, \tag{4.103}$$

where $\bar{u}_{h,k}$ is defined by (4.64). Moreover, (u, Ξ) is a solution of (P).

b) $\theta \in [0, 1/2]$: If, moreover, the pairs (h, k) satisfy the stability condition (4.24) and

$$1 - \frac{(1-2\theta)}{2}\tilde{M}ks(h)^2 \geq c > 0, \tag{4.104}$$

where c is a positive constant, then the conclusions of the case a) except (4.100) hold true.

Proof: The (first) *a priori* estimates derived when proving Theorem 4.1 remain valid. Due to the symmetry of $\{a_h\}$ we are able to obtain the second *a priori* estimates. Again the proof is divided into two cases: $\theta \in [\frac{1}{2}, 1]$ and $\theta \in [0, \frac{1}{2})$.

The case $\frac{1}{2} \leq \theta \leq 1$:

First, we substitute $v = u_{h,k}^{i+1} - u_{h,k}^i$ into $(P)_{h,k}^\theta$ and sum it from $i = 0$ to $i = n - 1$, $1 \leq n \leq r$. Then we estimate term by term. We see that

$$\sum_{i=1}^{n-1}\left(\frac{u_{h,k}^{i+1} - u_{h,k}^i}{k}, u_{h,k}^{i+1} - u_{h,k}^i\right) = \sum_{i=0}^{n-1}k\left|\frac{u_{h,k}^{i+1} - u_{h,k}^i}{k}\right|^2. \tag{4.105}$$

Using again (4.25), the symmetry of a_h, (3.25) and the fact that $\{u_{0h}\}$ is bounded in V, we obtain:

$$\sum_{i=0}^{n-1} a_h(u_{h,k}^{i+\theta}, u_{h,k}^{i+1} - u_{h,k}^i) \tag{4.106}$$

$$= \sum_{i=0}^{n-1}\left\{\frac{1}{2}\left(a_h(u_{h,k}^{i+1}, u_{h,k}^{i+1}) - a_h(u_{h,k}^i, u_{h,k}^i)\right)\right.$$

$$\left. + \frac{2\theta - 1}{2}a_h(u_{h,k}^{i+1} - u_{h,k}^i, u_{h,k}^{i+1} - u_{h,k}^i)\right\}$$

$$\geq \frac{\tilde{\alpha}}{2}\|u_{h,k}^n\|^2 + \frac{(2\theta - 1)\tilde{\alpha}}{2}\sum_{i=0}^{n-1}\|u_{h,k}^{i+1} - u_{h,k}^i\|^2 - C.$$

Next, the Cauchy–Schwarz inequality, (4.26) and (4.95) yield

$$\sum_{i=0}^{n-1}(f_{h,k}^{i+\theta}, u_{h,k}^{i+1} - u_{h,k}^i)_h \leq \sum_{i=0}^{n-1}C|u_{h,k}^{i+1} - u_{h,k}^i| \tag{4.107}$$

$$= \sum_{i=0}^{n-1}\left(Ck^{\frac{1}{2}}\right)\left(k^{\frac{1}{2}}\left|\frac{u_{h,k}^{i+1} - u_{h,k}^i}{k}\right|\right) \leq C(\varepsilon) + \varepsilon\sum_{i=0}^{n-1}k\left|\frac{u_{h,k}^{i+1} - u_{h,k}^i}{k}\right|^2.$$

Similarly, we deduce from (4.1),(4.16),(4.87) and from the first *a priori* estimates (4.34) that

$$\sum_{i=0}^{n-1} \int_\omega \Xi_{h,k}^{i+\theta} P_h(\Pi(u_{h,k}^{i+1} - u_{h,k}^i))dx \tag{4.108}$$

$$\leq \sum_{i=0}^{n-1} |\Xi_{h,k}^{i+\theta}|_{0,\omega} |P_h(\Pi(u_{h,k}^{i+1} - u_{h,k}^i))|_{0,\omega}$$

$$\leq \sum_{i=0}^{n-1} C(1 + |P_h(\Pi u_{h,k}^{i+\theta})|_{0,\omega})|u_{h,k}^{i+1} - u_{h,k}^i|$$

$$\leq C\Big(\sum_{i=0}^{n-1} k(1 + |u_{h,k}^{i+\theta}|)^2\Big)^{\frac{1}{2}} \Big(\sum_{i=0}^{n-1} k\Big|\frac{u_{h,k}^{i+1} - u_{h,k}^i}{k}\Big|^2\Big)^{\frac{1}{2}}$$

$$\leq C(\varepsilon)\Big(1 + \sum_{i=0}^{n-1} k|u_{h,k}^{i+\theta}|^2\Big) + \varepsilon \sum_{i=0}^{n-1} k\Big|\frac{u_{h,k}^{i+1} - u_{h,k}^i}{k}\Big|^2$$

$$\leq C(\varepsilon) + \varepsilon \sum_{i=0}^{n-1} k\Big|\frac{u_{h,k}^{i+1} - u_{h,k}^i}{k}\Big|^2.$$

From (4.105)–(4.108) it follows that

$$(1 - 2\varepsilon)\sum_{i=0}^{n-1} k\Big|\frac{u_{h,k}^{i+1} - u_{h,k}^i}{k}\Big|^2 + \frac{\bar{\alpha}}{2}\|u_{h,k}^n\|^2 \tag{4.109}$$

$$+\frac{(2\theta - 1)\tilde{\alpha}}{2} \sum_{i=0}^{n-1} \|u_{h,k}^{i+1} - u_{h,k}^i\|^2 \leq C(\varepsilon).$$

This implies the second *a priori* estimates:

$$\sum_{i=0}^{r-1} k\Big|\frac{u_{h,k}^{i+1} - u_{h,k}^i}{k}\Big|^2 \leq C, \tag{4.110}$$

$$\max_{1\leq i\leq r} \|u_{h,k}^i\|, \|u_{h,k}^{(i-1)+\theta}\| \leq C, \tag{4.111}$$

$$(2\theta - 1)\sum_{i=0}^{r-1} \|u_{h,k}^{i+1} - u_{h,k}^i\|^2 \leq C \tag{4.112}$$

provided that h, k are enough small (see step I in the proof of Theorem 4.1).

From (4.111) it follows that $\{u_{h,k}^\theta\}$, $\{\bar{u}_{h,k}\}$ are bounded in $L^\infty(0,T;V)$ and (4.110) together with (4.111) yield the boundedness of $\{\bar{u}_{h,k}\}$ in $H^1(0,T;H)$. From this the following convergence results follow:

$$u_{h,k}^\theta \overset{*}{\rightharpoonup} u \quad \text{in } L^\infty(0,T;V); \tag{4.113}$$

$$u_{h,k}^\theta \to u \quad \text{in } L^2(0,T;H); \tag{4.114}$$

$$\bar{u}_{h,k} \overset{*}{\rightharpoonup} \bar{u} \quad \text{in } L^\infty(0,T;V); \tag{4.115}$$

$$\bar{u}_{h,k} \rightharpoonup \bar{u} \quad \text{in } H^1(0,T;H), \tag{4.116}$$

as $h, k \to 0+$ with u being the solution of (P). Recall that (4.114) has been already proven in Theorem 4.1. Moreover, (4.115),(4.116) and the fact that $H^1(0, T; H) \hookrightarrow C([0, T]; H)$ yield that $\bar{u}_{h,k}(t) \to \bar{u}(t)$ in $L^2(\Omega)$ a.e. in $[0, T]$. This and Proposition 1.7 imply that

$$\bar{u}_{h,k} \to \bar{u} \quad \text{in } L^2(0, T; H). \tag{4.117}$$

The fact that $u = \bar{u}$ follows from Remark 4.9. The assertions (4.98) and (4.101) are proven.

By virtue of (4.1),(4.16) and (4.111) we easily get that

$$\max_{1 \le i \le r} |\Xi_{h,k}^{i+\theta}|_{0,\omega} \le C$$

so that

$$\Xi_{h,k}^\theta \stackrel{*}{\rightharpoonup} \Xi \quad \text{in } L^\infty(0, T; L^2(\omega)). \tag{4.118}$$

Thus, (4.103) holds.

In order to prove (4.102), i.e.,

$$\bar{u}_{h,k} \to u \quad \text{in } C([0, T]; H) \tag{4.119}$$

we use the relation

$$\frac{1}{2}|\bar{u}_{h,k}(t) - u(t)|^2 - \frac{1}{2}|u_{0h} - u(0)|^2 \tag{4.120}$$

$$= \int_0^t (\bar{u}_{h,k}' - u', \bar{u}_{h,k} - u)ds.$$

From (4.120) we easily obtain that (see (4.18),(4.116),(4.117)):

$$\frac{1}{2}|\bar{u}_{h,k}(t) - u(t)|^2 \le \frac{1}{2}|u_{0h} - u_0|^2$$

$$+ \|\bar{u}_{h,k}' - u'\|_{L^2(0,T;H)} \|\bar{u}_{h,k} - u\|_{L^2(0,T;H)} \to 0,$$

uniformly with respect to $t \in (0, T]$ as $h, k \to 0+$. Thus (4.102) is also verified.

It remains to establish the strong convergence of $\{u_{h,k}^\theta\}$ and $\{\bar{u}_{h,k}\}$ to u in the $L^2(0, T; V)$-norm.

We start with the sequence $\{u_{h,k}^\theta\}$. Due to (4.13) and the density result for polynomials of the form $a_0 + a_1 t + \ldots + a_n t^n$, $a_i \in V$ and $n \in \mathbf{N}$, in $L^2(0, T; V)$ it also holds that $\cup_h L^2(0, T; V_h)$ is dense in $L^2(0, T; V)$. Hence, there exists a sequence $\{\tilde{u}_{h,k}\}$, $\tilde{u}_{h,k} \in L^2(0, T; V_h)$, converging to u strongly in $L^2(0, T; V)$.

Now, from (3.25) we see that

$$\tilde{\alpha}\|\tilde{u}_{h,k} - u_{h,k}^\theta\|_{L^2(0,T;V)}^2 \tag{4.121}$$

$$\le \int_0^T a_h(\tilde{u}_{h,k}(t) - u_{h,k}^\theta(t), \tilde{u}_{h,k}(t) - u_{h,k}^\theta(t))dt$$

$$= \int_0^T a_h(\tilde{u}_{h,k}(t), \tilde{u}_{h,k}(t))dt - 2\int_0^T a_h(u_{h,k}^\theta(t), \tilde{u}_{h,k}(t))dt$$

$$+ \int_0^T a_h(u_{h,k}^\theta(t), u_{h,k}^\theta(t))dt.$$

Then, recalling (4.20),(4.113) we have:

$$\int_0^T a_h(\tilde{u}_{h,k}(t), \tilde{u}_{h,k}(t))dt \to \int_0^T a(u(t), u(t))dt, \qquad (4.122)$$

$$2\int_0^T a_h(u_{h,k}^\theta(t), \tilde{u}_{h,k}(t))dt \to 2\int_0^T a(u(t), u(t))dt.$$

Since $u_{h,k}^\theta, u$ solves (P)$_{h,k}^\theta$,(P), respectively, and (4.15),(4.96),(4.114), (4.116), (4.118) hold, we find that

$$\int_0^T a_h(u_{h,k}^\theta(t), u_{h,k}^\theta(t))dt = -\int_0^T ((\tilde{u}_{h,k})'(t), u_{h,k}^\theta(t))dt \qquad (4.123)$$

$$-\int_0^T (\Xi_{h,k}^\theta(t), P_h(\Pi u_{h,k}^\theta)(t))_{0,\omega} dt + \int_0^T (f_{h,k}^\theta(t), u_{h,k}^\theta(t))_h dt$$

$$\to -\int_0^T (u'(t), u(t))dt - \int_0^T (\Xi(t), \Pi u(t))_{0,\omega} dt$$

$$+\int_0^T (f(t), u(t))dt = \int_0^T a(u(t), u(t))dt.$$

Combining (4.122) with (4.123) we conclude from (4.121) that

$$\|\tilde{u}_{h,k} - u_{h,k}^\theta\|_{L^2(0,T;V)} \to 0 \qquad (4.124)$$

as $h, k \to 0+$, implying (4.99).

On the other hand, if $\theta \in (\frac{1}{2}, 1]$ we deduce from (4.112) (cf. (4.67)) that also $\{\tilde{u}_{h,k}\}$ converges to u strongly in $L^2(0, T; V)$, i.e., (4.100). Thus, the proof in the case of $\theta \in [1/2, 1]$ is complete.

The case $0 \le \theta < \frac{1}{2}$:

The only difference is the treatment of the term $\sum_{i=0}^{n-1} a_h(u_{h,k}^{i+\theta}, u_{h,k}^{i+1} - u_{h,k}^i)$: Using (4.14) we obtain that

$$\sum_{i=0}^{n-1} a_h(u_{h,k}^{i+\theta}, u_{h,k}^{i+1} - u_{h,k}^i) \qquad (4.125)$$

$$= \sum_{i=0}^{n-1} \left\{ \frac{1}{2}\left(a_h(u_{h,k}^{i+1}, u_{h,k}^{i+1}) - a_h(u_{h,k}^i, u_{h,k}^i)\right) \right.$$

$$\left. +\frac{2\theta-1}{2} a_h(u_{h,k}^{i+1} - u_{h,k}^i, u_{h,k}^{i+1} - u_{h,k}^i) \right\}$$

$$\ge \frac{\tilde{\alpha}}{2}\|u_{h,k}^n\|^2 - \frac{(1-2\theta)\tilde{M}}{2} \sum_{i=0}^{n-1}\|u_{h,k}^{i+1} - u_{h,k}^i\|^2 - C.$$

$$\ge \frac{\tilde{\alpha}}{2}\|u_{h,k}^n\|^2 - \frac{(1-2\theta)\tilde{M}}{2} s(h)^2 k \sum_{i=0}^{n-1} k\left|\frac{u_{h,k}^{i+1} - u_{h,k}^i}{k}\right|^2 - C.$$

Therefore, the inequality (4.109) is now replaced by

$$(1 - \frac{(1-2\theta)}{2}\tilde{M}ks(h)^2 - 2\varepsilon)\sum_{i=0}^{n-1}k\left|\frac{u_{h,k}^{i+1} - u_{h,k}^i}{k}\right|^2 \qquad (4.126)$$

$$+\frac{\tilde{\alpha}}{2}\|u_{h,k}^n\|^2 \le C(\varepsilon).$$

Taking into account the stability assumption (4.104) we can make the same conclusions as for the case $\theta \in [\frac{1}{2}, 1]$ except the a priori estimate (4.112). $\square$

Remark 4.10 *The results of Theorems 4.1 and 4.2 cannot be extended to the case when $\omega \subseteq \partial\Omega$. Note that the first a priori estimates can be proven by assuming (3.5) (cf. the treatment of the nonmonotone term in Section 3.2). But in order to guarantee the strong convergence of $P_h(\Pi u_{h,k}^\theta)$ to Πu in $L^2(\omega_T)$ (which is needed to establish the inclusion $\Xi(x,t) \in \hat{b}(x, \Pi u(x,t))$ in ω_T, cf. (4.79)) we need to prove also higher order a priori estimates (see the discussion in Miettinen and Panagiotopoulos, 1999). Now the difficulty arises from (4.16), in which one cannot anymore assume that $c(h) \to 0$ as $h \to 0+$. This fact was necessary for the derivation of the second a priori estimates (cf. (4.108)). Recall that if b is monotone (with respect to the second variable) it is possible to solve the above problems (see Trémoliéres, 1972, Glowinski et al., 1981).*

We close this section by showing how these abstract results can be applied to the approximation of Example 4.1 (with $\omega = \Omega$). First, we recall its formulation:

$$\begin{cases} \text{Find } (u, \Xi) \in W(V) \times L^2(\Omega_T) \text{ such that} \\ \langle u'(t), v\rangle + \int_\Omega \nabla u(t) \cdot \nabla v dx + + \int_\Omega \Xi(t)v dx \\ = \int_\Omega f(t)v dx \quad \forall v \in V, \text{ for a.a. } t \in (0,T), \\ \Xi(x,t) \in \hat{b}(x, u(x,t)) \quad \text{for a.a. } (x,t) \in \Omega_T \equiv \Omega \times (0,T) \\ \text{and } u(0) = u_0, \end{cases} \qquad (4.127)$$

where

Ω is a polygonal (polyhedral) domain in $\mathbf{R}^2$ ($\mathbf{R}^3$),

$V = H_0^1(\Omega)$, $\quad H = L^2(\Omega)$, $\quad \Pi v = v|_\Omega$, $u_0 \in H^2(\Omega)$, $f \in C(\overline{\Omega}_T)$.

Let $\{\mathcal{D}_h\}$ be a *strongly regular* family of triangulations of $\overline{\Omega}$ and $\{\mathcal{T}_h\}$ a family of partitions made of Type II elements. Then the approximations of the spaces V, $W \equiv \Pi(V) = V$ and $Y \equiv L^2(\Omega)$ are defined as follows:

$$V_h \equiv W_h = \{v_h \in C(\overline{\Omega}) \mid v_h|_D \in P_1(D) \quad \forall D \in \mathcal{D}_h, \ v_h = 0 \text{ on } \partial\Omega\},$$
$$Y_h = \{\mu_h \in L^\infty(\Omega) \mid \mu_h|_K \in P_0(K) \quad \forall K \in \mathcal{T}_h\}.$$

Now recalling the proof of Theorem 1.52 and Lemma 1.5 we see that the family $\{V_h\}$ satisfies (4.13) and (4.14) with $s(h) = \hat{c}h^{-1}$, $\hat{c}$ a positive constant. Setting $u_{0h} = r_h u_0$ with r_h being the piecewise linear lagrange interpolate operator, we can easily verify that $\{u_{0h}\}$ satisfies (4.18) and (4.97) as follows from (1.254) and the regularity assumption on u_0. The bilinear form a can be evaluated exactly (the integrand is piecewise constant) and, therefore, the assumptions (3.24), (3.25), (3.38) are trivially satisfied.

The approximation of the function f is defined by means of the integral mean value operator:

$$f_{h,k}^{\theta} = \sum_{i=0}^{r-1} ((1-\theta)f_{h,k}^i + \theta f_{h,k}^{i+1})\lambda^{i+1}, \tag{4.128}$$

where

$$f_{h,k}^i = \frac{1}{k}\int_{k(i-\frac{1}{2})}^{k(i+\frac{1}{2})} f(t)dt, \quad i = 1,...,r-1,$$

$$f_{h,k}^0 = \frac{2}{k}\int_0^{\frac{1}{2}k} f(t)dt, \quad f_{h,k}^r = \frac{2}{k}\int_{T-\frac{1}{2}k}^{T} f(t)dt.$$

Denote by $\tilde{\Delta}_i$ the intervals, over which the mean value is computed. The integral $\int_\Omega f_{h,k}^{\theta}(t)v_h dx$ will be approximated by the following quadrature formula

$$\int_\Omega f_{h,k}^{\theta}(t)v_h dx \approx \sum_D \text{meas } D(f_{h,k}^{\theta}(t)v_h)(Q_D) \equiv (f_{h,k}^{\theta}(t), v_h)_h,$$

in which Q_D is the barycentre of D. Let us check condition (4.95). It holds that

$$|(f_{h,k}^i, v_h)_h| = \frac{1}{k}\left|(\int_{\tilde{\Delta}_i} f(s)ds, v_h)_h\right| \tag{4.129}$$

$$\leq \frac{1}{k}\int_{\tilde{\Delta}_i} |(f(s), v_h)_h|ds = \frac{1}{k}\int_{\tilde{\Delta}_i}\left|\sum_{D\in\mathcal{D}_h}(f(s)v_h)(Q_D)\text{ meas }D\right|ds$$

$$\leq \frac{c}{k}\int_{\tilde{\Delta}_i}\sum_{D\in\mathcal{D}_h}|v_h(Q_D)|\text{ meas }Dds \leq c\sum_{D\in\mathcal{D}_h}\text{ meas }D|v_h(Q_D)|$$

$$\leq c(\text{meas }\Omega)^{\frac{1}{2}}\|v_h\|_{0,\Omega}.$$

Here we used the boundedness of f in $\overline{\Omega}_T$, and the last inequality can be proven in a similar way as in Section 3.4.

For the verification of (4.96) we employ a similar approach as in (3.85)-(3.87). Let $v_h \to v$ in $L^2(0, T; H)$, $v_h \in L^2(0, T; V_h)$. First, we write

$$\int_0^T (f_{h,k}^\theta(t), v_h(t))_h - (f(t), v(t))dt \qquad (4.130)$$

$$= \int_0^T (f_{h,k}^\theta(t), v_h(t))_h - (f(t), v_h(t))dt$$

$$+ \int_0^T (f(t), v_h(t)) - (f(t), v(t))dt.$$

The second term on the right hand side of (4.130) obviously tends to zero as $h \to 0+$. The analysis of the first term is more detailed. From the definition of $(\cdot, \cdot)_h$ it follows that

$$(f_{h,k}^\theta(t), v_h(t))_h - (f(t), v_h(t)) \qquad (4.131)$$

$$= \sum_{D \in \mathcal{D}_h} \{ \text{meas } D(f_{h,k}^\theta(t)v_h(t))(Q_D) - \int_D f(x, t)v_h(x, t)dx \}$$

holds for any $t \in [0, T]$. Let $t \in [(i-1)k, ik] \cap \tilde{\Delta}_i$ be fixed (the case $t \in [(i-1)k, ik] \cap \tilde{\Delta}_{i-1}$ can be done in a similar way). We shall estimate each term behind the sum:

$$\text{meas } D(f_{h,k}^\theta(t)v_h(t))(Q_D) - \int_D f(x, t)v_h(x, t)dx \qquad (4.132)$$

$$= (1 - \theta)\{ \text{meas } D(f_{h,k}^{i-1}v_h(t))(Q_D) - \int_D f(x, t)v_h(x, t)dx \}$$

$$+ \theta\{ \text{meas } D(f_{h,k}^i v_h(t))(Q_D) - \int_D f(x, t)v_h(x, t)dx \}$$

From this we see that it is sufficient to estimate the term of the form:

$$| \text{meas } D(f_{h,k}^i v_h(t))(Q_D) - \int_D f(x, t)v_h(x, t)dx| \qquad (4.133)$$

$$= |\frac{1}{k} \int_{\tilde{\Delta}_i} f(Q_D, s)ds \int_D v_h(x, t)dx - \int_D f(x, t)v_h(x, t)dx|$$

$$= |f(Q_D, \tilde{s}) \int_D v_h(x, t)dx - \int_D f(x, t)v_h(x, t)dx|$$

$$\leq \int_D |f(Q_D, \tilde{s}) - f(x, t)||v_h(x, t)|dx$$

$$\leq \max_{x \in D, \tilde{t} \in [(i-1)k, (i+1)k]} |f(Q_D, \tilde{s}) - f(x, \tilde{t})|(\text{meas } D)^{\frac{1}{2}}\|v_h(t)\|_{0, D},$$

for some $\tilde{s} \in \tilde{\Delta}_i$ as follows from the integral mean value theorem applied to $f \in C(\overline{\Omega}_T)$. Let $\varepsilon > 0$ be given. Then for h, k sufficiently small one has:

$$\max_{x \in D, \tilde{t} \in [(i-1)k, (i+1)k]} |f(Q_D, \tilde{s}) - f(x, \tilde{t})| \leq \varepsilon$$

because of the uniform continuity of f. From this (4.131)-(4.133) we conclude that (see also (3.87))

$$|(f^\theta_{h,k}(t), v_h(t))_h - (f(t), v_h(t))_h|$$
$$\leq c(\theta)\varepsilon(\text{meas }\Omega)^{\frac{1}{2}}\|v_h(t)\|_{0,\Omega}.$$

Thus, the first term on the right hand side of (4.130) can be estimated as follows:

$$|\int_0^T (f^\theta_{h,k}(t), v_h(t))_h - (f(t), v_h(t))dt|$$
$$\leq c(\theta)\varepsilon\|v_h\|_{L^2(0,T;H)} \leq c\varepsilon.$$

This completes the proof of (4.96).

The linear mapping $P_h : W_h \to Y_h$ will be defined by means of the piecewise constant lagrange interpolation operator (3.59). We check the assumptions (4.15)-(4.17). First, recall the error estimate

$$\|P_h w_h - w_h\|_{L^p(\Omega)} \leq ch\|\nabla w_h\|_{L^p(\Omega)} \quad \forall w_h \in W_h, \ \forall p \in [1,\infty],$$

which was proven in Section 3.4. Then from the definition of Π it follows that

$$\begin{aligned}\|P_h(\Pi v_h)\|_{0,\Omega} &\leq \|P_h v_h - v_h\|_{0,\Omega} + \|v_h\|_{0,\Omega}\\ &\leq ch\|\nabla v_h\|_{0,\Omega} + \|v_h\|_{0,\Omega},\end{aligned}$$

i.e., (4.16) holds with $c(h) = ch$ and $C = 1$.

Let $\{v_h\}$, $v_h \in L^2(0,T;V_h)$, be such that $v_h \rightharpoonup v$ in $L^2(0,T;V)$ and $v_h \to v$ in $L^2(0,T;H)$. Then

$$\begin{aligned}\|P_h(\Pi v_h) - \Pi v\|_{0,\Omega_T} &\leq \|P_h v_h - v_h\|_{0,\Omega_T} + \|v_h - v\|_{0,\Omega_T}\\ &= \left(\int_0^T \|P_h v_h(t) - v_h(t)\|^2_{0,\Omega}dt\right)^{\frac{1}{2}} + \|v_h - v\|_{0,\Omega_T}\\ &\leq \left(c^2 h^2 \int_0^T \|\nabla v_h(t)\|^2_{0,\Omega}dt\right)^{\frac{1}{2}} + \|v_h - v\|_{0,\Omega_T}\\ &\leq ch\|v_h\|_{L^2(0,T;V)} + \|v_h - v\|_{0,\Omega_T} \to 0 \quad \text{as } h \to 0+\end{aligned}$$

implying (4.15). Finally, we note that $c(h)s(h) = (ch)(\hat{c}h^{-1}) = c\hat{c}$, i.e., (4.17) holds as well. Thus all the assumptions of Theorem 4.2 are satisfied and approximate solutions $(u^\theta_{h,k}, \Xi^\theta_{h,k})$ of $(P)^\theta_{h,k}$ are close on subsequences to a solution (u, Ξ) of (P) in the sense of (4.98)-(4.103).

Remark 4.11 *Below we formulate two hemivariational inequalities of a hyperbolic type and introduce possible approximation schemes. We shall not discuss their convergence properties (as far as we know this is not still done and proofs could be rather technical).*

First, let us recall the problem studied by Pop and Panagiotopoulos, 1998:
Find $u \in L^2(0, T; V)$, *with* $u' \in L^2(0, T; V)$, $u'' \in L^2(0, T; V^*)$ *and* $\Xi \in L^2(\omega_T)$
$(\omega \subseteq \Omega)$ *such that*

$$
\begin{cases}
\langle u''(t), v \rangle + a(u(t), v) + \displaystyle\int_\omega \Xi(t)\Pi v \, dx \\
\quad = \langle f(t), v \rangle \quad \forall v \in V, \text{ for a.a. } t \in (0, T), \\
\Xi(x, t) \in \hat{b}(x, \Pi u(x, t)) \quad \text{for a.a. } (x, t) \in \omega_T \\
\text{and } u(0) = u_0, \ u'(0) = u_1.
\end{cases} \qquad (P)_1
$$

Above we assume that:

- *a* : $V \times V \to \mathbf{R}$ *is a bounded, symmetric and V-elliptic bilinear form;*

- *b satisfies (3.2), (3.3), (3.16), (4.1);*

- $f \in L^2(0, T; V^*)$;

- $u_0 \in V$, $u_1 \in H$;

- Π *is a continuous linear mapping from V to* $L^2(\omega)$.

Then it is possible to prove that problem $(P)_1$ *has at least one solution (see Pop and Panagiotopoulos, 1998). Using the (classical) approximation theory for hyperbolic problems (see, e.g., Trémoliéres, 1972; Glowinski et al., 1981) and the theory for parabolic hemivariational inequalities developed here we can define the following approximation scheme for* $(P)_1$:

$$
\begin{cases}
\text{Find } (u_{h,k}^{i+\theta}, \Xi_{h,k}^{i+\theta}) \in V_h \times Y_h \text{ for all } i = 1, ..., r - 1 \text{ such that} \\
\left(\dfrac{u_{h,k}^{i+1} - 2u_{h,k}^i + u_{h,k}^{i-1}}{k^2}, v_h \right) + a_h(u_{h,k}^{i+\theta}, v_h) + \displaystyle\int_\omega \Xi_{h,k}^{i+\theta} P_h(\Pi v_h) dx \\
\quad = \left\langle f_{h,k}^{i+\theta}, v_h \right\rangle_h \quad \forall v_h \in V_h, \\
\Xi_{h,k}^{i+\theta}(x) \in \hat{b}(\displaystyle\sum_{j=1}^m \mathcal{X}_{\text{int}_w K_j}(x) x_h^j, P_h(\Pi u_{h,k}^{i+\theta})(x)) \quad \text{for a.a. } x \in \omega, \\
u_{h,k}^0 = u_{0h} \text{ and } u_{h,k}^1 = u_{0h} + k u_{1h}.
\end{cases}
$$

By $\{u_{1h}\}$, $u_{1h} \in V_h$, *we denoted a sequence approximating* u_1. *The meaning of the other symbols remains the same as in the parabolic case.*
The second example is taken from Goeleven et al., 1999: Find $u \in L^2(0, T; V)$, *with* $u' \in L^2(0, T; V)$, $u'' \in L^2(0, T; V^*)$ *and* $\Xi \in L^2(\omega_T)$ $(\omega \subseteq \Omega)$ *such that*

$$
\begin{cases}
\langle u''(t), v \rangle + a_1(u'(t), v) + a_2(u(t), v) + \displaystyle\int_\omega \Xi(t)\Pi v \, dx \\
\quad = \langle f(t), v \rangle \quad \forall v \in V, \text{ for a.a. } t \in (0, T), \\
\Xi(x, t) \in \hat{b}(x, \Pi u'(x, t)) \quad \text{for a.a. } (x, t) \in \omega_T \\
\text{and } u(0) = u_0, \ u'(0) = u_1.
\end{cases} \qquad (P)_2
$$

Now, besides of the assumptions imposed on $(P)_1$ *we suppose:*

- both $a_1, a_2 : V \times V \to \mathbf{R}$ are bounded, symmetric and V-elliptic bilinear forms;

- $f \in L^2(0, T; H)$.

Then $(P)_2$ has a solution (see Goeleven et al., 1999). Below, we present two possible approximation schemes for the numerical realization of $(P)_2$:

$$
\begin{cases}
\text{Find } (u_{h,k}^{i+\theta}, \Xi_{h,k}^i) \in V_h \times Y_h \text{ for all } i = 1, ..., r - 1 \text{ such that} \\
\left(\dfrac{u_{h,k}^{i+1} - 2u_{h,k}^i + u_{h,k}^{i-1}}{k^2}, v_h \right) + a_{1h}\left(\dfrac{u_{h,k}^{i+1} - u_{h,k}^{i-1}}{2k}, v_h \right) + a_{2h}(u_{h,k}^{i+\theta}, v_h) \\
\quad + \displaystyle\int_\omega \Xi_{h,k}^i P_h(\Pi v_h) dx = \left\langle f_{h,k}^{i+\theta}, v_h \right\rangle_h, \quad \forall v_h \in V_h, \\
\Xi_{h,k}^i(x) \in \hat{b}\left(\displaystyle\sum_{j=1}^m \mathcal{X}_{\text{int}_\omega K_j}(x) x_h^j, P_h(\Pi(\dfrac{u_{h,k}^{i+1} - u_{h,k}^{i-1}}{2k}))(x)) \right) \quad \text{for a.a. } x \in \omega, \\
u_{h,k}^0 = u_{0h} \text{ and } u_{h,k}^1 = u_{0h} + k u_{1h}
\end{cases}
$$

or

$$
\begin{cases}
\text{Find } (u_{h,k}^{i+\theta}, \Xi_{h,k}^i) \in V_h \times Y_h \text{ for all } i = 1, ..., r - 1 \text{ such that} \\
\left(\dfrac{u_{h,k}^{i+1} - 2u_{h,k}^i + u_{h,k}^{i-1}}{k^2}, v_h \right) + a_{1h}\left(\dfrac{u_{h,k}^{i+1} - u_{h,k}^i}{k}, v_h \right) + a_{2h}(u_{h,k}^{i+\theta}, v_h) \\
\quad + \displaystyle\int_\omega \Xi_{h,k}^i P_h(\Pi v_h) dx = \left\langle f_{h,k}^{i+\theta}, v_h \right\rangle_h, \quad \forall v_h \in V_h, \\
\Xi_{h,k}^i(x) \in \hat{b}\left(\displaystyle\sum_{j=1}^m \mathcal{X}_{\text{int}_\omega K_j}(x) x_h^j, P_h(\Pi(\dfrac{u_{h,k}^{i+1} - u_{h,k}^i}{k}))(x)) \right) \quad \text{for a.a. } x \in \omega, \\
u_{h,k}^0 = u_{0h} \text{ and } u_{h,k}^1 = u_{0h} + k u_{1h}.
\end{cases}
$$

4.3 ALGEBRAIC REPRESENTATION

Following the presentation in Section 3.5 we give the algebraic form of the discrete parabolic hemivariational inequality $(P)_{h,k}^\theta$. Let the parameters $h > 0$, $k > 0$ and $\theta \in [0, 1]$ be fixed (dim $V = n$, dim $Y = m$). Below, if possible, we skip these symbols from our notations.

Recalling the conventions used in the static case we can write the discrete problem as follows:

$$
\begin{cases}
\text{Find } (u^{i+\theta}, \Xi^{i+\theta}) \in V \times Y \text{ for all } i = 0, ..., r - 1 \text{ such that} \\
\left(\dfrac{u^{i+1} - u^i}{k}, v \right) + a(u^{i+\theta}, v) + \displaystyle\int_\omega \Xi^{i+\theta} P(\Pi v) dx \\
\quad = \langle f^{i+\theta}, v \rangle \quad \forall v \in V, \\
\Xi^{i+\theta}(x^j) \in \hat{b}(x^j, P(\Pi u^{i+\theta})(x^j)) \quad \forall j = 1, ..., m \\
\text{and } u^0 = u_0.
\end{cases} \tag{P}
$$

Similarly, the *algebraic* representation of (P) takes the following form:

$$\begin{cases} \text{Find } (\vec{u}^{i+\theta}, \vec{\Xi}^{i+\theta}) \in \mathbf{R}^n \times \mathbf{R}^m \text{ for all } i = 0, ..., r-1 \text{ such that} \\ \left(\mathbf{M}\left(\dfrac{\vec{u}^{i+1} - \vec{u}^i}{k}\right), \vec{v}\right)_{\mathbf{R}^n} + \left(\mathbf{A}\vec{u}^{i+\theta}, \vec{v}\right)_{\mathbf{R}^n} + \left(\vec{\Xi}^{i+\theta}, \mathbf{P}(\Pi\vec{v})\right)_{\mathbf{R}^m} \\ = (\vec{f}^{i+\theta}, \vec{v})_{\mathbf{R}^n} \quad \forall \vec{v} \in \mathbf{R}^n \\ \Xi_j^{i+\theta} \in c_j \hat{b}(x^j, (\mathbf{P}(\Pi\vec{u}^{i+\theta})_j)) \quad \forall j = 1, ..., m \\ \text{and} \quad \vec{u}^0 = \vec{u}_0, \end{cases} \qquad (\vec{P})$$

where $\mathbf{M}$ is the standard mass matrix, $\vec{u}_0$ is the vector whose components are defined by $u_{0j} = (u_{0h}, \varphi_j)$, $j = 1, ..., n$, and the other symbols have the same meaning as in Section 3.5. From this we see that at each time level $i = 0, ...r-1$ we solve the following static discrete hemivariational inequality ($\theta \in (0, 1]$):

$$\begin{cases} \text{Find } (\vec{u}^{i+\theta}, \vec{\Xi}^{i+\theta}) \in \mathbf{R}^n \times \mathbf{R}^m \text{ such that} \\ \left((k\mathbf{A} + \dfrac{1}{\theta}\mathbf{M})\vec{u}^{i+\theta}, \vec{v}\right)_{\mathbf{R}^n} + \left(k\vec{\Xi}^{i+\theta}, \mathbf{P}(\Pi\vec{v})\right)_{\mathbf{R}^m} \\ = (k\vec{f}^{i+\theta} + \frac{1}{\theta}\mathbf{M}\vec{u}^i, \vec{v})_{\mathbf{R}^n} \quad \forall \vec{v} \in \mathbf{R}^n \\ \Xi_j^{i+\theta} \in c_j \hat{b}(x^j, (\mathbf{P}(\Pi\vec{u}^{i+\theta})_j) \quad \forall j = 1, ..., m. \end{cases} \qquad (\vec{P})_i$$

Notice that for $\theta = 0$, which corresponds to the explicit scheme, the formulation $(\vec{P})_i$ reduces to the following simple (linear) problem:

$$\begin{cases} \text{Find } (\vec{u}^{i+1}, \vec{\Xi}^i) \in \mathbf{R}^n \times \mathbf{R}^m \text{ such that} \\ (\mathbf{M}\vec{u}^{i+1}, \vec{v})_{\mathbf{R}^n} = -(k\vec{\Xi}^i, \mathbf{P}(\Pi\vec{v}))_{\mathbf{R}^m} \\ + (k\vec{f}^i + \mathbf{M}\vec{u}^i - k\mathbf{A}\vec{u}^i, \vec{v})_{\mathbf{R}^n} \quad \forall \vec{v} \in \mathbf{R}^n \\ \Xi_j^i \in c_j \hat{b}(x^j, (\mathbf{P}(\Pi\vec{u}^i)_j) \quad \forall j = 1, ..., m. \end{cases}$$

In order to solve $(\vec{P})_i$ numerically we have to find the corresponding superpotential $\mathcal{L}_i : \mathbf{R}^n \to \mathbf{R}$ (the trivial case $\theta = 0$ is excluded): First, we define

$$\mathbf{B} = k\mathbf{A} + \frac{1}{\theta}\mathbf{M};$$

$$\Psi(\vec{v}) = \sum_{j=1}^m kc_j \int_0^{(\Lambda\vec{v})_j} b(x^j, \xi)d\xi;$$

$$\vec{g}^i = k\vec{f}^{i+\theta} + \frac{1}{\theta}\mathbf{M}\vec{u}^i,$$

where the coefficients c_j are the same as in Section 3.5. Then, the superpotential $\mathcal{L}_i$ takes the form

$$\mathcal{L}_i(\vec{v}) = \frac{1}{2}(\mathbf{B}\vec{v}, \vec{v})_{\mathbf{R}^n} - (\vec{g}^i, \vec{v})_{\mathbf{R}^n} + \Psi(\vec{v}). \qquad (4.134)$$

Finally, recall that the relation between the discrete (static) hemivariational problem $(\vec{P})_i$ and the problem of finding substationary points of $\mathcal{L}_i$ has already been analysed in Theorems 3.5 and 3.6.

4.4   CONSTRAINED HEMIVARIATIONAL INEQUALITIES

The last section is devoted to extensions of results from Section 3.6 to a parabolic case: we shall study the approximation of a class of constrained parabolic inequalities.

Let K be a nonempty, closed and convex subset of V. We shall consider the following parabolic hemivariational inequalities (P):

$$
\begin{cases}
\text{Find } (u, \Xi) \in W(V) \cap L^2(0, T; K) \times L^2(\omega_T) \text{ such that} \\[4pt]
\displaystyle\int_0^T \langle u'(t), v(t) - u(t) \rangle dt + \int_0^T a(u(t), v(t) - u(t)) dt \\[8pt]
\displaystyle \quad + \int_0^T \int_\omega \Xi(t) (\Pi v(t) - \Pi u(t)) dx dt \\[8pt]
\displaystyle \geq \int_0^T \langle f(t), v(t) - u(t) \rangle dt \qquad \forall v \in L^2(0, T; K), \\[8pt]
\Xi(x, t) \in \hat{b}(x, \Pi u(x, t)) \qquad \text{for a.a. } (x, t) \in \omega_T \\[4pt]
\text{and } u(0) = u_0.
\end{cases}
$$

Here, we assume that a is *symmetric*, $u_0 \in K$ and $f \in L^2(0, T; H)$. The other assumptions remain the same as in the unconstrained case and are collected below. Also the definition of the corresponding approximation scheme $(P)^\theta_{h,k}$ is very straightforward. Indeed, a pair of functions $(u^\theta_{h,k}, \Xi^\theta_{h,k})$ is called to be a solution of the approximation scheme $(P)^\theta_{h,k}$, if their time level values $\{(u^{i+\theta}_{h,k}, \Xi^{i+\theta}_{h,k})\}_{i=0}^{r-1}$ solve the problem:

$$
\begin{cases}
\text{Find } (u^{i+\theta}_{h,k}, \Xi^{i+\theta}_{h,k}) \in K_h \times Y_h \text{ such that} \\[6pt]
\displaystyle \left(\frac{u^{i+1}_{h,k} - u^i_{h,k}}{k}, v_h - u^{i+\theta}_{h,k} \right) + a_h(u^{i+\theta}_{h,k}, v_h - u^{i+\theta}_{h,k}) \\[10pt]
\displaystyle \quad + \int_\omega \Xi^{i+\theta}_{h,k} P_h(\Pi v_h - \Pi u^{i+\theta}_{h,k}) dx \\[10pt]
\displaystyle \geq (f^{i+\theta}_{h,k}, v_h - u^{i+\theta}_{h,k})_h \quad \forall v_h \in K_h, \\[8pt]
\displaystyle \Xi^{i+\theta}_{h,k}(x) \in \hat{b}(\sum_{j=1}^m \chi_{\text{int}_\omega K_j}(x) x^j_h, P_h(\Pi u^{i+\theta}_{h,k})(x)) \quad \text{for a.a. } x \in \omega.
\end{cases}
\qquad (P)^\theta_{h,k}
$$

Below we prove the solvability of discrete problem $(P)^\theta_{h,k}$ and analyse its relation to (P) as $h, k \to 0+$. In fact, we prove the convergence only for $\theta = 1$, i.e., for the implicit scheme (for the other schemes the convergence is still open, see Remark 4.12).

For clarity, we again collect the assumptions which will be needed in what follows:

(i) the function b satisfies (3.2), (3.3), (3.16), (4.1);

(ii) the system of approximated symmetric bilinear forms $\{a_h\}$ satisfies (3.24), (3.25), (3.52);

(iii) the system of approximated linear forms $\{f^\theta_{h,k}\}$ satisfies (4.95), (4.96) and $f \in L^2(0,T;H)$;

(iv) the sequence of approximated initial conditions $\{u_{0h}\}$, $u_{0h} \in K_h$, satisfies (4.18) and is bounded in V;

(v) the mapping Π is continuous from V to $L^2(\omega)$;

(vi) the mapping P_h satisfies (4.15)-(4.17);

(vii) the system $\{V_h\}$ satisfies (4.14);

(viii) the system $\{K_h\}$ is such that $0 \in K_h$ for any $h > 0$ and the conditions (1.172) and (1.173) are satisfied.

Lemma 4.3 *Let all the assumptions (i)-(viii) be satisfied. Then $(P)^\theta_{h,k}$ has at least one solution for any h, $k > 0$ and $\theta \in (0,1]$ (k sufficiently small).*

Proof: It is enough to transform $(P)^\theta_{h,k}$ at each time level $i = 0, ..., r-1$ to a constrained static problem:

$$
\begin{cases}
\text{Find } (u^{i+\theta}_{h,k}, \Xi^{i+\theta}_{h,k}) \in K_h \times Y_h \text{ such that} \\[4pt]
ka_h(u^{i+\theta}_{h,k}, v_h - u^{i+\theta}_{h,k}) + \dfrac{1}{\theta}(u^{i+\theta}_{h,k}, v_h - u^{i+\theta}_{h,k}) \\[6pt]
\quad + k \displaystyle\int_\omega \Xi^{i+\theta}_{h,k} P_h(\Pi v_h - \Pi u^{i+\theta}_{h,k})\,dx \\[6pt]
\quad \geq k(f^{i+\theta}_{h,k}, v_h - u^{i+\theta}_{h,k})_h + \dfrac{1}{\theta}(u^i_{h,k}, v_h - u^{i+\theta}_{h,k}) \qquad \forall v_h \in K_h, \\[8pt]
\Xi^{i+\theta}_{h,k}(x) \in \hat{b}\Big(\displaystyle\sum_{j=1}^m \mathcal{X}_{\mathrm{int}_w K_j}(x) x^j_h, P_h(\Pi u^{i+\theta}_{h,k})(x)\Big) \quad \text{for a.a. } x \in \omega,
\end{cases}
$$

and then to apply Theorem 3.8. Note that the lack of the assumption (3.5) can be treated as in the proof of Lemma 4.1. $\qquad\square$

As we have mentioned above we prove the convergence of the fully implicit scheme, corresponding to $\theta = 1$. For simplicity of notations, the upper index $\theta = 1$ will be omitted, in what follows.

Theorem 4.3 *Let all the assumptions (i)-(viii) be satisfied. Let $\{(u_{h,k}, \Xi_{h,k})\}$ be a sequence of solutions to fully implicit scheme $(P)_{h,k}$. Then there exist: a subsequence of $\{(u_{h,k}, \Xi_{h,k})\}$ and an element $(u, \Xi) \in L^\infty(0,T;K) \cap H^1(0,T;H) \times L^\infty(0,T;L^2(\omega))$ such that*

$$u_{h,k}, \bar{u}_{h,k} \overset{*}{\rightharpoonup} u \quad \text{in } L^\infty(0,T;V), \tag{4.135}$$

$$u_{h,k}, \bar{u}_{h,k} \rightarrow u \quad \text{in } L^2(0,T;V), \tag{4.136}$$

$$\bar{u}_{h,k} \rightharpoonup u \quad \text{in } H^1(0,T;H), \tag{4.137}$$

$$\bar{u}_{h,k} \rightarrow u \quad \text{in } C([0,T];H), \tag{4.138}$$

$$\Xi_{h,k} \overset{*}{\rightharpoonup} \Xi \quad \text{in } L^\infty(0,T;L^2(\omega)), \ h,k \rightarrow 0+, \tag{4.139}$$

where $\bar{u}_{h,k}$ is the piecewise linear interpolate of $u_{h,k}$ defined by (4.64). More-over, (u, Ξ) is a solution of (P).

Proof: Step I: A priori estimates. First we insert $v_h = 0 \in K_h$ into (P)$_{h,k}$. We get:

$$\left(\frac{u_{h,k}^{i+1} - u_{h,k}^i}{k}, u_{h,k}^{i+1}\right) + a_h(u_{h,k}^{i+1}, u_{h,k}^{i+1})$$
$$+ \int_\omega \Xi_{h,k}^{i+1} P_h(\Pi u_{h,k}^{i+1}) dx \leq (f_{h,k}^{i+1}, u_{h,k}^{i+1})_h.$$

Proceeding in a similar way as in the proof of Theorem 4.1 (the case $\theta \in [\frac{1}{2}, 1]$) we get the first *a priori* estimates:

$$\max_{1 \leq i \leq r} |u_{h,k}^i| \leq C,$$
$$\|u_{h,k}\|_{L^2(0,T;V)} \leq C.$$

Next we substitute $v_h = u_{h,k}^i$ into (P)$_{h,k}$. Then we obtain:

$$\left(\frac{u_{h,k}^{i+1} - u_{h,k}^i}{k}, u_{h,k}^{i+1} - u_{h,k}^i\right) + a_h(u_{h,k}^{i+1}, u_{h,k}^{i+1} - u_{h,k}^i)$$
$$+ \int_\omega \Xi_{h,k}^{i+1} P_h(\Pi(u_{h,k}^{i+1} - u_{h,k}^i)) dx \leq (f_{h,k}^{i+1}, u_{h,k}^{i+1} - u_{h,k}^i)_h.$$

Using the same approach as in the proof of Theorem 4.2 (the case $[\frac{1}{2}, 1]$) we obtain the second *a priori* estimates:

$$\sum_{i=0}^{r-1} k \left|\frac{u_{h,k}^{i+1} - u_{h,k}^i}{k}\right|^2 \leq C, \tag{4.140}$$

$$\max_{1 \leq i \leq r} \|u_{h,k}^i\| \leq C, \tag{4.141}$$

$$\sum_{i=0}^{r-1} \|u_{h,k}^{i+1} - u_{h,k}^i\|^2 \leq C, \tag{4.142}$$

$$\max_{1 \leq i \leq r} |\Xi_{h,k}^i|_{0,\omega} \leq C. \tag{4.143}$$

Step II: Convergence of subsequences. From (4.140),(4.141),(4.143) we can con-clude that there exist subsequences of $\{u_{h,k}\}$, $\{\bar{u}_{h,k}\}$ and $\{\Xi_{h,k}\}$ such that

$$u_{h,k} \overset{*}{\rightharpoonup} u \quad \text{in } L^\infty(0,T;V); \tag{4.144}$$

$$\bar{u}_{h,k} \overset{*}{\rightharpoonup} \bar{u} \quad \text{in } L^\infty(0,T;V); \tag{4.145}$$

$$\bar{u}_{h,k} \rightharpoonup \bar{u} \quad \text{in } H^1(0,T;H); \tag{4.146}$$

$$\Xi_{h,k} \overset{*}{\rightharpoonup} \Xi \quad \text{in } L^\infty(0,T;L^2(\omega)). \tag{4.147}$$

Then from Proposition 1.6 it follows (see also (4.117)) that

$$\bar{u}_{h,k} \to \bar{u} \quad \text{in } L^2(0,T;H), \tag{4.148}$$

and, from (4.142),(4.144) and (4.67) we obtain that

$$u_{h,k} \to \bar{u} \quad \text{in } L^2(0,T;H). \tag{4.149}$$

Thus, also $u = \bar{u}$. Therefore, (4.135), (4.137) and (4.139) hold. Further, (4.138) follows from (4.149) (see (4.120)).

It remains to prove (4.136). Let $\{\tilde{u}_{h,k}\}$, $\tilde{u}_{h,k} \in L^2(0,T;K_h)$, be a sequence converging to u strongly in $L^2(0,T;V)$. According to (3.25) we have

$$\tilde{\alpha}\|u_{h,k} - \tilde{u}_{h,k}\|^2_{L^2(0,T;V)} \tag{4.150}$$

$$\leq \int_0^T a_h(u_{h,k} - \tilde{u}_{h,k}, u_{h,k} - \tilde{u}_{h,k})dt$$

$$= \int_0^T a_h(u_{h,k}, u_{h,k} - \tilde{u}_{h,k})dt - \int_0^T a_h(\tilde{u}_{h,k}, u_{h,k} - \tilde{u}_{h,k})dt.$$

Because of (4.20),(4.144) the second term above tends to zero as $h, k \to 0+$. On the other hand, the first term can be estimated as follows:

$$\int_0^T a_h(u_{h,k}, u_{h,k} - \tilde{u}_{h,k})dt \tag{4.151}$$

$$\leq -\int_0^T ((\bar{u}_{h,k})', u_{h,k} - \tilde{u}_{h,k})dt - \int_0^T (\Xi_{h,k}, P_h(\Pi u_{h,k} - \Pi \tilde{u}_{h,k}))_{0,\omega}dt$$

$$+ \int_0^T (f_{h,k}, u_{h,k} - \tilde{u}_{h,k})_h dt \to 0, \quad \text{as } h, k \to 0+,$$

which is due to the fact that $u_{h,k}$ solves $(P)_{h,k}$ and (4.15),(4.96),(4.146),(4.147), (4.149). Combining (4.150) and (4.151) we get the strong convergence

$$u_{h,k} \to u \quad \text{in } L^2(0,T;V). \tag{4.152}$$

Finally, we remind that the strong convergence of $\{\tilde{u}_{h,k}\}$ to u in $L^2(0,T;V)$ can be established as in the proof of Theorem 4.2.

Step III: Limit procedure. We first prove that $u(t) \in K$ for a.a. $t \in (0,T)$. Indeed, from (4.152) it follows (passing to a subsequence, if necessary) that

$$u_{h,k}(t) \to u(t) \quad \text{in } V \text{ for a.a. } t \in (0,T).$$

Recalling (1.173) we get that $u(t) \in K$ for a.a. $t \in (0,T)$.

Now we show that u solves (P). Let $\phi \in C_0^\infty((0,T))$ and $v \in K$ be given. Then by virtue of (1.172) there exists a sequence $\{v_h\}$, $v_h \in K_h$, converging to v strongly in V. Thus, it holds that $\Phi_h = v_h\phi$ belongs to $L^2(0,T;K_h)$ and

$$\Phi_h \to \Phi = v\phi \quad \text{in } L^2(0,T;V), \tag{4.153}$$

as $h \to 0+$. Substituting Φ_h into $(P)_{h,k}$ and integrating over $(0,T)$ we obtain:

$$\int_0^T (\bar{u}'_{h,k}, \Phi_h - u_{h,k})dt + \int_0^T a_h(u_{h,k}, \Phi_h - u_{h,k})dt \tag{4.154}$$

$$+ \int_0^T (\Xi_{h,k}, P_h(\Pi\Phi_h - \Pi u_{h,k}))_{0,\omega}dt \geq \int_0^T (f_{h,k}, \Phi_h - u_{h,k})_h dt.$$

Now, we shall pass to the limit with $h, k \to 0+$. Taking into account (4.15), (4.20),(4.96),(4.146),(4.147),(4.152),(4.153) we see that (4.154) converges to

$$\int_0^T (u', \Phi - u)dt + \int_0^T a(u, \Phi - u)dt \qquad (4.155)$$

$$+ \int_0^T (\Xi, (\Pi\Phi - \Pi u))_{0,\omega} dt \geq \int_0^T (f, \Phi - u)dt.$$

The initial condition $u(0) = u_0$ follows from the fact that $\{\bar{u}_{h,k}\}$ converges to u strongly in $C([0, T]; H)$ and (4.18).

Finally, the inclusion

$$\Xi(x, t) \in \hat{b}(x, \Pi u(x, t)) \qquad \text{for a.a. } (x, t) \in \omega_T \qquad (4.156)$$

can be proven as Theorem 4.1. $\qquad\qquad\qquad\qquad\qquad\qquad\qquad\qquad\square$

Remark 4.12 *Let us discuss the case $\theta \neq 1$. It is easy to see that the first a priori estimates can be derived as in the unconstrained case (cf. (4.27),(4.81)). Recalling the proof of Theorem 4.1 especially how the strong convergence of $\{u_{h,k}^\theta\}$ to u in $L^2(0, T; H)$ was established, it is obvious that we cannot apply this approach to constrained problems. Therefore, it is necessary to prove the second a priori estimates. But $u_{h,k}^i$, $i = 1, ..., r$, are not necessarily elements of K_h (only what we know is $u_{h,k}^0, u_{h,k}^{i+\theta}$, $i = 1, ..., r - 1$, belong to K_h). Therefore, we cannot repeat the proof of Theorem 4.2 (the substitution $v_h = u_{h,k}^{i+1}$ is not allowed). Recall again from Trémoliéres, 1972, Glowinski et al., 1981 how this problem is solved if b is monotone (with respect to the second variable).*

References

Fairweather, G. (1978). *Finite Element Galerkin Methods for Differential Equations.* Marcel Dekker, New York.

Glowinski, R. (1984). *Numerical Methods for Nonlinear Variational Problems.* Springer-Verlag, New York.

Glowinski, R., Lions, J. L., and Trémoliéres, R. (1981). *Numerical analysis of variational inequalities*, volume 8 of *Studies in Mathematics and its Applications*. North Holland, Amsterdam, New York.

Goeleven, D., Miettinen, M., and Panagiotopoulos, P. D. (1999). Dynamic hemivariational inequalities and their applications. *to appear in J. Opt. Theory Appl.*

Miettinen, M. (1996). A parabolic hemivariational inequality. *Nonlinear Analysis*, 26:725–734.

Miettinen, M. and Haslinger, J. (1998). Finite element approximation of parabolic hemivariational inequalities. *Numer. Funct. Anal. Optimiz.*, 19:565–585.

Miettinen, M. and Panagiotopoulos, P. D. (1999). On parabolic hemivariational inequalities and applications. *Nonlinear Analysis*, 35:885–915.

Pop, G. and Panagiotopoulos, P. D. (1998). On a type of hyperbolic variational-hemivariational inequalities. *preprint*.

Thomée, V. (1984). *Galerkin finite element methods for parabolic problems*. Springer-Verlag, Berlin, Heidelberg, New York.

Trémoliéres, R. (1972). *Inéquations variationnelles: existence, approximations, résolution*. PhD thesis, Université de Paris VI.

III NONSMOOTH OPTIMIZATION METHODS

5 NONSMOOTH OPTIMIZATION METHODS

5.1 CONVEX CASE

From the previous chapters we know that after the discretization, elliptic and parabolic hemivariational inequalities can be transformed into substationary point type problems for locally Lipschitz superpotentials and as such will be solved. There is a class of mathematical programming methods especially developed for this type of problems. The aim of this chapter is to give an overview of nonsmooth optimization techniques with special emphasis on the first and the second order bundle methods. We present their basic ideas in the convex case and necessary modifications for nonconvex optimization. We shall use them in the next chapter for the numerical realization of several model examples. Let us mention also that bundle type methods can be applied to a large class of locally Lipschitz superpotentials without any special requirement on their properties (such as the difference of two convex functions, e.g.).

In this section we consider an unconstrained nonsmooth optimization problem of the form

$$\begin{cases} \text{minimize} & f(x) \\ \text{subject to} & x \in \mathbf{R}^n, \end{cases} \tag{CP}$$

where the objective function $f : \mathbf{R}^n \to \mathbf{R}$ is supposed to be convex (not necessarily differentiable). Recall from Chapter 1, that f is convex in $\mathbf{R}^n$ iff

$$f(\lambda x + (1 - \lambda)y) \le \lambda f(x) + (1 - \lambda)f(y) \tag{5.1}$$

holds for all $x, y \in \mathbf{R}^n$ and $\lambda \in (0, 1)$. We suppose, that at every point $x \in \mathbf{R}^n$ we can calculate the function value $f(x)$ and find a subgradient $\xi(x)$ from the subdifferential (cf. Definition 1.2)

$$\partial f(x) = \{\xi(x) \in \mathbf{R}^n \mid f(y) \geq f(x) + \xi(x)^T(y - x) \quad \text{for all } y \in \mathbf{R}^n\}. \quad (5.2)$$

Note, that if f is continuously differentiable, then $\partial f(x) = \{\nabla f(x)\}$.

For a convex function we recall the following necessary and sufficient optimality condition (cf. Theorem 1.23).

Theorem 5.1 *A convex function* $f : \mathbf{R}^n \to \mathbf{R}$ *attains its global minimum at* x^*, *if and only if*

$$0 \in \partial f(x^*). \quad (5.3)$$

Typically, numerical optimization methods are iterative: starting from a given initial point $x_1 \in \mathbf{R}^n$, they construct a sequence $\{x_k\}_{k=1}^{\infty} \subset \mathbf{R}^n$ which is intended to converge to a global minimum x^*. A general iterative algorithm for solving problem (CP) can be stated as follows:

General Algorithm

Step 0: (Initialization). Find a starting point $x_1 \in \mathbf{R}^n$ and set $k := 1$.

Step 1: (Stopping criterion). If x_k is "close enough" to x^*, then STOP.

Step 2: (Direction finding). Find a *descent direction* $d_k \in \mathbf{R}^n$ such that

$$f(x_k + t d_k) < f(x_k) \quad \text{for some} \quad t > 0.$$

Step 3: (Line search). Find a step size $t_k > 0$ such that

$$t_k \approx \operatorname*{argmin}_{t > 0}\{f(x_k + t d_k)\}.$$

Step 4: (Updating). Set $x_{k+1} := x_k + t_k d_k$, $k := k + 1$ and go to Step 1.

Most of the optimization algorithms are constructed for smooth versions of problem (CP). If the objective function f is differentiable, then a descent direction in Step 2 may be generated by exploiting the fact that $-\nabla f(x)$ is locally the steepest descent direction. Due to the optimality condition (Theorem 5.1) for a optimum point we have $\nabla f(x^*) = 0$, and by continuity $\nabla f(x) \to 0$, whenever $x \to x^*$. This fact may yield a stopping criterion in Step 1; we can stop the iterations whenever $\|\nabla f(x_k)\| < \varepsilon_s$ for some tolerance $\varepsilon_s > 0$. In methods for smooth problems the line search procedures (Step 3) usually employ some efficient univariate smooth optimization method or some polynomial interpolation.

Nonsmoothness creates a lot of difficulties and requires an additional work at almost every step. The first and hardest problem is Step 2, since the direction

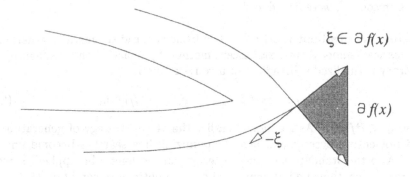

Figure 5.1.   The nondescent opposite direction of an arbitrary subgradient.

$-\xi(x)$ for a subgradient $\xi(x) \in \partial f(x)$ need not be one of descent (see Fig.5.1).

This fact forces us to modify the classical line search operations as well. Also, the stopping criterion is not clear any more, since we do not require the knowledge of the whole subdifferential $\partial f(x)$ but only one representative $\xi(x)$. The simplest counter example in the nonsmooth case is the absolute-value function $f(x) = |x|$ in **R**, which attains its global minimum at $x^* = 0$ although, for example, $0 \neq 1 \in \partial f(0)$ is a suitable nonzero subgradient at that minimum point.

In what follows we shortly describe several methods for convex nonsmooth optimization, which can be divided into two main classes: subgradient methods and bundle methods. The direction finding (Step 2) can be considered as the heart of the optimization algorithm and the point where the most remarkable differences among different methods exist. For this reason in the sequel we concentrate ourselves on the ways they generate descent directions.

Subgradient Methods

The subgradient methods were mainly developed in the Soviet Union (for an excellent overview see Shor, 1985). The history of subgradient methods starts in the 60s. Their basic idea is to generalize the smooth optimization methods by replacing the gradient by an arbitrary subgradient. This simple structure makes them widely used, although they suffer from some serious drawbacks. Nondescent search direction may occur and standard line search operation can not be applied, which means that the step sizes have to be chosen a priori. Also the lack of an implementable stopping criterion and the poor rate of the convergence speed are disadvantages of the subgradient methods. To overcome the last handicap the variable metric ideas were adopted in this context (see Shor, 1985) by introducing the space dilation methods (along the gradients and along the difference of gradients). Also some modified ideas have been proposed in Uryas'ev, 1991, where two adaptive variable metric methods, deviating in step size control, were introduced.

Original subgradient method. The basic subgradient method generalizes the classical (smooth) steepest descent method by replacing the gradient by an arbitrary normalized subgradient of a convex function i.e.

$$x_{k+1} := x_k + t_k d_k, \qquad d_k = -\xi_k/\|\xi_k\|, \qquad (5.4)$$

where $\xi_k \in \partial f(x_k)$. As mentioned earlier the above strategy of generating d_k need not ensure descent and hence minimizing line searches become unrealistic. Also the standard stopping criterion can no longer be applied, since a subgradient contains no information about the optimality condition (5.3).

Due to these facts we are forced to use an a priori choice of step sizes t_k to avoid line searches and the stopping criterion. However, it is easy to prove the following results of Lemaréchal, 1989, which justify some choices of step sizes.

Proposition 5.1 *Let x^* be and x_k not be a solution of (CP). Then*

$$\|x_{k+1} - x^*\| < \|x_k - x^*\|, \qquad (5.5)$$

whenever

$$0 < t_k < 2[f(x_k) - f(x^*)]/\|\xi_k\|, \qquad (5.6)$$

where $\xi_k \in \partial f(x_k)$.

Proposition 5.2 *At each k-th iteration we have*

$$\|x_1 - x_k\| \le \sum_{j=1}^{k} t_j \qquad (5.7)$$

In view of the above results, in order to guarantee the global convergence, we require that

$$t_k \downarrow 0 \quad \text{when} \quad k \to \infty \quad \text{and} \quad \sum_{j=1}^{\infty} t_j = \infty. \qquad (5.8)$$

Variable metric methods. In order to accelerate the rate of convergence we may try to generalize more efficient smooth methods than the steepest descent method. Quasi-Newton (variable metric) methods represent a possible candidate for that. Replacing the gradient by the normalized subgradient in a quasi-Newton method we obtain

$$x_{k+1} := x_k + t_k d_k, \quad \text{where} \quad d_k = -H_k \xi_k/(\xi_k^T H_k \xi_k)^{1/2}, \qquad (5.9)$$

where $H_k \in \mathcal{L}(\mathbf{R}^n, \mathbf{R}^n)$ is an approximate of inverse Hessian matrix. However, due to Lemaréchal, 1982, the direct generalization by employing the standard

DFP (Davidon-Fletcher-Powell) or BFGS (Broyden-Fletcher-Goldfarb-Shanno) updating formulas (see, for example Fletcher, 1987) for H_k gives very poor numerical results. These difficulties have led to the appearance of modified ideas in the construction of generalized variable metric methods. At the moment, space dilation methods by Shor, 1985 are the most efficient ones. The space dilation along the gradients (SDG), recently known as the ellipsoid method, starts with $H_1 := I$ and updates H_k by

$$H_{k+1} := H_k + \beta_k \frac{H_k \xi_k \xi_k^T H_k}{\xi_k^T H_k \xi_k}, \tag{5.10}$$

where β_k is the so called space dilation coefficient. However, better numerical results were obtained by the space dilation along the difference of subgradients (also known as r-algorithm), where

$$H_{k+1} := H_k + \beta_k \frac{H_k(\xi_{k+1} - \xi_k)(\xi_{k+1} - \xi_k)^T H_k}{(\xi_{k+1} - \xi_k)^T H_k(\xi_{k+1} - \xi_k)}. \tag{5.11}$$

The more recent modified ideas have been proposed in Uryas'ev, 1991, where two adaptive variable metric methods, deviating in controlling the step size t_k, were introduced. Both of them employ the following updating rule for H_k:

$$H_{k+1} := H_k + \beta_k \frac{H_k \xi_{k+1} \xi_k^T + \xi_k \xi_{k+1}^T}{\|\xi_{k+1}\| \|\xi_k\|}. \tag{5.12}$$

The crucial point in the efficiency of these methods is the choice of the step size parameter t_k and the space dilation parameter β_k. As a conclusion: subgradient methods suffer from quite poor theoretical convergence results, but may be applied quite efficiently in certain special cases.

Bundle Methods

In this section we concentrate ourselves on the bundle methods, which are at the moment the most efficient and promising methods for nonsmooth optimization. We shall give a short review concerning their history and development.

General bundle algorithm. Firstly we describe a general bundle method that produces a sequence $\{x_k\}_{k=1}^{\infty} \subset \mathbf{R}^n$ converging to a global minimum of f, if this exists. We suppose that in addition to the current iteration point x_k we have some trial points $y_j \in \mathbf{R}^n$ (from past iterations) and subgradients $\xi_j \in \partial f(y_j)$ for $j \in J_k$, where the index set J_k is a nonempty subset of $\{1, \ldots, k\}$.

The idea behind bundle methods is to approximate f from below by a piecewise linear function, in other words, we replace f by the so-called *cutting-plane model* (see Fig.5.2)

$$\hat{f}_k(x) := \max_{j \in J_k}\{f(y_j) + \xi_j^T(x - y_j)\}, \tag{5.13}$$

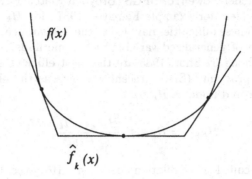

Figure 5.2.   The cutting-plane model.

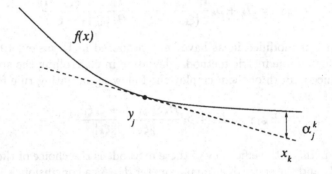

Figure 5.3.   The linearization error.

which can be written in the equivalent form

$$\hat{f}_k(x) = \max_{j \in J_k}\{f(x_k) + \xi_j^T(x - x_k) - \alpha_j^k\},$$

with the *linearization error* (see Fig.5.3).

$$\alpha_j^k := f(x_k) - f(y_j) - \xi_j^T(x_k - y_j) \quad \text{for all} \quad j \in J_k. \tag{5.14}$$

The next iterate candidate is then defined by

$$y_{k+1} := x_k + d_k,$$

where the search direction d_k is calculated by

$$d_k := \operatorname*{argmin}_{d \in \mathbb{R}^n}\{\hat{f}_k(x_k + d) + \frac{1}{2}d^T M_k d\}. \tag{5.15}$$

The role of the stabilizing term $\frac{1}{2}d^T M_k d$ is to guarantee the existence of the solution d_k and to keep the approximation local enough. The $n \times n$ matrix M_k

is chosen to be regular and symmetric, and it serves to accumulate information about the curvature of f around x_k.

A *serious step* $x_{k+1} := y_{k+1}$ is taken if y_{k+1} is significantly better than x_k, in other words if

$$f(y_{k+1}) \leq f(x_k) + m_L v_k,$$

where $m_L \in (0, \frac{1}{2})$ is a line search parameter and

$$v_k = \hat{f}_k(y_{k+1}) - f(x_k)$$

is the predicted descent of f at x_k.

Otherwise, a *null step* $x_{k+1} := x_k$ is taken and to improve the cutting plane model we set $J_{k+1} := J_k \cup \{k + 1\}$ and create $\hat{f}_{k+1}$ using $\xi_{k+1} \in \partial f(y_{k+1})$. The iteration terminates if $v_k \geq -\varepsilon_s$, where $\varepsilon_s > 0$ is a final accuracy tolerance supplied by the user.

Notice, that the problem (5.15) still is a nonsmooth optimization problem. However, due to its piecewise linear nature it can be rewritten as a (smooth) quadratic programming subproblem

$$\begin{cases} \text{minimize} & v + \frac{1}{2} d^T M_k d \\ \text{subject to} & -\alpha_j^k + \xi_j^T d \leq v \quad \text{for all} \quad j \in J_k. \end{cases} \tag{QP}$$

For computational reasons it might be, in some cases, more efficient to solve the dual problem of (QP), i.e., to find multipliers λ_j^k for $j \in J_k$ solving the problem

$$\begin{cases} \text{minimize} & \frac{1}{2} \left[\sum_{j \in J_k} \lambda_j \xi_j \right]^T M_k^{-1} \left[\sum_{j \in J_k} \lambda_j \xi_j \right] + \sum_{j \in J_k} \lambda_j \alpha_j^k \\ \text{subject to} & \sum_{j \in J_k} \lambda_j = 1 \\ \text{and} & \lambda_j \geq 0. \end{cases} \tag{DP}$$

Theorem 5.2 *Problems (QP) and (DP) are equivalent: they have unique solutions (d_k, v_k) and λ_j^k for $j \in J_k$, respectively, related by*

$$d_k = -\sum_{j \in J_k} \lambda_j^k M_k^{-1} \xi_j, \tag{5.16}$$

$$v_k = -d_k^T M_k d_k - \sum_{j \in J_k} \lambda_j^k \alpha_j^k. \tag{5.17}$$

In what follows we shortly present several versions of bundle methods, which all conform the above ideas. To avoid technical details we focus on their main differences in the choice of the cutting plane approximation $\hat{f}_k$, the linearization error α_j^k or the stabilizing matrix M_k.

Cutting plane methods. The history of bundle methods originates from the cutting plane idea, which was developed independently by Cheney and Goldstein, 1959, and Kelley, 1960. In the original cutting plane method the curvature matrix was chosen to be

$$M_k \equiv 0.$$

For this reason the direction finding subproblems (QP) and (DP) are linear programming problems and thus simpler to solve. It is also obvious, that if the original objective function f is piecewise linear or almost piecewise linear, then the cutting plane method may converge in a reliable way and rapidly to the exact global optimum.

However, the absence of the stabilizing quadratic term affects that (QP) or (DP) may not have a finite optimum. In addition, the convergence of the cutting plane algorithm can, in some cases, be really slow. This is mainly due to the fact that $\|d_k\|$ may be so large that the new trial point y_{k+1} is far from the previous ones y_j, $j = 1, \ldots, k$. Then y_{k+1} lies in the region where the model $\hat{f}_k$ poorly approximates the original objective f. The method also stores all the previous subgradients, in other words $J_k = \{1, \ldots, k\}$, which may cause serious computational difficulties in the form of unbounded storage requirement.

Conjugate subgradient methods. The next step towards bundle methods was made by the method of conjugate subgradients, which was developed in two journal papers by Wolfe, 1975 and Lemaréchal, 1975 in the same issue. The main idea of the conjugate subgradient method was to treat each previous subgradient $\xi_j \in \partial f(y_j)$ for $j \in J_k$ as it was a subgradient at the current point x_k, in other words the linearization error (5.14) was neglected. On the other hand, no curvature of the objective function was taken account, thus the search direction finding problems (QP) and (DP) were applied with

$$M_k \equiv I \quad \text{and} \quad \alpha_j^k \equiv 0.$$

With such a choice of M_k and α_j^k the opposite direction of d_k can be found by projecting the origin onto the set

$$\overline{\text{conv}}\{\xi_j \mid j \in J_k\}$$

approximating the set $\partial f(x_k)$. The neglection of α_j^k affects that the cutting-plane model $\hat{f}_k$ is an usable approximation to f only if the trial points y_j are close enough to x_k. For this reason the choice of the index set $J_k \subset \{1, \ldots, k\}$ is a crucial point of the algorithm. Several *subgradient selection strategies* to choose J_k were proposed, for instance, in Wolfe, 1975, Lemaréchal, 1975, Mifflin, 1977 and Polak et al., 1983 to keep the approximation local enough and limit the number of stored subgradients. In Wolfe, 1975 and Lemaréchal, 1975 the authors introduce for the first time the *subgradient aggregation strategy*, which requires storing only the finite number of subgradients. However, the numerical experiments have shown, that the convergence of the conjugate subgradient method is rather slow in practice (see Lemaréchal, 1982).

ε-steepest descent methods. The first method really called the "bundle method" was the ε-steepest descent method introduced by Lemaréchal, 1976. The method was further developed in Lemaréchal et al., 1981, where the subgradient aggregation strategy was utilized to limit the number of stored subgradients, and more lately by Strodiot et al., 1983. Contrary to the conjugate subgradient method some attention to the linearization error (5.14) was paid. In contrast to the standard treatment of the linearization error as a term appearing in the objective function of (DP), this error was controlled by an extra constraint. Since no curvature of the objective function was still taken account, in other words

$$M_k \equiv I,$$

the dualized search direction finding problem (DP) has the form

$$\begin{cases} \text{minimize} & \frac{1}{2}\|\sum_{j\in J_k} \lambda_j \xi_j\|^2 \\ \text{subject to} & \sum_{j\in J_k} \lambda_j = 1 \\ & \lambda_j \geq 0 \\ \text{and} & \sum_{j\in J_k} \lambda_j \alpha_j^k \leq \varepsilon_k, \end{cases} \tag{DP'}$$

where $\varepsilon_k > 0$ is a controlling parameter. This tolerance controls the radius of the ball in which the cutting-plane model is thought to be a good approximation of the objective function. The main difficulty in the ε-steepest descent method is the design of convergent rules for automatic updating schemas of ε_k. However, due to the numerical tests of Lemaréchal, 1982 this method works better in general than the conjugate subgradient method.

Generalized cutting plane methods. To avoid the difficulties of above methods when handling the linearization error, the idea of the generalized cutting plane method was introduced by Lemaréchal, 1978 and further developed by Kiwiel, 1985. The starting point was the classical cutting plane algorithm. The stabilizing term $\frac{1}{2}\|d\|^2$ was added to the objective function to guarantee the existence of the solution and to keep the approximation local enough. Thus the search direction was obtained as a solution of (QP) or (DP) with the first order choice

$$M_k \equiv I.$$

In spite of different backgrounds, the ε-steepest descent method and the generalized cutting plane method have the following connection: if λ_j^k for $j \in J_k$ are optimal multipliers of (DP) with $M_k \equiv I$, then they solve also the problem (DP') with the choice

$$\varepsilon_k = \sum_{j\in J_k} \lambda_j^k \alpha_j^k.$$

The storage saving subgradient selection and aggregation strategies were introduced in Kiwiel, 1985. Especially the aggregation strategy has been successfully applied in forthcoming methods. When compared with the ε-steepest descent method the difficulty of selecting the approximation tolerance ε_k was avoided in the generalized cutting plane method. Also the numerical tests of Mäkelä, 1989 support the superiority of the generalized cutting plane method. However, the tests also disclosed its sensitivity to the scaling of the objective function (i.e. multiplication of f by a positive constant).

Diagonal variable metric bundle methods. The diagonal variable metric idea was the next improvement in the history of the bundle methods. A weighting parameter was added to the quadratic term of the objective functions in (QP) and (DP) in order to accumulate some second order information about the curvature of f around x_k. Thus the variable metric matrix M_k took the diagonal form

$$M_k = u_k I,$$

with a weighting parameter $u_k > 0$. Based on the proximal point algorithm of Rockafellar, 1976 and the work of Auslender, 1987 the *proximal bundle* method was derived by Kiwiel, 1990, where also an adaptive safeguarded quadratic interpolation technique for updating u_k was introduced. Somewhat similar outcome was concluded in Schramm and Zowe, 1992, where the *bundle trust region* method was developed combining the bundle idea with the classical trust region method of Levenberg, 1944 and Marquardt, 1963. Furthermore, the *diagonal quasi-Newton* method of Lemaréchal and Sagastizábal, 1994 based on the Moreau-Yosida regularizations (see Moreau, 1965 and Yosida, 1964) belongs to this class of bundle methods. The methods deviate mainly in the strategies for updating the weights u_k: the bundle trust region method employs the same safeguarded quadratic interpolation technique than the proximal bundle method, while the diagonal quasi-Newton method relies on the curved search technique. The numerical tests in Kiwiel, 1990, Mäkelä and Neittaanmäki, 1992 and Schramm and Zowe, 1992 demonstrate the obvious progress in the convergence speed of the diagonal variable metric bundle methods when compared with the earlier methods.

Variable metric bundle methods. The development of the second order methods fascinated the researchers in nonsmooth optimization during its whole history. Although the tools of nonsmooth analysis already existed, their influence in the numerical methodology has not been as fruitful as in the first order case. However, several attempts to employ

$$M_k \text{ as a full matrix}$$

with some updating scheme have been proposed by various authors. Already in his pioneering work Lemaréchal, 1978 derived a version of the variable metric bundle method utilizing the classical BFGS secant updating formula from smooth optimization. Due to the disappointing numerical results in Lemaréchal, 1982 this idea was buried nearly for two decades. In Gaudioso and

Monaco, 1992 the space dilation updating scheme of Shor, 1985 was adopted from the subgradient method context. More lately, based on the Moreau-Yosida regularization and BFGS update, variants of the variable metric bundle idea were proposed in Bonnans et al., 1995, Lemaréchal and Sagastizábal, 1994 and Mifflin, 1996. According to very limited numerical experiments (see for example Gaudioso and Monaco, 1992) it seems that the variable metric bundle methods work fairly well. However, when compare the results to the extra computational efforts, which are needed with the full matrix algebra, they do not offer a substantial advancement in numerical solution process.

Translated bundle methods. Another strategy how to exploit higher order information about f is to keep the stabilizing matrix constant, in other words

$$M_k \equiv I,$$

but to move the supporting hyperplanes by varying the linearization error (5.14). First attempts in this direction were made in Mifflin, 1982, where the so-called α-function was introduced, and in Gaudioso and Monaco, 1982, where the linearization error α_j^k was replaced by the quantity

$$\beta_j^k := f(y_j) - f(x_k). \tag{5.18}$$

This idea was extended by Gaudioso and Monaco, 1992, where the combination

$$\rho_j^k := \max\left\{(1 - \theta)\alpha_j^k, \min\{\alpha_j^k, \beta_j^k\}\right\},$$

with $\theta \in (0, 1)$ was utilized. In the same paper also the quadratically based translation approach was proposed in the form

$$\rho_j^k := \xi_k^T (x_k - y_j).$$

In numerical tests none of the above methods behave uniformly better than the others. Due to these limited results the convergence rate of the translated bundle methods places itself between the ε-steepest descent method and diagonal variable metric bundle methods.

Tilted bundle methods. Based on the work Tarasov and Popova, 1984 the tilted bundle method was developed by Kiwiel, 1991 in order to accumulate some second-order information and some interior point features to the proximal bundle method of Kiwiel, 1990. The matrix M_k was handled in the original diagonal form

$$M_k = u_k I,$$

with the safeguarded quadratic interpolation technique for updating u_k. The cutting-plane model (5.13) was replaced by the so-called *tilted cutting-plane model*

$$\check{f}_k(x) := \max_{j \in J_k}\{f(x_k) + (1 - \theta_j^k)\xi_j^T (x - x_k) - \rho_j^k\}$$

with some tilting parameters $\theta_j^k \in [0, 1 - \kappa]$ for $j \in J_k$ and $\kappa \in (0, 1]$. The linearization error (5.14) was replaced by the combination

$$\rho_j^k := \max\{\breve{\alpha}_j^k, \kappa\alpha_j^k, \},$$

where the tilted linearization error is given by

$$\breve{\alpha}_j^k := f(x_k) - f(y_j) - (1 - \theta_j^k)\xi_j^T(x_k - y_j).$$

Note, that if $\kappa = 1$, then $\theta_j^k = 0$ for all $j \in J_k$ and the method restores to the proximal bundle method. The tilted cutting plane is a real "cutting plane", since it cuts parts of the epigraph of f (see Fig.5.4), while the standard cutting planes do not. The numerical tests in Kiwiel, 1991 appear rather promising, but the question how to choose the tilting parameters κ and θ_j^k is still open.

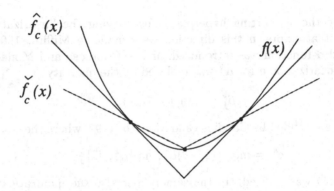

Figure 5.4.   The tilted cutting plane model.

Level bundle methods.   Since the polyhedral level-sets of the cutting plane model (5.13) are easy to calculate, a new bundle method variant based on the minimization of the stabilizing quadratic term subject to some level set constraint of $\hat{f}_k$ was proposed in Lemaréchal et al., 1995. In other words, the search direction finding problem (5.15) was replaced by

$$d_k := \operatorname*{argmin}_{d \in \mathbf{R}^n}\{\tfrac{1}{2}d^T M_k d \mid \hat{f}_k(x_k + d) \le f_k^{\mathrm{lev}}\},$$

where the target level $f_k^{\mathrm{lev}} < f(x_k)$ is chosen to ensure $f_k^{\mathrm{lev}} \to \inf f$ as $k \to \infty$. The stabilizing matrix was chosen to be constant

$$M_k \equiv I.$$

Bounded storage versions of the level bundle idea were introduced by Brännlund, 1994, Kiwiel, 1995 and more lately by Brännlund et al., 1996, where the global convergence was proven without any compactness assumptions as in Lemaréchal et al., 1995. In numerical tests the level bundle methods worked in a reliable way but they lost, for example, when compared with diagonal variable metric bundle methods.

Bundle-Newton method. The recent advance in the development of the second-order bundle method was made by Lukšan and Vlček, 1995, where the bundle-Newton method was derived. Instead of the piecewise linear cutting-plane model (5.13) they introduced a quadratic model of the form

$$\tilde{f}_k(x) := \max_{j \in J_k}\{f(y_j) + \xi_j^T(x - y_j) + \frac{1}{2}\varrho_j(x - y_j)^T M_j(x - y_j)\},$$

where $\varrho_j \in [0, 1]$ is a damping parameter. The search direction finding problem (5.15) was then replaced by the problem

$$d_k := \operatorname*{argmin}_{d \in \mathbf{R}^n}\{\tilde{f}_k(x_k + d)\}. \tag{5.19}$$

When compared with the earlier variable metric bundle methods we can state that the bundle-Newton method is the "real" second-order method, since every part of the model contains the second-order information in the form of the stabilizing matrix M_j. For the approximation

$$M_j \approx \nabla^2 f(y_j)$$

the authors proposed optionally analytic or finite difference approximations. The numerical experiments in Lukšan and Vlček, 1995 seem to be very promising, especially in piecewise quadratic test cases (see Mäkelä et al., 1999).

5.2 NONCONVEX CASE

In this part we consider the following nonsmooth and nonconvex optimization problem

$$\begin{cases} \text{minimize} & f(x) \\ \text{subject to} & x \in K, \end{cases} \tag{NP}$$

where an objective function $f : \mathbf{R}^n \to \mathbf{R}$ is supposed to be locally Lipschitz in $\mathbf{R}^n$. Note, that f needs not to be convex nor differentiable. Due to our further applications the feasible set K has the following specific structure

$$K = \{x \in \mathbf{R}^n \mid x_l \le x \le x_u\}, \tag{5.20}$$

where x_l and x_u are the lower and upper bounds for variables, respectively. The generalized gradient f is now defined (cf. Remark 1.4) by

$$\bar{\partial}f(x) = \overline{\operatorname{conv}}\{\lim_{i \to \infty} \nabla f(x_i) \mid x_i \to x \text{ and } \nabla f(x^i) \text{ and } \lim \text{ exist}\}.$$

Note, that if f is convex, then $\bar{\partial}f(x) = \partial f(x)$.

For a locally Lipschitz function f the following necessary optimality condition holds (cf. Theorem 1.24).

Theorem 5.3 *If a locally Lipschitz function $f : \mathbf{R}^n \to \mathbf{R}$ attains its local minimum at x^* on a nonempty, closed and convex set K, then x^* is a substationary point of f on K, i.e.*

$$0 \in \bar{\partial}f(x^*) + N_K(x^*), \tag{5.21}$$

where $N_K(x^*)$ is the normal cone of K at x^*.

Taking into account the special structure of K in (5.20), we can rewrite (5.21) in the following form:

Theorem 5.4 *If a locally Lipschitz function $f : \mathbf{R}^n \to \mathbf{R}$ attains its local minimum at x^* on K, then there exist vectors of multipliers $\mu_u, \mu_l \in \mathbf{R}^n$ such that $\mu_u, \mu_l \geq 0$, $\mu_u^T(x_u - x^*) = 0$, $\mu_l^T(x_l - x^*) = 0$ and*

$$0 \in \bar{\partial} f(x^*) + \mu_u - \mu_l. \tag{5.22}$$

Since the optimality condition is not sufficient without the convexity assumption, the methods can not guarantee even the local optimality of the solutions. *Only some candidates, substationary points, satisfying (5.21) are to be looked for.*

In the convex case the cutting-plane model (5.13) was an underestimate for the objective function and the nonnegative linearization error (5.14) measured how well the model approximates the original problem (see (5.40)). In the nonconvex case these facts are not valid anymore: α_j^k may take a tiny (or even negative!) value, although the trial point y_j lies far away from the current iteration point x_k and thus the corresponding subgradient ξ_j is useless (see Fig.5.5).

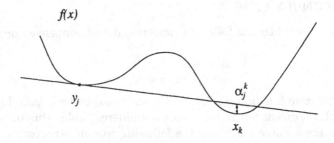

Figure 5.5.

For the reasons mentioned above, the main modifications in the methods needed in the nonconvex case concern of the linearization error α_j^k. Since the problem is now more complicated, some of them include the so-called resetting strategies, and also some modifications in the line search procedure have to be done in order to guarantee the convergence properties (see Kiwiel, 1985). In what follows, we concentrate on the changes concerning of the form of the linearization error. At the end we shall describe two methods, namely the proximal bundle method and the bundle-Newton method, in more details.

Handling Nonconvexity

In this section we introduce two different strategies to handle the nonconvexity: subgradient deletion rules and subgradient locality measures.

Subgradient deletion rules. As mentioned before, the conjugate subgradient methods neglected the linearization error (5.14) and some deletion rules were needed to reduce the past subgradient information in order to localize the approximations. For this reason they were used also for nonconvex problems in Mifflin, 1977 and Polak et al., 1983, where, for example, the following type of deletion rules was proposed

$$J_k := \{1 \le j \le k \mid \|x_k - y_j\| \le \delta_k\}, \tag{5.23}$$

where $\delta_k > 0$ tends to zero. Without any special deletion rules the translated bundle method introduced in Gaudioso and Monaco, 1982 was proposed to be suitable also for nonconvex case in Gaudioso and Monaco, 1988 since the substitute linearization error (5.18) is always nonnegative (see Section 5.1 the general bundle algorithm), in other words

$$\beta_j^k = f(y_j) - f(x_k) \ge 0. \tag{5.24}$$

More complicated deletion rules were derived in Kiwiel, 1985 for generalized cutting plane methods. The linearization error (5.14) was replaced by its absolute value, in other words

$$\beta_j^k := |\alpha_j^k| = |f(x_k) - f(y_j) - \xi_j^T(x_k - y_j)| \quad \text{for all} \quad j \in J_k \tag{5.25}$$

and the algorithm was reset (deleting the old subgradient information) whenever

$$\|d_k\| \le m_S \cdot \max_{j \in J_k}\{s_j^k\}, \tag{5.26}$$

where $m_S > 0$ is a reset tolerance supplied by the user and

$$s_j^k := \|x_j - y_j\| + \sum_{i=j}^{k-1} \|x_{i+1} - x_i\| \tag{5.27}$$

is the distance measure estimating

$$\|x_k - y_j\| \tag{5.28}$$

without the need to store the trial points y_j.

More lately in Kiwiel, 1996, the author introduced his *restricted step* proximal bundle method, where the search direction finding problem (5.15) was replaced by

$$d_k := \arg \min_{d \in \mathbf{R}^n} \left\{ \hat{f}_k(x_k + d) + \frac{1}{2} d^T M_k d \mid \|d\| \le \delta_k \right\}, \tag{5.29}$$

where $\delta_k > 0$ tends to zero and

$$M_k = u_k I, \tag{5.30}$$

with a weighting parameter $u_k > 0$. The resetting test (5.26) was now replaced by

$$u_k\|d_k\| \leq m_S \cdot \max_{j \in J_k}\{s_j^k\} - \sum_{j \in J_k} \lambda_j^k \beta_j^k, \tag{5.31}$$

where λ_j^k for $j \in J_k$ are the optimal multipliers of (DP) and β_j^k are defined by (5.25).

Subgradient locality measures. In order to add some localizing information to the model, the linearization error (5.14) was replaced by the so-called subgradient locality measure

$$\beta_j^k =: \max\{\alpha_j^k, \gamma\|x_k - y_j\|^2\} \tag{5.32}$$

in Lemaréchal et al., 1981 for the ε-steepest descent method. The distance measure parameter $\gamma \geq 0$ can be set to zero when f is convex. The authors also proposed to use the distance measure (5.42) avoiding to store the trial points y_j, in other words to replace (5.32) by

$$\beta_j^k =: \max\{\alpha_j^k, \gamma(s_j^k)^2\}. \tag{5.33}$$

In Mifflin, 1982 the subgradient locality measure (5.32) and

$$\beta_j^k =: \max\{|\alpha_j^k|, \gamma(s_j^k)^2\} \tag{5.34}$$

used by Kiwiel, 1985 were introduced for the generalized cutting plane method. Furthermore, in Schramm and Zowe, 1992 (bundle trust region method) the subgradient locality measure in the form (5.32) and in Mäkelä and Neittaanmäki, 1992 (proximal bundle method) in the form (5.34) were proposed for a diagonal variable metric bundle method. In Lukšan and Vlček, 1995 the generalization of (5.34) was introduced by

$$\beta_j^k =: \max\{|\alpha_j^k|, \gamma(s_j^k)^\omega\} \tag{5.35}$$

with $\omega \geq 1$ for the bundle-Newton method.

The latest modification of the subgradient locality measure was proposed in Kiwiel, 1996:

$$\beta_j^k =: \max\{|\alpha_j^k|, \gamma_1(s_j^k)^2, \gamma_2\|\xi_j\|s_j^k\} \tag{5.36}$$

with $\gamma_1, \gamma_2 \geq 0$. The numerical tests in Kiwiel, 1996 comparing this Levenberg-Marquardt type method with the restricted step method (5.29) show the slight superiority of the latter subgradient deletion rules based method. Especially in the worst cases the restricted step method did not need as much objective function and subgradient evaluations as the Levenberg-Marquardt type method.

Proximal Bundle Method

In this part we shortly describe the ideas of the proximal bundle method for nonsmooth and nonconvex minimization. For more details we refer to Kiwiel, 1990, Schramm and Zowe, 1992 and Mäkelä and Neittaanmäki, 1992.

Direction finding. Our aim is to produce a sequence $\{x_k\}_{k=1}^{\infty}$, $x_k \in \mathbf{R}^n$ converging to a substationary point of problem (NP). Suppose that the starting point x_1 is feasible and at the k-th iteration of the algorithm we have at our disposal the current iteration point x_k, some trial points $y_j \in \mathbf{R}^n$ (from previous iterations) and subgradients $\xi_j \in \bar{\partial} f(y_j)$ for $j \in J_k$, where the index set J_k is a nonempty subset of $\{1, \ldots, k\}$.

The idea behind the proximal bundle method is the same as in the general bundle method, in other words, we approximate the objective function by a piecewise linear cutting-plane model (cf. (5.13))

$$\hat{f}_k(x) := \max_{j \in J_k}\{f(y_j) + \xi_j^T(x - y_j)\}, \tag{5.37}$$

which equivalently can be written in the form

$$\hat{f}_k(x) = \max_{j \in J_k}\{f(x_k) + \xi_j^T(x - x_k) - \alpha_j^k\}, \tag{5.38}$$

with the linearization error

$$\alpha_j^k := f(x_k) - f(y_j) - \xi_j^T(x_k - y_j) \quad \text{for all} \quad j \in J_k. \tag{5.39}$$

Note, that in the convex case we have that

$$\hat{f}_k(x) \le f(x) \quad \text{for all } x \in \mathbf{R}^n \quad \text{and} \quad \alpha_j^k \ge 0 \quad \text{for all } j \in J_k. \tag{5.40}$$

In other words, if f is convex, then the cutting-plane model $\hat{f}_k$ is an under estimate for f and the nonnegative linearization error α_j^k measures how well the model approximates the original problem as we have already said. In the nonconvex case these facts are not valid anymore. For these reasons the linearization error α_j^k is replaced by the so-called *subgradient locality measure* introduced by Kiwiel, 1985:

$$\beta_j^k := \max\{|\alpha_j^k|, \gamma(s_j^k)^2\}, \tag{5.41}$$

where $\gamma \ge 0$ is the *distance measure parameter* ($\gamma = 0$ if f is convex) and

$$s_j^k := \|x_j - y_j\| + \sum_{i=j}^{k-1} \|x_{i+1} - x_i\| \tag{5.42}$$

is the *distance measure* estimating

$$\|x_k - y_j\| \tag{5.43}$$

without the need to store the trial points y_j. Then obviously $\beta_j^k \ge 0$ for all $j \in J_k$ and $\min_{x \in K} \hat{f}_k(x) \le f(x_k)$, since

$$\min_{x \in K} \hat{f}_k(x) \le \hat{f}_k(x_k) = f(x_k) - \max_{j \in J_k} \beta_j^k \le f(x_k). \tag{5.44}$$

In order to calculate a search direction $d_k \in \mathbf{R}^n$ we replace the original problem (NP) by the cutting plane model

$$\begin{cases} \text{minimize} \quad \hat{f}_k(x_k + d) + \frac{1}{2}u_k d^T d \\ \text{subject to} \quad x_k + d \in K, \end{cases} \tag{CPP}$$

where the regularizing quadratic penalty term $\frac{1}{2}u_k d^T d$ is added to guarantee the existence of the solution d_k and to keep the approximation local enough. The weighting parameter $u_k > 0$ was added to improve the convergence rate and to accumulate some second order information about the curvature of f around x_k. It was adapted from the proximal point algorithm by Rockafellar, 1976 and Auslender, 1987 and was used for the first time by Kiwiel, 1990 and Schramm and Zowe, 1992 (see also Outrata et al., 1998).

Notice, that problem (CPP) is still a nonsmooth optimization problem. However, due to its piecewise linear nature it can be rewritten as a (smooth) quadratic programming subproblem of finding the solution $(d_k, v_k) \in \mathbf{R}^{n+1}$ to

$$\begin{cases} \text{minimize} \quad v + \frac{1}{2}u_k d^T d \\ \text{subject to} \quad -\beta_j^k + \xi_j^T d \leq v \quad \text{for all} \quad j \in J_k \\ \text{and} \qquad x_k + d \in K. \end{cases} \tag{QP}$$

Line search. In the previous section we found a search direction d_k. Next we consider the problem of determining the step size along this direction. We assume that $m_L \in (0, \frac{1}{2})$, $m_R \in (m_L, 1)$ and $\bar{t} \in (0, 1]$ are fixed line search parameters. First we shall search for the largest number $t_L^k \in [0, 1]$ such that $t_L^k \geq \bar{t}$ and

$$f(x_k + t_L^k d_k) \leq f(x_k) + m_L t_L^k v_k, \tag{5.45}$$

where v_k is the predicted amount of the descent. If such t_L^k exists we take a *long serious step*

$$x_{k+1} := x_k + t_L^k d_k \quad \text{and} \quad y_{k+1} := x_{k+1}.$$

Otherwise, if (5.45) holds but $0 < t_L^k < \bar{t}$ then a *short serious step*

$$x_{k+1} := x_k + t_L^k d_k \quad \text{and} \quad y_{k+1} := x_k + t_R^k d_k$$

is taken. Finally, if $t_L^k = 0$ we take a *null step*

$$x_{k+1} := x_k \quad \text{and} \quad y_{k+1} := x_k + t_R^k d_k,$$

where $t_R^k > t_L^k$ is such that

$$-\beta_{k+1}^{k+1} + \xi_{k+1}^T d_k \geq m_R v_k. \tag{5.46}$$

In the long serious step a significant decrease of the value of the objective function occurs. Thus there is no need for detecting discontinuities of the gradient of f, and so we set $\xi_{k+1} \in \bar{\partial} f(x_{k+1})$. In short serious and null steps there exists the discontinuity of the gradient of f. Then the requirement (5.46) ensures that x_k and y_{k+1} lie on the opposite sides of this discontinuity and the new subgradient $\xi_{k+1} \in \bar{\partial}f(y_{k+1})$ will force a remarkable modification of the next search direction finding problem. We use the line search algorithm presented in Mäkelä and Neittaanmäki, 1992. The convergence proof of the algorithm requires f to be upper semismooth (see (5.48)).

The iteration terminates if

$$v_k \geq -\varepsilon_s, \tag{5.47}$$

where $\varepsilon_s > 0$ is a final accuracy tolerance supplied by the user.

Weight updating. One of the most important questions concerning the proximal bundle method is the choice of the weight u_k. The simplest strategy might be to keep it constant $u_k \equiv u_{fix}$. This, however, leads to several difficulties. Due to Theorem 5.2 we observe the following:

- If u_{fix} is very large, then $|v_k|$ and $\|d_k\|$, will be small and almost all steps are serious and we have slow descent;

- If u_{fix} is very small, then $|v_k|$ and $\|d_k\|$, will be large and each serious step will be followed by many null steps.

Therefore, we keep the weight as a variable and update it when necessary. For updating u_k we use the safeguarded quadratic interpolation algorithm due to Kiwiel, 1990.

The following semismoothness assumption due to Bihain, 1984 is needed for the convergence of the method.

Definition 5.1 *A function $f : \mathbf{R}^n \to \mathbf{R}$ is said to be upper semismooth, if for any $x \in \mathbf{R}^n$, $d \in \mathbf{R}^n$ and sequences $\{g_i\}$, $g_i \in \mathbf{R}^n$ and $\{t_i\}$, $t_i \in (0, \infty)$ satisfying $g_i \in \bar{\partial}f(x + t_id)$ and $t_i \to 0_+$, one has*

$$\limsup_{i \to \infty} g_i^T d \geq \liminf_{i \to \infty} [f(x + t_id) - f(x)]/t_i. \tag{5.48}$$

For locally Lipschitz and upper semismooth objective functions one can prove the following global convergence result (see Kiwiel, 1985).

Theorem 5.5 *Let f be locally Lipschitz and upper semismooth, and the sequence $\{x_k\}_{k=1}^{\infty}$ be bounded. Then every accumulation point of $\{x_k\}$ is substationary.*

Bundle-Newton Method

Next we describe the main ideas of the second order bundle-Newton method. For more details we refer to Lukšan and Vlček, 1995.

Direction finding. We suppose that at each $x \in K$ we can evaluate, in addition to the function value $f(x)$ and a subgradient $\xi(x) \in \bar{\partial}f(x)$, also an $n \times n$ symmetric matrix $G(x)$ approximating the Hessian matrix $\nabla^2 f(x)$. For example, at the kink point y of a piecewise twice differentiable function we can take $G(y) = \nabla^2 f(x)$, where x is "infinitely close" to y.

Instead of the piecewise linear cutting-pane model (5.13) we introduce a piecewise quadratic model of the form

$$\tilde{f}_k(x) := \max_{j \in J_k}\{f(y_j) + \xi_j^T(x - y_j) + \frac{1}{2}\varrho_j(x - y_j)^T G_j(x - y_j)\}, \qquad (5.49)$$

where $G_j = G(y_j)$ and $\varrho_j \in [0, 1]$ is a damping parameter. The model (5.49) can again be equivalently written as

$$\tilde{f}_k(x) = \max_{j \in J_k}\{f(x_k) + \xi_j^T(x - x_k) + \frac{1}{2}\varrho_j(x - x_k)^T G_j(x - x_k) - \alpha_j^k\} \quad (5.50)$$

and for all $j \in J_k$ the error takes now the form

$$\alpha_j^k := f(x_k) - f(y_j) - g_j^T(x_k - y_j) - \frac{1}{2}\varrho_j(x_k - y_j)^T G_j(x_k - y_j). \qquad (5.51)$$

Now, even in the convex case, α_j^k might be negative. Therefore we replace (5.51) again by the subgradient locality measure (5.41) and we keep the property (see Lukšan and Vlček, 1995)

$$\min_{x \in K} \tilde{f}_k(x) \leq f(x_k). \qquad (5.52)$$

A search direction $d_k \in \mathbf{R}^n$ is now calculated as the solution of

$$\begin{cases} \text{minimize} & \tilde{f}_k(x_k + d) \\ \text{subject to} & x_k + d \in K. \end{cases} \qquad (\text{CN})$$

Note, that since the model already has the second order information, no regularizing quadratic terms are needed like in (CPP). Problem (CN) is transformed into a nonlinear programming problem, which is then solved by a recursive quadratic programming method (see Lukšan and Vlček, 1995). If we denote

$$\xi_j^k := \xi_j + \varrho_j G_j(x_k - y_j),$$

this procedure leads to a quadratic programming subproblem of finding the solution $(d_k, v_k) \in \mathbf{R}^{n+1}$ of

$$\begin{cases} \text{minimize} & v + \frac{1}{2}d^T W_k d \\ \text{subject to} & -\beta_j^k + (\xi_j^k)^T d \leq v \quad \text{for all} \quad j \in J_k \\ \text{and} & x_k + d \in K, \end{cases} \qquad (\text{QN})$$

where

$$W_k := \sum_{j \in J_{k-1}} \lambda_j^{k-1} \varrho_j G_j$$

and λ_j^{k-1} for $j \in J_{k-1}$ are the Lagrange multipliers of (QN) from the previous $(k - 1)$-th iteration. In calculations W_k is replaced by some positive definite modification, if necessary (see Lukšan and Vlček, 1995).

Line search.   The line search operation of the bundle-Newton method follows the same principles than the proximal bundle method presented in Section 5.5. The only remarkable difference occurs in the termination condition for short and null steps, in other words (5.46) is replaced by two conditions

$$-\beta_{k+1}^{k+1} + \xi_{k+1}^T d_k \geq m_R v_k \tag{5.53}$$

and

$$\|x_{k+1} - y_{k+1}\| \leq C_S, \tag{5.54}$$

where $C_S > 0$ is a parameter supplied by the user.

The bundle-Newton method uses the line search algorithm presented in Lukšan and Vlček, 1995. For a locally Lipschitz and upper semismooth objective function one can prove the following global convergence result (see Lukšan and Vlček, 1995).

Theorem 5.6 *Let f be locally Lipschitz and upper semismooth, and the sequences $\{x_k\}_{k=1}^\infty$ and $\{H_k\}_{k=1}^\infty$ be bounded. Then every accumulation point of $\{x_k\}$ is substationary.*

Remark 5.1 *In order to find all the substationary points we can change the initial point x_1 in above methods. Then the local minimas and saddle points can be found, at least in theory. Furthermore, by minimizing $-f$ also all the local maximas are, in principle, reachable.*

References

Auslender, A. (1987). Numerical methods for nondifferentiable convex optimization. *Mathematical Programming Study*, 30:102–126.

Bihain, A. (1984). Optimization of upper semidifferentiable functions. *Journal of Optimization Theory and Applications*, 4:545–568.

Bonnans, J. F., Gilbert, J. C., Lemaréchal, C., and Sagastizábal, C. (1995). A family of variable metric proximal methods. *Mathematical Programming*, 68:15–47.

Brännlund, U. (1994). A descent method with relaxation type step. In Henry, J. and Yvon, J. P., editors, *Lecture Notes in Control and Information Sciences*, volume 197, pages 177–186, New York. Springer-Verlag.

Brännlund, U., Kiwiel, K. C., and Lindberg, P. O. (1996). Preliminary computational experience with a descent level method for convex nondifferentiable optimization. In zal, J. D. and Fidler, J., editors, *System Modelling and Optimization*, pages 387–394, London. Chapman & Hall.

Cheney, E. W. and Goldstein, A. A. (1959). Newton's method for convex programming and tchebycheff approximation. *Numerische Mathematik*, 1:253–268.

Fletcher, R. (1987). *Practical Methods of Optimization*. John Wiley, Chichester.

Gaudioso, M. and Monaco, M. F. (1982). A bundle type approach to the unconstrained minimization of convex nonsmooth functions. *Mathematical Programming*, 23:216–226.

Gaudioso, M. and Monaco, M. F. (1988). Some techniques for finding the search direction in nonsmooth minimization problems. Technical Report 75, Dipartimento di Sistemi, Universita' della Calabria.

Gaudioso, M. and Monaco, M. F. (1992). Variants to the cutting plane approach for convex nondifferentiable optimization. *Optimization*, 25:65–75.

Kelley, J. E. (1960). The cutting plane method for solving convex programs. *SIAM J*, 8:703–712.

Kiwiel, K. C. (1985). *Methods of Descent for Nondifferentiable Optimization*. Springer-Verlag, Berlin.

Kiwiel, K. C. (1990). Proximity control in bundle methods for convex nondifferentiable optimization. *Mathematical Programming*, 46:105–122.

Kiwiel, K. C. (1991). A tilted cutting plane proximal bundle method for convex nondifferentiable optimization. *Operations Research Letters*, 10:75–81.

Kiwiel, K. C. (1995). Proximal level bundle methods for convex nondifferentiable optimization, saddle-point problems and variational inequlities. *Mathematical Programming*, 69:89–109.

Kiwiel, K. C. (1996). Restricted step and levenberg-marquardt techniques in proximal bundle methods for nonconvex nondifferentiable optimization. *SIAM Journal on Optimization*, 6:227–249.

Lemaréchal, C. (1975). An extension of davidon methods to non differentiable problems. *Mathematical Programming Study*, 31:95–109.

Lemaréchal, C. (1976). Combining kelley's and conjugate gradient methods. In *Abstracts of IX International Symposium on Mathematical Programming*, Budapest, Hungary.

Lemaréchal, C. (1978). Nonsmooth optimization and descent methods. Technical report, IIASA-report, Laxemburg, Austria.

Lemaréchal, C. (1982). Numerical experiments in nonsmooth optimization. In Nurminski, E. A., editor, *Progress in Nondifferentiable Optimization*, IIASA-report, pages 61–84, Laxemburg, Austria.

Lemaréchal, C. (1989). Nondifferentiable optimization. In Nemhauser, G. L., Kan, A. H. G. R., and Todd, M. J., editors, *Optimization*, pages 529–572, Amsterdam. North-Holland.

Lemaréchal, C., Nemirovskii, A., and Nesterov, Y. (1995). New variants of bundle methods. *Mathematical Programming*, 69:111–147.

Lemaréchal, C. and Sagastizábal, C. (1994). An approach to variable metric bundle methods. In Henry, J. and Yvon, J. P., editors, *Lecture Notes in Control and Information Sciences*, volume 197, pages 144–162, New York. Springer-Verlag.

Lemaréchal, C., Strodiot, J.-J., and Bihain, A. (1981). On a bundle algorithm for nonsmooth optimization. In Mangasarian, O. L., Mayer, R. R., and Robinson, S. M., editors, *Nonlinear Programming*, volume 4, pages 245–281, New York. Academic Press.

Levenberg, K. (1944). A method for the solution of certain nonlinear problems in least squares. *Quart. Appl. Math.*, 2:164–166.

Lukšan, L. and Vlček, J. (1995). A bundle-newton method for nonsmooth unconstrained minimization. Technical report 654, Institute of Computer Science, Academy of Sciences of the Czech Republic.

Mäkelä, M. M. (1989). Methods and algorithms for nonsmooth optimization. Reports on Applied Mathematics and Computing 2, University of Jyväskylä, Department of Mathematics.

Mäkelä, M. M., Miettinen, M., Lukšan, L., and Vlček, J. (1999). Comparing nonsmooth nonconvex bundle methods in solving hemivariational inequalities. *J. Global Optim.*, 14:117–135.

Mäkelä, M. M. and Neittaanmäki, P. (1992). *Nonsmooth Optimization: Analysis and Algorithms with Applications to Optimal Control.* World Scientific Publishing Co., Singapore.

Marquardt, D. W. (1963). An algorithm for least-squares estimation of nonlinear parameters. *SIAM Journal on Applied Mathematics*, 11:431–441.

Mifflin, R. (1977). An algorithm for constrained optimization with semismooth functions. *Mathematics of Operations Research*, 2:191–207.

Mifflin, R. (1982). A modification and an extension of lemaréchal's algorithm for nonsmooth minimization. *Mathematical Programming Study*, 17:77–90.

Mifflin, R. (1996). A quasi-second-order proximal bundle algorithm. *Mathematical Programming*, 73:51–72.

Moreau, J. J. (1965). Proximité et dualité dans un espace hilbertien. *Bulletin de la Société Mathématique de France*, 93:273–299.

Outrata, J., Kočvara, M., and Zowe, J. (1998). *Nonsmooth Approach to Optimization Problems with Equilibrium Constraints. Theory, Applications and Numerical Results*, volume 28 of *Nonconvex Optimization and its Applications*. Kluwer Academic Publishers, Dordrecht.

Polak, E., Mayne, D. Q., and Wardi, Y. (1983). On the extension of constrained optimization algorithms from differentiable to nondifferentiable problems. *SIAM Journal on Optimal Control and Optimization*, 21:179–203.

Rockafellar, R. T. (1976). Monotone operators and the proximal point algorithm. *SIAM Journal on Optimal Control and Optimization*, 14:877–898.

Schramm, H. and Zowe, J. (1992). A version of the bundle idea for minimizing a nonsmooth functions: Conceptual idea, convergence analysis, numerical results. *SIAM Journal on Optimization*, 2:121–152.

Shor, N. Z. (1985). *Minimization Methods for Non-differentiable Functions.* Springer-Verlag, Berlin.

Strodiot, J.-J., Nguyen, V. H., and Heukemes, N. (1983). ε-optimal solutions in nondifferentiable convex programming and some related questions. *Mathematical Programming*, 25:307–328.

Tarasov, V. N. and Popova, N. K. (1984). A modification of the cutting-plane method with accelerated convergence. In Demyanov, V. F. and Pallaschke, D., editors, *Nondifferentiable Optimization: Motivations and Applications*, pages 284–290, Berlin. Springer-Verlag.

Uryas'ev, S. P. (1991). New variable metric algorithms for nondifferentiable optimization problems. *Journal of Optimization Theory and Applications*, 71:359–388.

Wolfe, P. (1975). A method of conjugate subgradients for minimizing nondifferentiable functions. *Mathematical Programming Study*, 3:145–173.

Yosida, K. (1964). *Functional Analysis*. Springer-Verlag.

IV Numerical Examples

6 NUMERICAL EXAMPLES

The aim of this chapter is to illustrate how previous theoretical results can be used for the numerical realization of several model examples. Our strategy is to transform the discrete hemivariational inequality to a problem of finding substationary points of the corresponding superpotential, and then to solve this by using nonsmooth and nonconvex optimization methods introduced in Chapter 5 (see Miettinen and Haslinger, 1995, Miettinen et al., 1995, Mäkelä et al., 1999 for some earlier numerical tests). The advantage of this strategy is that it is mathematically justified and is applicable to a large class of hemivariational inequalities. Other possibilities are: either to use some heuristic methods or to impose some additional restrictions on the nonconvexity (a difference of two convex functions, e.g.) and then to use some special methods (see Panagiotopoulos, 1993, Dem'yanov et al., 1996, Mistakidis and Stavroulakis, 1998).

We restrict ourselves to static hemivariational inequalities of scalar type. First three examples analyse the behaviour of an elastic structure supported by a foundation assuming nonmonotone multivalued responses on the contact part corresponding to nonmonotone friction and contact conditions. In the fourth example we study a simple laminated composite structure under loading when the binding material between the laminae obeys a nonmonotone multivalued law.

In the numerical tests we have used two proximal bundle codes: PB by L. Lukšan and J. Vlček and PBM by M. Mäkelä (used only in Example 6.4) and one bundle-Newton code by L. Lukšan and J. Vlček. We refer to Mäkelä and

Neittaanmäki, 1992, Lukšan and Vlček, 1995, Lukšan and Vlček, 1998 for their details. The numerical experiments have been performed by using Pentium II PC (240 MHz, 256 MB) in Examples 6.1-6.3 and HP9000/J280 (180 MHz) in Example 6.4.

6.1 NONMONOTONE FRICTION AND CONTACT PROBLEMS

Physical data for the first set of examples are the same: the body, represented by the unit square $1m \times 1m$ is made of an elastic isotropic material, characterized by the modulus of elasticity $E = 2.15 \cdot 10^{11} N/m^2$ and Poisson's ratio $\nu = 0.29$ (steel). Assuming the plane stress case, the linear Hooke's law is expressed by

$$\tau_{ij} = \frac{E\nu}{1-\nu^2}\delta_{ij}\vartheta + \frac{E}{1+\nu}\varepsilon_{ij}, \quad i,j = 1,2, \tag{6.1}$$

where $\vartheta = \varepsilon_{ii}$ is the trace of the strain tensor ε and δ_{ij} is the Kronecker symbol. The boundary $\partial\Omega$ of Ω is decomposed into Γ_u, Γ_c and $\Gamma_P^1 \cup \Gamma_P^2$ (as follows from Fig. 6.1).

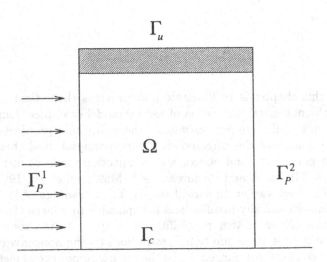

Figure 6.1.

On Γ_u the body is fixed:

$$u_i = 0 \quad \text{on } \Gamma_u, \ i = 1,2, \tag{6.2}$$

while on $\Gamma_P^1 \cup \Gamma_P^2$ surface tractions T will be applied. We take

$$T = (P,0) \quad \text{on } \Gamma_P^1, \text{where } P = 1.0 \times 10^6 N/m^2, \tag{6.3}$$
$$T = (0,0) \quad \text{on } \Gamma_P^2.$$

The examples differ by conditions prescribed on the contact part Γ_c.

Example 6.1 (*bilateral contact with nonmonotone friction*) We assume that

$$\begin{cases} u_2 = 0 & \text{on } \Gamma_c \\ -T_1(x) \in \hat{b}(u_1(x)) & \text{for a.a. } x \in \Gamma_c, \end{cases} \tag{6.4}$$

where the multifunction $\hat{b}$ is depicted in Fig. 6.2 with the following values of the parameters: $\delta_1 = 9.0 \times 10^{-6} m$, $\gamma_1 = 1.0 \times 10^3 N/m^2$, $\gamma_2 = 0.5 \times 10^3 N/m^2$.

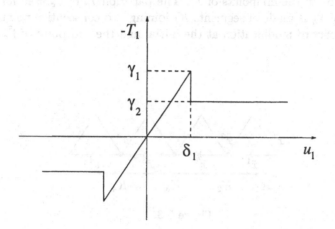

Figure 6.2.

Our aim is to find an equilibrium state of Ω which is characterized by the following hemivariational inequality:

$$\begin{cases} \text{Find } (u, \Xi) \in V \times L^2(\Gamma_c) \text{ such that} \\ a(u, v) + \displaystyle\int_{\Gamma_c} \Xi v_1 dx_1 = P \int_{\Gamma_P^1} v_1 dx_2 \quad \forall v \in V \\ \Xi(x) \in \hat{b}(u_1(x)) \quad \text{for a.a. } x \in \Gamma_c, \end{cases} \tag{$\mathcal{P}$}_1$$

where

$$V = \{v \in H^1(\Omega; \mathbf{R}^2) \mid v_i = 0 \text{ on } \Gamma_u, \ i = 1, 2, \ v_2 = 0 \text{ on } \Gamma_c\}$$

and

$$a(u, v) = \int_{\Omega} \tau_{ij}(u)\varepsilon_{ij}(v)dx \tag{6.5}$$

with $\tau(u)$, $\varepsilon(u)$ related by means of (6.1).

The approximation of $(\mathcal{P})_1$ will be realized by the finite element method as described in Chapter 3. The triangulation $\mathcal{D}_h$ of $\overline{\Omega}$ is constructed as follows: $\overline{\Omega}$ is partitioned into small squares of the size h and then each square is divided by its diagonal into two triangles. Four different triangulations $\mathcal{D}_h$ corresponding to $h = 1/8, 1/16, 1/32$ and $1/64$ will be used in the sequel.

We define

$$V_h = \{v_h = (v_{h1}, v_{h2}) \in C(\overline{\Omega}; \mathbf{R}^2) \mid v_h|_T \in (P_1(T))^2 \; \forall T \in \mathcal{D}_h,$$
$$v_{hi} = 0 \text{ on } \Gamma_u, \; i = 1, 2, \; v_{h2} = 0 \text{ on } \Gamma_c\}.$$

Type IV elements will be used for the construction of Y_h being the approximation of $Y = L^2(\Gamma_c)$. Let $\{T_i'\}$ be the system of all edges of the boundary elements $T_i \in \mathcal{D}_h$ such that $T_i' \subseteq \overline{\Gamma}_c$. Denote by x^i the nodes of $\mathcal{D}_h$ placed on $\overline{\Gamma}_c$ and by $\bar{x}^i$ the midpoints of T_i'. The partition $\mathcal{T}_h$ of $\overline{\Gamma}_c$ used for the construction of Y_h is made of segments K_i joining two consecutive points $\bar{x}^i$ with a straightforward modification at the initial and the end point of $\overline{\Gamma}_c$ (see Fig. 6.3).

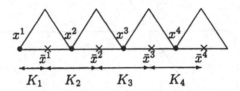

Figure 6.3.

Then we set

$$Y_h = \{\mu_h \in L^2(\Gamma_c) \mid \mu_h|_{K_i} \in P_0(K_i) \quad \forall K_i \in \mathcal{T}_h\}.$$

Since the mapping $\Pi : V \to L^2(\Gamma_c)$ is defined by $\Pi v = v_1$, $v \in V$ it is readily seen that

$$W_h = \Pi(V_h) = \{w_h \in C(\overline{\Gamma}_c) \mid w_h|_{T_i'} \in P_1(T_i') \quad \forall T_i'\}.$$

Let $P_h : W_h \to Y_h$ be defined by (3.76), i.e., P_h is the piecewise constant lagrange operator at $\{x^i\}$:

$$P_h w_h(x) = \sum_i w_h(x^i) \mathcal{X}_{\text{int}_{\Gamma_c} K_i}(x).$$

The approximation of $(\mathcal{P})_1$ now reads as follows:

$$\left\{ \begin{array}{l} \text{Find } (u_h, \Xi_h) \in V_h \times Y_h \text{ such that} \\[2mm] a(u_h, v_h) + \displaystyle\int_{\Gamma_c} \Xi_h P_h v_{h1} \, dx_1 = P \int_{\Gamma_P^1} v_{h1} \, dx_2 \quad \forall v_h \in V_h \\[4mm] \Xi_h(x^i) \in \hat{b}(u_{h1}(x^i)) \quad \forall i. \end{array} \right. \qquad (\mathcal{P})_1^h$$

Since the bilinear form a, the linear form f contains piecewise constant, piecewise linear integrands, respectively, all integrals can be evaluated exactly, i.e., no numerical integration is necessary. From Theorem 3.4 it follows that $(\mathcal{P})_1$ and $(\mathcal{P})_1^h$ are close on subsequences. More precisely it holds that:

$$u_h \to u \quad \text{in } V$$
$$\Xi_h \rightharpoonup \Xi \quad \text{in } L^2(\Gamma_c), \ h \to 0+$$

for an appropriate subsequence of solutions to $(\mathcal{P})_1^h$.

For the numerical realization of $(\mathcal{P})_1^h$ its algebraic representation will used (see Section 3.5):

$$
\begin{cases}
\text{Find } (\vec{u}, \vec{\Xi}) \in \mathbf{R}^n \times \mathbf{R}^m \text{ such that} \\
(\mathbf{A}\vec{u}, \vec{v})_{\mathbf{R}^n} + (\vec{\Xi}, \mathbf{P}\mathbf{\Pi}\vec{v})_{\mathbf{R}^m} = (\vec{f}, \vec{v})_{\mathbf{R}^n} \quad \forall \vec{v} \in \mathbf{R}^n, \\
\Xi_i \in c_i \hat{b}((\mathbf{P}\mathbf{\Pi}\vec{u})_i) \quad \forall i,
\end{cases}
\tag{6.6}
$$

where $n = \dim V_h$, $m = \dim Y_h$, $c_1 = c_m = h/2$, $c_i = h$, $2 \le i \le m-1$. The remaining symbols have the same meaning as in Section 3.5.

In the case of nonmonotone relations prescribed on $\partial\Omega$ (or its part) one can considerably simplify the problem by eliminating those components of $\vec{u}$ which do not appear on the right hand side of the inclusion. We shall illustrate this procedure in the case of (6.6).

Let $\vec{v} \in \mathbf{R}^n$ be the nodal value vector of $v_h = (v_{h1}, v_{h2}) \in V_h$. By $\vec{v}_t$ we denote a subvector of $\vec{v}$, containing the nodal values of v_{h1} at the nodes $x^i \in \overline{\Gamma}_c$. Assuming that the components of $\vec{v}_t$ are listed last in $\vec{v}$, we may write $\vec{v} = (\vec{v}_i, \vec{v}_t) \in \mathbf{R}^{n-d} \times \mathbf{R}^d$. In our case it is easy to see that $d = m$. The decomposition of $\vec{v}$ into $\vec{v}_i$ and $\vec{v}_t$ yields the following block structure of $\mathbf{A}$:

$$
\mathbf{A} = \begin{pmatrix} \mathbf{A}_{ii} & \mathbf{A}_{it} \\ \mathbf{A}_{ti} & \mathbf{A}_{tt} \end{pmatrix}.
$$

The substitution of vectors of the form $(\vec{v}_i, \vec{0}_t)$, $(\vec{0}_i, \vec{v}_t)$, where $\vec{0}$ stands for the zero vector in $\mathbf{R}^n$, leads to the following equivalent formulation of (6.6):

$$
\begin{cases}
\text{Find } (\vec{u}_i, \vec{u}_t, \vec{\Xi}) \in \mathbf{R}^{n-m} \times \mathbf{R}^m \times \mathbf{R}^m \text{ such that} \\
\mathbf{A}_{ii}\vec{u}_i + \mathbf{A}_{it}\vec{u}_t = \vec{f}_i \\
\mathbf{A}_{ti}\vec{u}_i + \mathbf{A}_{tt}\vec{u}_t + \vec{\Xi} = \vec{f}_t \\
\Xi_i \in c_i \hat{b}((\vec{u}_t)_i) \quad \forall i.
\end{cases}
\tag{6.7}
$$

Here we used the fact that $\mathbf{\Pi}\vec{v} = \vec{v}_t$ and $\mathbf{P} = \mathrm{id}$.

From the first equation in (6.7) we can compute $\vec{u}_i$:

$$\vec{u}_i = \mathbf{A}_{ii}^{-1}(\vec{f}_i - \mathbf{A}_{it}\vec{u}_t). \tag{6.8}$$

Inserting (6.8) into the second equation in (6.7) we can completely eliminate $\vec{u}_i$. The resulting problem reads as follows:

$$
\begin{cases}
\text{Find } (\vec{u}_t, \vec{\Xi}) \in \mathbf{R}^m \times \mathbf{R}^m \text{ such that} \\
\underset{\sim}{\mathbf{A}}\vec{u}_t + \vec{\Xi} = \underset{\sim}{\vec{f}} \\
\Xi_i \in c_i \hat{b}((\vec{u}_t)_i) \quad \forall i,
\end{cases}
\tag{6.9}
$$

where $\underset{\sim}{\mathbf{A}} = \mathbf{A}_{tt} - \mathbf{A}_{ti}\mathbf{A}_{ii}^{-1}\mathbf{A}_{it}$ is the Schur complement and $\underset{\sim}{\vec{f}} = \vec{f}_t - \mathbf{A}_{ti}\mathbf{A}_{ii}^{-1}\vec{f}_i$.

Therefore it is sufficient to find a solution $(\vec{u}_t, \vec{\Xi})$ of (6.9) and from (6.8) the subvector $\vec{u}_i$ can be recovered.

As we have already mentioned in Section 3.5, no mathematically justified methods for the direct realization of (6.9) are available at present. For this reason, the hemivariational inequality (6.9) will be replaced by a substationary type problem for an appropriate superpotential $\mathcal{L}$ as we have shown in Section 3.5. Let Ψ_t be defined by

$$\Psi_t(\vec{v}_t) = \sum_i c_i \int_0^{(\vec{v}_t)_i} b(\xi)d\xi,$$

where $c_i \in \mathbf{R}$ are the same as in (6.9), $(\vec{v}_t)_i$ stands for the i-th component of $\vec{v}_t$ and $b \in L^\infty(\mathbf{R})$ is the function determining $\hat{b}$ (see (3.6)-(3.8)). Then

$$\mathcal{L}(\vec{v}_t) = \frac{1}{2}(\underset{\sim}{\mathbf{A}}\vec{v}_t, \vec{v}_t)_{\mathbf{R}^m} - (\underset{\sim}{\vec{f}}, \vec{v}_t)_{\mathbf{R}^m} + \Psi_t(\vec{v}_t)$$

is the respective superpotential. Instead of (6.9) we shall solve:

$$\text{Find } \vec{u}_t \in \mathbf{R}^m : \vec{0}_t \in \bar{\partial}\mathcal{L}(\vec{u}_t) \equiv \underset{\sim}{\mathbf{A}}\vec{u}_t - \underset{\sim}{\vec{f}} + \bar{\partial}\Psi_t(\vec{u}_t). \tag{6.10}$$

From the properties of b and $P_h : W_h \to Y_h$ it follows that both problems (6.9) and (6.10) are equivalent in the sense of Theorem 3.6.

For the numerical realization of (6.10) nonsmooth minimization methods of Chapter 5 can be used. This and next two examples are solved by using *the proximal bundle first order* and *the bundle-Newton second order method*. The convergence of these methods to a substationary point of $\mathcal{L}$ is guaranteed by the upper semismoothness of $\mathcal{L}$ and Theorems 5.5 and 5.6 (this holds also for Examples 6.2-6.4).

Remark 6.1 *Let us check that the function $\mathcal{L}$ is upper semismooth. Let $\vec{v}_t, \vec{d}_t \in \mathbf{R}^m$ be given. From Definition 5.1 of the upper semismoothness we see that it is sufficient to study the behaviour of $\mathcal{L}$ on the straight line $S = \{\vec{v}_t + r\vec{d}_t : r \in \mathbf{R}\}$. Due to the character of the function $\hat{b}$, the superpotential $\mathcal{L}|_S$ is smooth except a finite number of points (discontinuity points of b). Now if $\mathcal{L}|_S$ is smooth at $\vec{v}_t$, it is smooth also in some small (one-dimensional) neighbourhood of $\vec{v}_t$ and the upper semismoothness condition*

$$\limsup_{k\to\infty} \vec{g}_k^T \vec{d}_t \geq \liminf_{k\to\infty} [\mathcal{L}(\vec{v}_t + r_k\vec{d}_t) - \mathcal{L}(\vec{v}_t)]/r_k, \tag{6.11}$$

in which $\vec{g}_k \in \bar{\partial}\mathcal{L}(\vec{v}_t + r_k\vec{d}_t)$ and $r_k \downarrow 0$, is satisfied trivially. On the other hand, if $\vec{v}_t$ is a nonsmooth point of $\mathcal{L}$ there exists a small neighbourhood of $\vec{v}_t$ in which $\mathcal{L}|_S$ is smooth except $\vec{v}_t$ and the classical one-sided continuous derivatives exist at $\vec{v}_t$. This implies that (6.11) is reduced to

$$\limsup_{k\to\infty} \mathcal{L}'(\vec{v}_t + r_k\vec{d}_t; \vec{d}_t) \geq \mathcal{L}'(\vec{v}_t; \vec{d}_t) \tag{6.12}$$

which is satisfied (in (6.12) there is actually the equality).

As usually, all minimization methods provide us with a numerical solution $\vec{u}_t^{comp}$ which can be more or less accurate. To verify if $\vec{u}_t^{comp}$ is good enough we go back to (6.9): the vector $\vec{\Xi} \in \mathbf{R}^m$ is the solution of

$$\vec{\Xi} = \vec{f} - \mathbf{A}\vec{u}_t. \tag{6.13}$$

Let $\vec{\Xi}^{comp}$ be the solution of (6.13) with the right hand side $\vec{f} - \mathbf{A}\vec{u}_t^{comp}$. If $\vec{u}_t^{comp}$ was good enough then the components Ξ_i^{comp} of $\vec{\Xi}^{comp}$ should satisfy the inclusions

$$\Xi_i^{comp} \in c_i \hat{b}((\vec{u}_t^{comp})_i) \quad \forall i \tag{6.14}$$

as follows from (6.9). The violation of any such inclusion simply means that $\vec{u}_t^{comp}$ has to be found with a higher accuracy. The computation of $\vec{\Xi}$ from (6.13) together with the verification of (6.14) serve as a reliable stopping criterion of the computational process. Moreover, from the results of Chapter 3 we know that the respective $\Xi_h \in Y_h$ tends weakly in $L^2(\Gamma_c)$ to an element Ξ having a nice mechanical meaning, namely $\Xi = -T_1$ (friction force) on Γ_c.

The stopping criterion based solely on the use of $\vec{u}_t^{comp}$ could be rather deceptive. Indeed, after 200 iterations of the bundle-Newton method, the first component of $\vec{u}_t^{comp}$ (i.e., the tangential displacements of the first contact node x^1) is equal to $9.8568 \times 10^{-6}m$ while the respective Ξ_1 is less that $200N/m^2$ being very far from the prescribed value $1000N/m^2$. After 1200 iterations the value of $\vec{u}_t^{comp}$ at x^1 is equal to $9.8465 \times 10^{-6}m$ but Ξ_1 equals to $999.7N/m^2$ (a dramatic improvement).

Let us briefly comment the following figures representing the numerical results:

Fig 6.4. shows the behaviour of the tangential displacements along Γ_c for the different number of the contact nodes ($m = 9, 17, 33, 65$ corresponding to $\mathcal{D}_h$ with $h = 1/8, 1/16, 1/32, 1/64$);

Fig 6.5. shows the distribution of the tangential component $-T_1$ of the stress vector along Γ_c after 1200 iterations of the bundle-Newton method for the number of the contact nodes $m = 65$. The graph is obtained by connecting the points (i, Ξ_i), $i = 1, ..., 65$, where Ξ_i is the i-th component of $\vec{\Xi}$ computed from (6.13);

Fig 6.6. the same as in the previous figure but using the proximal bundle method PB;

Fig 6.7. shows the history of $\vec{\Xi}$ computed from (6.13) after 200,400,...,1200 iterations of the bundle-Newton method for the number of the contact nodes $m = 65$;

Fig 6.8. compares the tangential displacements at the first ten contact nodes for (i) *a frictionless case* (ii) *a monotone model of friction* characterized by the multifunction $\hat{b}$ given by Fig. 2.1 with the given

slip stress bound $g = 1000N/m^2$ (iii) *a nonmonotone friction law*
depicted in Fig.6.2.

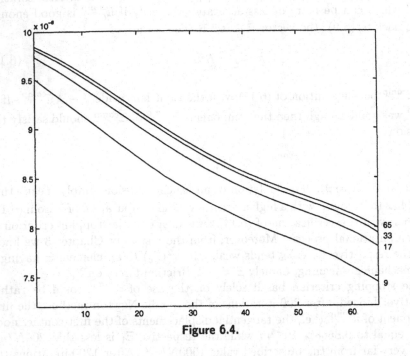

Figure 6.4.

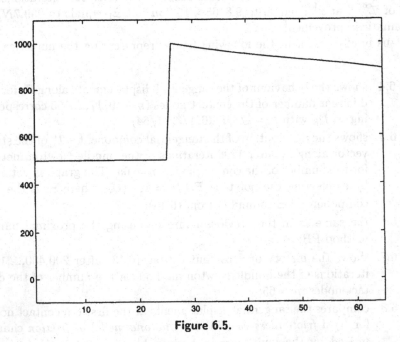

Figure 6.5.

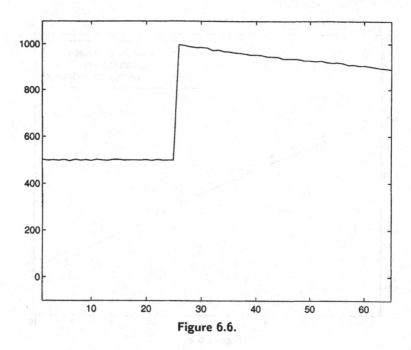

Figure 6.6.

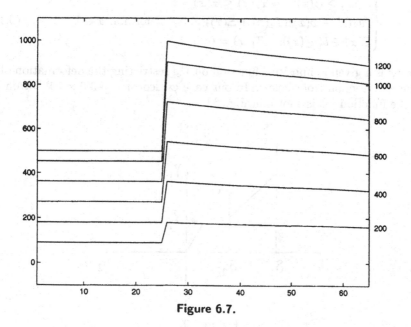

Figure 6.7.

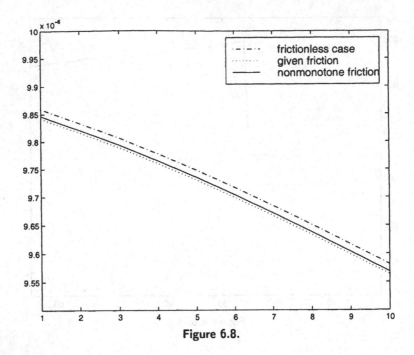

Figure 6.8.

Example 6.2 (*nonmonotone unilateral contact without friction*) In this case, the boundary conditions on Γ_c read as follows:

$$\begin{cases} u_2(x) \geq \varphi(x), \quad -T_2(x) \leq \Xi(x) \\ (u_2(x) - \varphi(x))(T_2(x) + \Xi(x)) = 0 \quad \text{for a.a. } x \in \Gamma_c \\ \Xi(x) \in \hat{b}(u_2(x)), \quad T_1(x) = 0 \end{cases} \quad (6.15)$$

where φ is a given nonpositive function on Γ_c, restricting the deformation of Ω in the x_2-direction from below. In our case $\varphi \equiv \text{const.} = -3.0 \times 10^{-6} m$ on Γ_c and the function $-\hat{b}$ is shown in Fig. 6.9,

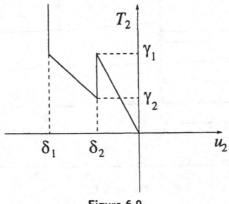

Figure 6.9.

with $\gamma_1 = 1.0 \times 10^3 N/m^2$, $\gamma_2 = 0.5 \times 10^3 N/m^2$, $\delta_1 = -3.0 \times 10^{-6} m$, $\delta_2 = -1.5 \times 10^{-6} m$. The weak form of this problem leads to the following constrained hemivariational inequality:

$$\begin{cases} \text{Find } (u, \Xi) \in K \times L^2(\Gamma_c) \text{ such that} \\ a(u, v - u) + \int_{\Gamma_c} \Xi(v_2 - u_2) dx_1 \geq P \int_{\Gamma_P^1} (v_1 - u_1) dx_2 \quad \forall v \in K \qquad (\mathcal{P})_2 \\ \Xi(x) \in \hat{b}(u_2(x)) \quad \text{for a.a. } x \in \Gamma_c, \end{cases}$$

where

$$K = \{ v \in H^1(\Omega; \mathbf{R}^2) \mid v_i = 0 \text{ on } \Gamma_u, \ i = 1, 2, \ v_2 \geq \varphi \text{ a.e. on } \Gamma_c \}$$

is the set of kinematically admissible displacements and the other symbols have the same meaning as before.

Next, we describe the approximation of $(\mathcal{P})_2$. Let $\mathcal{D}_h$ be the triangulation from Example 6.1. Denote by

$$K_h = \{ v_h = (v_{h1}, v_{h2}) \in C(\overline{\Omega}; \mathbf{R}^2) \mid v_h|_T \in (P_1(T))^2 \ \forall T \in \mathcal{D}_h,$$
$$v_{hi} = 0 \text{ on } \Gamma_u, \ i = 1, 2, \ v_{h2}(x^i) \geq \varphi(x^i) \ \forall i \},$$

i.e., K_h is the convex subset of all continuous, piecewise linear functions on $\mathcal{D}_h$, vanishing on Γ_u and satisfying the unilateral boundary condition at all the contact nodes x^i. Since $\varphi \equiv \text{const.} < 0$ on Γ_c we have that $K_h \subset K$ and $0 \in K_h$ $\forall h > 0$. The space Y_h is defined as in Example 6.1.

The approximation of $(\mathcal{P})_2$ reads as follows:

$$\begin{cases} \text{Find } (u_h, \Xi_h) \in K_h \times Y_h \text{ such that} \\ a(u_h, v_h - u_h) + \int_{\Gamma_c} \Xi_h P_h(v_{h2} - u_{h2}) dx_1 \\ \geq P \int_{\Gamma_P^1} (v_{h1} - u_{h1}) dx_2 \quad \forall v_h \in K_h \qquad (\mathcal{P})_2^h \\ \Xi(x^i) \in \hat{b}(u_{h2}(x^i)) \quad \forall i, \end{cases}$$

with P_h given again by (3.76). From Theorem 3.9 it follows that solutions of $(\mathcal{P})_2$ and $(\mathcal{P})_2^h$ are close on subsequences:

$$u_h \to u \quad \text{in } V,$$
$$\Xi_h \rightharpoonup \Xi \quad \text{in } L^2(\Gamma_c), \ h \to 0+$$

for an appropriate sequence of solutions to $(\mathcal{P})_2^h$.

The algebraic form of $(\mathcal{P})_2^h$ is obvious:

$$\begin{cases} \text{Find } (\vec{u}, \vec{\Xi}) \in \mathcal{K} \times \mathbf{R}^m \text{ such that} \\ (\mathbf{A}\vec{u}, \vec{v} - \vec{u})_{\mathbf{R}^n} + (\vec{\Xi}, \mathbf{P}\Pi(\vec{v} - \vec{u}))_{\mathbf{R}^m} \geq (\vec{f}, \vec{v} - \vec{u})_{\mathbf{R}^n} \quad \forall \vec{v} \in \mathcal{K} \qquad (6.16) \\ \Xi_i \in c_i \hat{b}((\mathbf{P}\Pi\vec{u})_i) \quad \forall i. \end{cases}$$

The symbol $\mathcal{K}$ stands for the convex subset of $\mathbf{R}^n$ defined by

$$\mathcal{K} = \{\vec{x} \in \mathbf{R}^n \mid \vec{x}_\nu \geq -3.0 \times 10^{-6}\},$$

where $\vec{x}_\nu$ is the subvector of $\vec{x}$, whose components are the normal displacements at the contact nodes $x^i \in \bar{\Gamma}_c$. The other symbols keep the meaning from the previous example. As before, problem (6.16) can be transformed to a simpler problem by eliminating all unconstrained variables. To this end write $\vec{x} = (\vec{x}_i, \vec{x}_\nu)$, where $\vec{x}_i$ is the subvector of $\vec{x}$ containing the displacements at the inner nodes of $\mathcal{D}_h$ and the tangential components of displacements at the contact nodes. The stiffness matrix $\mathbf{A}$ can be decomposed as well:

$$\mathbf{A} = \begin{pmatrix} \mathbf{A}_{ii} & \mathbf{A}_{i\nu} \\ \mathbf{A}_{\nu i} & \mathbf{A}_{\nu\nu} \end{pmatrix}.$$

Then (6.16) is equivalent to

$$\begin{cases} \text{Find } (\vec{u}_i, \vec{u}_\nu, \vec{\Xi}) \in \mathbf{R}^{n-m} \times \mathcal{K}_\nu \times \mathbf{R}^m \text{ such that} \\ \mathbf{A}_{ii}\vec{u}_i + \mathbf{A}_{i\nu}\vec{u}_\nu = \vec{f}_i \\ (\underset{\sim}{\mathbf{A}}\vec{u}_\nu, \vec{v}_\nu - \vec{u}_\nu)_{\mathbf{R}^m} + (\vec{\Xi}, \vec{v}_\nu - \vec{u}_\nu)_{\mathbf{R}^m} \\ \geq (\underset{\sim}{\vec{f}}, \vec{v}_\nu - \vec{u}_\nu)_{\mathbf{R}^m} \quad \forall \vec{v}_\nu \in \mathcal{K}_\nu \\ \Xi_i \in c_i \hat{b}((\vec{u}_\nu)_i) \quad \forall i, \end{cases} \tag{6.17}$$

where

$$\mathcal{K}_\nu = \{\vec{v}_\nu \in \mathbf{R}^m \mid \vec{v}_\nu \geq -3.0 \times 10^{-6}\},$$
$$\underset{\sim}{\mathbf{A}} = \mathbf{A}_{\nu\nu} - \mathbf{A}_{\nu i}\mathbf{A}_{ii}^{-1}\mathbf{A}_{i\nu},$$
$$\underset{\sim}{\vec{f}} = \vec{f}_\nu - \mathbf{A}_{\nu i}\mathbf{A}_{ii}^{-1}\vec{f}_i,$$

taking into account that $\Pi\vec{v} = \vec{v}_t$ and $\mathbf{P} = \text{id}$.

First we find solutions to the constrained hemivariational inequality for $\vec{u}_\nu$. We replace the inequality in (6.17) by an equivalent problem of finding substationary points of a superpotential $\mathcal{L}$ with respect to $\mathcal{K}_\nu$:

$$\text{Find } \vec{u}_\nu \in \mathbf{R}^m \text{ such that } \vec{0}_\nu \in \bar{\partial}\mathcal{L}(\vec{u}_\nu) + N_{\mathcal{K}_\nu}(\vec{u}_\nu), \tag{6.18}$$

where

$$\mathcal{L}(\vec{v}_\nu) = \frac{1}{2}(\underset{\sim}{\mathbf{A}}\vec{v}_\nu, \vec{v}_\nu)_{\mathbf{R}^m} - (\underset{\sim}{\vec{f}}, \vec{v}_\nu) + \Psi_\nu(\vec{v}_\nu),$$

Ψ_ν is a Lipschitz function derived from $\hat{b}$ and $N_{\mathcal{K}_\nu}(\vec{u}_\nu)$ stands for the normal cone of $\mathcal{K}_\nu$ at $\vec{u}_\nu$. Problem (6.18) can be realized by nonsmooth bundle type methods. Having $\vec{u}_\nu$ at our disposal we recover $\vec{u}_i$ from the first equation in (6.17). Using the approach described in Section 3.6 one can also determine the components Ξ_i of $\vec{\Xi}$ at those $x^i \in \overline{\Gamma}_c$ which are not in contact with the foundation (see (3.120)).

The numerical results are depicted in the following figures:

Fig 6.10. shows the deformation of Γ_c for the different number of the contact nodes $(m = 9, 17, 33, 65)$;

Fig 6.11. shows the distribution of the normal component T_1 of the stress vector on Γ_c after 176 iterations of the bundle-Newton method for the number of the contact nodes $m = 65$. The graph is obtained by connecting the points (i, Ξ_i), where the index i corresponds to the nonactive constraint and Ξ_i is computed by (3.120);

Fig 6.12. shows the history of $\vec{\Xi}$ after $10, 30, 60, 176$ iterations of the bundle-Newton method;

Fig 6.13. compares the deformations of Γ_c obtained by solving the classical Signorini problem without friction defined by (1.83)-(1.86) and our nonmonotone unilateral problem, respectively.

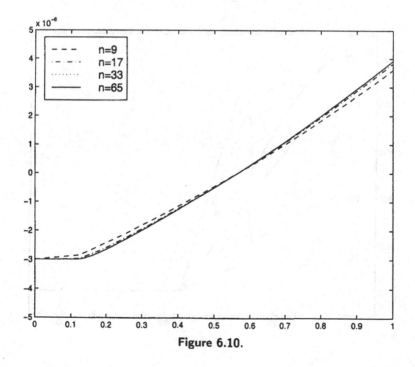

Figure 6.10.

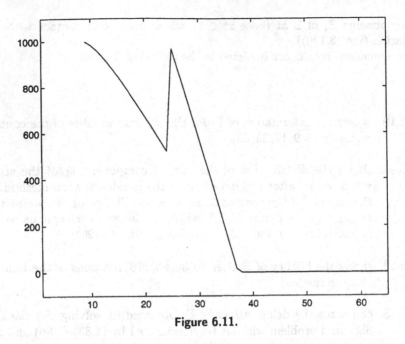

Figure 6.11.

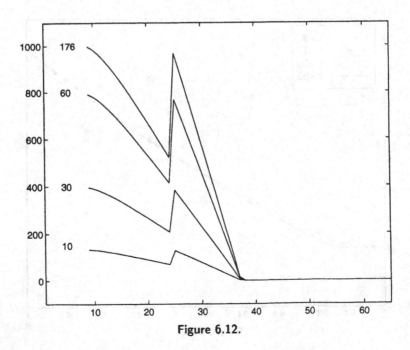

Figure 6.12.

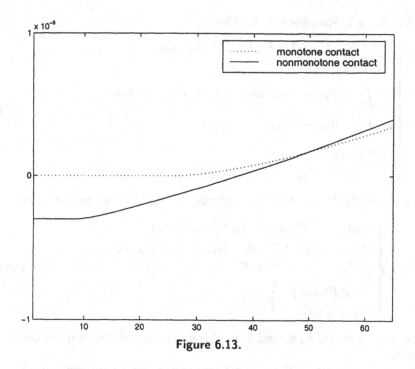

Figure 6.13.

Example 6.3 (*nonmonotone unilateral and friction problem*) This example combines both, the nonmonotone friction of Example 6.1 and the nonmonotone unilateral conditions of Example 6.2. The weak formulation of this problem reads as follows (see Remark 3.18):

$$
\begin{cases}
\text{Find } (u, \Xi', \Xi'') \in K \times (L^2(\Gamma_c))^2 \text{ such that} \\[2mm]
a(u, v - u) + \displaystyle\int_{\Gamma_c} \Xi'(v_1 - u_1)dx_1 + \int_{\Gamma_c} \Xi''(v_2 - u_2)dx_1 \\[2mm]
\geq P \displaystyle\int_{\Gamma_P^1} (v_1 - u_1)dx_2 \quad \forall v \in K \\[2mm]
\left.\begin{array}{l}
\Xi'(x) \in \hat{b}'(u_1(x)) \\[1mm]
\Xi''(x) \in \hat{b}''(u_2(x))
\end{array}\right\} \quad \text{for a.a. } x \in \Gamma_c,
\end{cases} \qquad (\mathcal{P})_3
$$

where the convex set K is the same as in Example 6.2, the multifunction $\hat{b}'$ is represented by Fig. 6.2 with $\gamma_1 = 1.0 \times 10^3 N/m^2$, $\gamma_2 = 0.8 \times 10^3 N/m^2$ and $\delta_1 = 1.25 \times 10^{-5}m$, while $-\hat{b}''$ by Fig. 6.9 with the following values of the parameters: $\gamma_1 = 1.0 \times 10^4 N/m^2$, $\gamma_2 = 0.5 \times 10^4 N/m^2$, $\delta_1 = -3.0 \times 10^{-6}m$, $\delta_2 = -1.5 \times 10^{-6}m$. For the approximation of $(\mathcal{P})_3$ we use exactly the same sets K_h, Y_h as in Example 6.2. We obtain:

$$\left\{\begin{array}{l} \text{Find } (u_h, \Xi_h', \Xi_h'') \in K_h \times (Y_h)^2 \text{ such that} \\ a(u_h, v_h - u_h) \\ + \displaystyle\int_{\Gamma_c} \Xi_h' P_h(v_{h1} - u_{h1}) dx_1 + \int_{\Gamma_c} \Xi_h'' P_h(v_{h2} - u_{h2}) dx_1 \\ \geq P \displaystyle\int_{\Gamma_P^1} (v_{h1} - u_{h1}) dx_2 \quad \forall v_h \in K_h \\ \left.\begin{array}{l} \Xi_h'(x^i) \in \hat{b}'(u_{h1}(x^i)) \\ \Xi_h''(x^i) \in \hat{b}''(u_{h2}(x^i)) \end{array}\right\} \quad \forall i \end{array}\right. \qquad (\mathcal{P})_3^h$$

with P_h given by (3.76). Also the algebraic form of $(\mathcal{P})_3^h$ can be easily derived:

$$\left\{\begin{array}{l} \text{Find } (\vec{u}, \vec{\Xi}', \vec{\Xi}'') \in \mathcal{K} \times (\mathbf{R}^m)^2 \text{ such that} \\ (\mathbf{A}\vec{u}, \vec{v} - \vec{u})_{\mathbf{R}^n} + (\vec{\Xi}', \vec{v}_t - \vec{u}_t)_{\mathbf{R}^m} + (\vec{\Xi}'', \vec{v}_\nu - \vec{u}_\nu)_{\mathbf{R}^m} \\ \geq (\vec{f}, \vec{v} - \vec{u})_{\mathbf{R}^n} \quad \forall \vec{v} \in \mathcal{K} \\ \left.\begin{array}{l} \Xi_i' \in \hat{b}'((\vec{u}_t)_i) \\ \Xi_i'' \in \hat{b}''((\vec{u}_\nu)_i) \end{array}\right\} \quad \forall i \end{array}\right. \qquad (6.19)$$

where the meaning of $\mathcal{K}$, $\vec{u}_\nu$ and $\vec{u}_t$ has been already introduced in the previous examples.

Denote by $\vec{v}_{t\nu} = (\vec{v}_t, \vec{v}_\nu) \in \mathbf{R}^{2m}$ the vector, whose components are the displacements at $x^i \in \overline{\Gamma}_c$. Then the displacement vector $\vec{v} \in \mathbf{R}^n$ can be written as $\vec{v} = (\vec{v}_i, \vec{v}_{t\nu})$ implying the following block structure of $\mathbf{A}$:

$$\mathbf{A} = \begin{pmatrix} \mathbf{A}_{ii} & \mathbf{A}_i^{t\nu} \\ \mathbf{A}_{t\nu}^i & \mathbf{A}_{t\nu}^{t\nu} \end{pmatrix}.$$

Then (6.19) is equivalent to

$$\left\{\begin{array}{l} \text{Find } (\vec{u}_i, \vec{u}_{t\nu}, \vec{\Xi}', \vec{\Xi}'') \in \mathbf{R}^{n-2m} \times \mathcal{K}_{t\nu} \times (\mathbf{R}^m)^2 \text{ such that} \\ \mathbf{A}_{ii}\vec{u}_i + \mathbf{A}_i^{t\nu}\vec{u}_{t\nu} = \vec{f}_i \\ (\mathbf{A}_{t\nu}^i \vec{u}_i, \vec{v}_{t\nu} - \vec{u}_{t\nu})_{\mathbf{R}^{2m}} + (\mathbf{A}_{t\nu}^{t\nu}\vec{u}_{t\nu}, \vec{v}_{t\nu} - \vec{u}_{t\nu})_{\mathbf{R}^{2m}} \\ +(\vec{\Xi}', \vec{v}_t - \vec{u}_t)_{\mathbf{R}^m} + (\vec{\Xi}'', \vec{v}_\nu - \vec{u}_\nu)_{\mathbf{R}^m} \\ \geq (\vec{f}_{t\nu}, \vec{v}_{t\nu} - \vec{u}_{t\nu})_{\mathbf{R}^{2m}} \quad \forall \vec{v}_{t\nu} \in \mathcal{K}_{t\nu} \\ \left.\begin{array}{l} \Xi_i' \in \hat{b}'((\vec{u}_t)_i) \\ \Xi_i'' \in \hat{b}''((\vec{u}_\nu)_i) \end{array}\right\} \quad \forall i \end{array}\right. \qquad (6.20)$$

where $\mathcal{K}_{t\nu} \equiv \mathbf{R}^m \times \mathcal{K}_\nu$ with the same $\mathcal{K}_\nu$ as in Example 6.2. The second inequality in (6.20) together with both inclusions define a constrained hemivariational inequality on $\overline{\Gamma}_c$ for the unknown vector $\vec{u}_{t\nu}$. This inequality will be again replaced by a substationary type problem:

$$\left\{\begin{array}{l} \text{Find } \vec{u}_{t\nu} \in \mathbf{R}^{2m} \text{ such that} \\ \vec{0}_{t\nu} \in \underset{\sim}{\mathbf{A}}\vec{u}_{t\nu} - \vec{f} + \bar{\partial}\Psi_t(\vec{u}_t) + \bar{\partial}\Psi_\nu(\vec{u}_\nu) + N_{\mathcal{K}_{t\nu}}(\vec{u}_{t\nu}), \end{array}\right. \qquad (6.21)$$

where Ψ_t, Ψ_ν are derived from $\hat{b}'$, $\hat{b}''$, respectively, $\mathbf{A} = \mathbf{A}_{t\nu}^{t\nu} - \mathbf{A}_{t\nu}^{i}\mathbf{A}_{ii}^{-1}\mathbf{A}_{i}^{t\nu}$
and $\vec{f} = \vec{f}_{t\nu} - \mathbf{A}_{t\nu}^{i}\mathbf{A}_{ii}^{-1}\vec{f}_{i}$. Problem (6.21) can be realized by using nonsmooth
bundle type methods. The remaining part $\vec{u}_i$ of $\vec{u}$ can be computed from the
first equation in (6.20).

Also the vectors $\vec{\Xi}'$, $\vec{\Xi}''$ can be recovered from (6.20). Indeed, let us write $\vec{u}$
in the form $\vec{u} = (\vec{u}_i, \vec{u}_t, \vec{u}_\nu)$ and similarly

$$\mathbf{A} = \begin{pmatrix} \mathbf{A}_{ii} & \mathbf{A}_{it} & \mathbf{A}_{i\nu} \\ \mathbf{A}_{ti} & \mathbf{A}_{tt} & \mathbf{A}_{t\nu} \\ \mathbf{A}_{\nu i} & \mathbf{A}_{\nu t} & \mathbf{A}_{\nu\nu} \end{pmatrix}.$$

Then (6.20) is equivalent to the following system:

$$\begin{cases} \text{Find } (\vec{u}_i, \vec{u}_t, \vec{u}_\nu, \vec{\Xi}', \vec{\Xi}'') \in \mathbf{R}^{n-2m} \times \mathbf{R}^m \times \mathcal{K}_\nu \times (\mathbf{R}^m)^2 \text{ such that} \\ \mathbf{A}_{ii}\vec{u}_i + \mathbf{A}_{it}\vec{u}_t + \mathbf{A}_{i\nu}\vec{u}_\nu = \vec{f}_i \\ \mathbf{A}_{ti}\vec{u}_i + \mathbf{A}_{tt}\vec{u}_t + \mathbf{A}_{t\nu}\vec{u}_\nu + \vec{\Xi}' = \vec{f}_t \\ (\mathbf{A}_{\nu i}\vec{u}_i, \vec{v}_\nu - \vec{u}_\nu)_{\mathbf{R}^m} + (\mathbf{A}_{\nu t}\vec{u}_t, \vec{v}_\nu - \vec{u}_\nu)_{\mathbf{R}^m} \\ +(\mathbf{A}_{\nu\nu}\vec{u}_\nu, \vec{v}_\nu - \vec{u}_\nu)_{\mathbf{R}^m} + (\vec{\Xi}'', \vec{v}_\nu - \vec{u}_\nu)_{\mathbf{R}^m} \\ \geq (\vec{f}_\nu, \vec{v}_\nu - \vec{u}_\nu)_{\mathbf{R}^m} \quad \forall \vec{v}_\nu \in \mathcal{K}_\nu \\ \left.\begin{array}{l} \Xi_i' \in \hat{b}'((\vec{u}_t)_i) \\ \Xi_i'' \in \hat{b}''((\vec{u}_\nu)_i) \end{array}\right\} \quad \forall i. \end{cases} \tag{6.22}$$

If $\vec{u}_i$, $\vec{u}_t$ and $\vec{u}_\nu$ are known, then $\vec{\Xi}'$ can be obtained by solving the second
equation in (6.22). Using the same approach as in Example 6.2 one can find
also those components Ξ_i'' of $\vec{\Xi}''$ corresponding to the nodes which are not in
contact with the foundation.

Fig 6.14. and 6.15. show the distribution of the tangential, the normal component of contact stresses, respectively, after 1000 iterations of the
bundle-Newton method;

Fig 6.16. compares the tangential displacements on Γ_c of this and the previous
example.

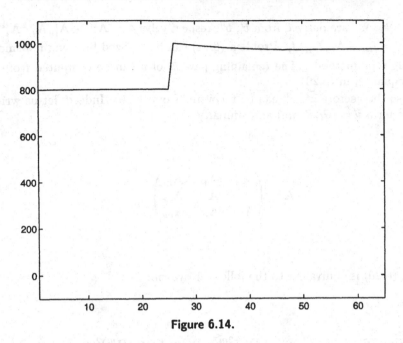

Figure 6.14.

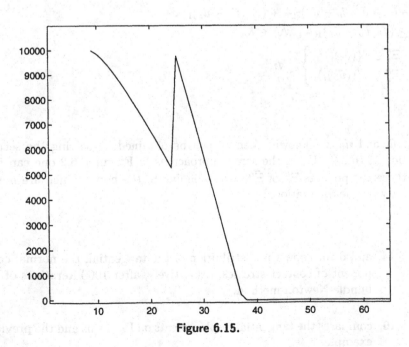

Figure 6.15.

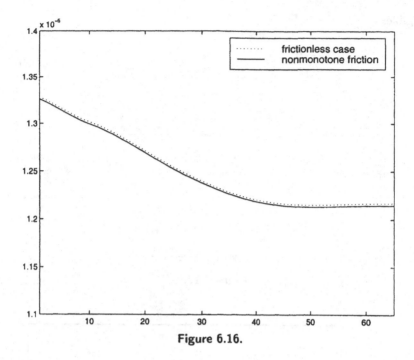

Figure 6.16.

6.2 DELAMINATION PROBLEM

Example 6.4 The studied two-dimensional laminated structure is depicted in Fig. 6.17. As in the previous examples we assume that the structure is made of a linear elastic isotropic material obeying the plane stress model (6.1). The used data are as follows: the modulus of elasticity $E = 1.7 \times 10^{11} N/m^2$, Poisson's ratio $\nu = 0.315$ (steel again) and the element thickness $t = 5mm$. Because of the symmetry of the structure and by assuming that the forces applied to the lower and the upper lamina are exactly the same it is sufficient to consider only the upper part.

Let $\Omega \subset \mathbf{R}^2$ be the upper lamina. The boundary $\partial\Omega$ of Ω consists of four subsets Γ_u, Γ_c, Γ_P^1 and Γ_P^2 (see Fig. 6.17). On Γ_u we have:

$$u_i = 0 \quad \text{on } \Gamma_u, \ i = 1, 2. \tag{6.23}$$

On the other hand, on $\Gamma_P^1 \cup \Gamma_P^2$ the boundary forces are prescribed:

$$T = (0, P) \quad \text{on } \Gamma_P^1, \tag{6.24}$$
$$T = (0, 0) \quad \text{on } \Gamma_P^2.$$

Finally we assume that the nonmonotone law $\hat{b}$ depicted in Fig. 6.18 describes the behaviour of the binding interlayer material in the normal direction on Γ_c, while the interaction between the laminae in the tangential direction s negligible:

$$\begin{cases} T_1 = 0 & \text{on } \Gamma_c \\ -T_2(x) \in \hat{b}(u_2(x)) & \text{for a.a. } x \in \Gamma_c. \end{cases} \quad (6.25)$$

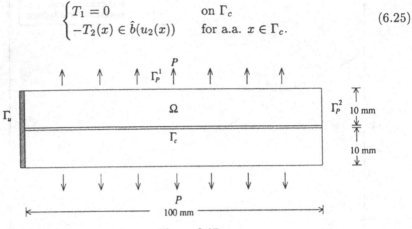

Figure 6.17.

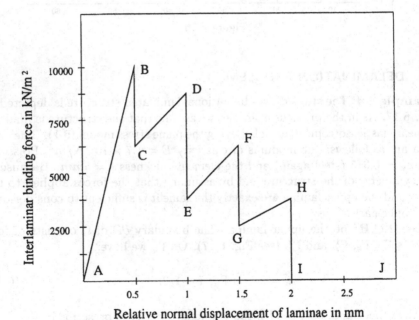

Relative normal displacement of laminae in mm

Figure 6.18.

Further, because of the nonpenetration of the laminae we have also the unilateral boundary condition on Γ_c:

$$u_2 \geq 0 \quad \text{a.e. on } \Gamma_c. \quad (6.26)$$

Then the equilibrium state of the body Ω is characterized by the following constrained hemivariational inequality:

$$\begin{cases} \text{Find } (u, \Xi) \in K \times L^2(\Gamma_c) \text{ such that} \\ a(u, v - u) + \displaystyle\int_{\Gamma_c} \Xi(v_2 - u_2)dx_1 \geq P \int_{\Gamma_P^1} (v_2 - u_2)dx_1 \quad \forall v \in K \\ \Xi(x) \in \hat{b}(u_2(x)) \quad \text{for a.a. } x \in \Gamma_c, \end{cases} \tag{$\mathcal{P}$}_4$$

where the bilinear form a is defined by (6.5) and

$$V = \{v \in H^1(\Omega; \mathbf{R}^2) \mid v = 0 \text{ a.e. on } \Gamma_u\},$$
$$K = \{v \in V \mid v_2 \geq 0 \text{ a.e. on } \Gamma_c\}.$$

For the approximation of $(\mathcal{P})_4$ we apply the finite element method presented in Chapter 3. The construction of the triangulation $\mathcal{D}_h$ of $\overline{\Omega}$ is depicted in Fig. 6.19.

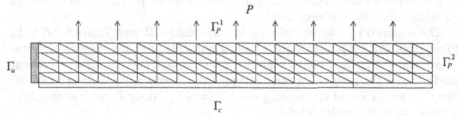

Figure 6.19. 20×4 triangulation $\mathcal{D}_h$ of $\overline{\Omega}$

We have used four different triangulations 10×2, 20×4, 40×8 and 80×16 corresponding to $h = 1, 1/2, 1/4, 1/8$, respectively. We define

$$K_h = \big\{v_h = (v_{h1}, v_{h2}) \in C(\overline{\Omega}; \mathbf{R}^2) \mid v_h|_T \in (P_1(T))^2 \quad \forall T \in \mathcal{D}_h,$$
$$v_{hi} = 0 \text{ on } \Gamma_u, \ i = 1, 2, \ v_{h2}(x^i) \geq 0 \ \forall i\big\},$$

in which x^i are the contact nodes on $\overline{\Gamma}_c \setminus \overline{\Gamma}_u$ and $K_h \subset K$ is a convex subset of all continuous, piecewise linear kinematically admissible functions on $\mathcal{D}_h$. For the approximation Y_h of $Y = L^2(\Gamma_c)$ we use Type IV elements as in Examples 6.1-6.3 with the necessary modification of one element $K \in \mathcal{T}_h$ due to nonempty intersection of $\overline{\Gamma}_c$ and $\overline{\Gamma}_u$ (see p.130). Also P_h is defined by (3.76). Then the approximation of $(\mathcal{P})_4$ is defined as follows:

$$\begin{cases} \text{Find } (u_h, \Xi_h) \in K_h \times Y_h \text{ such that} \\ a(u_h, v_h - u_h) + \displaystyle\int_{\Gamma_c} \Xi_h P_h(v_{h2} - u_{h2})dx_1 \\ \geq P \int_{\Gamma_P^1} (v_{h2} - u_{h2})dx_1 \quad \forall v_h \in K_h \\ \Xi(x^i) \in \hat{b}(u_{h2}(x^i)) \quad \forall i. \end{cases} \tag{$\mathcal{P}$}_4^h$$

Applying Theorem 3.9 we conclude that $(\mathcal{P})_4$ and $(\mathcal{P})_4^h$ are close on subsequences meaning that

$$u_h \to u \quad \text{in } V,$$
$$\Xi_h \rightharpoonup \Xi \quad \text{in } L^2(\Gamma_c), \ h \to 0+$$

for an appropriate subsequence of solutions to $(\mathcal{P})_4^h$.

The algebraic form of $(\mathcal{P})_4^h$ is the same as in Example 6.2, i.e., (6.16) except $\mathcal{K}$ which is now defined by

$$\mathcal{K} = \{\vec{x} \in \mathbf{R}^n \mid \vec{x}_\nu \geq 0\}.$$

In numerical calculations we will use the transformed algebraic form (6.17) and the corresponding superpotential $\mathcal{L}$ given by (6.18).

For the numerical realization of (6.18) two versions of the proximal bundle methods PB and PBM and the bundle-Newton code BN are applied. We have solved (6.18) with three constant loadings $P = 6000, 8000, 10000 kN/m^2$ and with four discretizations 10×2, 20×4, 40×8 and 80×16 (see Fig. 6.19) corresponding to the following number of the contact nodes $m = 10, 20, 40, 80$. In calculations the incremental procedure is used: the load applied to the structure is increased monotonically and the solution of the previous load is used as an initial guess for the next load.

The obtained results are collected in Figs. 6.20-6.25 and Tables 6.1-6.5. In Figs. 6.20-6.25 we illustrate the physical quantities like the normal displacements and the normal binding forces on the contact part Γ_c for the different loadings and the progress of the damage of the interlaminar binding material. For the calculation of the binding force Ξ $(= -T_2)$ along Γ_c we use the same strategy as in Examples 6.1-6.3.

In Tables 6.1-6.5 we have listed some details of the numerical experiments. Let us comment them shortly. We have tried to choose the same stopping criterions for the different codes: we require that the relative accuracy in the objective function value is about 6 digits. Note that this criterion is different from that one used in Examples 6.1-6.3.

Remark 6.2 *The different stopping criterions are one reason for the fact that the iteration numbers Ni of BN code are much higher in Examples 6.1-6.3 than in Example 6.4. Another reason is that we used the exact Hessian matrix in Example 6.4 while an approximation of this matrix was used in Examples 6.1-6.3. Further, we scaled the data of Example 6.4 in such a way that we avoided the calculation with very small and large numbers causing a numerical inaccuracy.*

From Tables 6.1-6.4 we see that all the codes found the same solutions for the loads 6000, $10000 kN/m^2$ and the objective function values differed only in the fifth digits (in some cases six digits were the same). Hence, we can conclude that all of them are reliable at least in this example. The differences in the solutions for the load $8000 kN/m^2$ are due to the fact that the damage of the binding interlaminar material starts close to this load. Therefore, the structure is very unstable and sensitive to the increase of the loading resulting in different solutions for different codes. If we compare a number of iterations Ni the bundle-Newton code BN is clearly superior. This is, of course, very expectable because of the piecewise quadratic nature of the objective function: BN as the second order method uses piecewise quadratic approximations for the objective function and, on the other hand, PB and PBM as the first order methods use piecewise polyhedral approximations. Further, what is very important that the iteration number of the bundle-Newton code does not seem to depend on the dimension of the problem. The results also indicate that PB is more efficient than PBM.

Finally in Table 6.5 we have listed the average error of the normal displacements on Γ_c $(= 1/m \sum_{i=1}^{m} |u_{h2}^{comp}(x^i) - u_{h2}^{exac}(x^i)|)$ for different discretizations. As an exact solution we have used the solution obtained with the finest grid 80×16. These results are due to the bundle-Newton code BN. From these we see that the convergence rate is approximately $\mathcal{O}(h^2)$.

Below we explain more precisely the details of Figs. 6.20-6.25 and of Tables 6.1-6.5:

Fig. 6.20. illustrates the development of the normal displacements u_2 of the upper lamina along Γ_c for the different applied loads 6000, 8000, $10000kN/m^2$ corresponding to the discretization 40×8 (number of the contact nodes is $m = 40$). Note that the relative displacements of the upper and the lower lamina are twice larger. The proximal bundle (PB) and the bundle-Newton (BN) gives different solutions with the load $8000kN/m^2$ (for more details see Fig. 6.21).

Fig. 6.21. shows the evolution of the delamination (corresponding to the solutions in Fig. 6.20). With the load $6000kN/m^2$ all the contact nodes are on the branch A-B of Fig. 6.18: no damage of the interlaminar binding material occurs. When the load is close to $8000kN/m^2$ the delamination starts and the structure is very sensitive on a small increase of the loading. This is illustrated by the fact that PB and BN gives qualitatively different solutions. There is no failure of the interface material for the solution by PB and, therefore, all the nodes are still on the branch A-B. On the other hand, for the solution by BN a partial damage of the binding material appears. Thus, some of the nodes are on the branch C-D. Finally with the load $10000kN/m^2$ the complete damage of the laminate structure has occurred: the most of the nodes are on the branch I-J. Only the fact that the laminate structure is fixed along Γ_u prevents that the upper and the lower lamina are not completely separated.

Fig. 6.22. shows the development of the normal binding force $-T_2$ along Γ_c for the loads 6000, 8000, $10000kN/m^2$ when $m = 40$. From these graphs it is easy to see which nodes are on branch A-B, C-D, D-E, F-G or I-J of Fig. 6.18. The results are due to BN.

Fig. 6.23. the same as the previous figure but for $m = 80$.

Fig. 6.24. illustrates the history of the binding force $-T_2$ for the loading 8000 kN/m^2 along Γ_c after 0 (initial guess given by the solution corresponding to the load $6000kN/m^2$), 5, 10 and 17 (the final solution) iterations of BN with $m = 40$.

Fig. 6.25. the same as the previous figure but for the load $10000kN/m^2$.

Tabs. 6.1.-6.4. show the final results obtained by different codes (BN, PB and PBM) for the differents loads (6000, 8000, $10000kN/m^2$) and for the different discretizations ($m = 10, 20, 40, 80$). Ni denotes the number of iterations and $\mathcal{L}$ the objective function value at termination.

Tab. 6.5. shows the average error of the normal displacements on Γ_c for the loads 6000, 8000, $10000kN/m^2$ when $m = 10, 20, 40$. As an exact solution we take the solution obtained for $m = 80$. These results are due to the bundle-Newton method (BN).

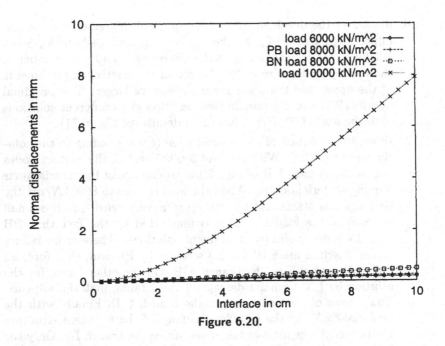

Figure 6.20.

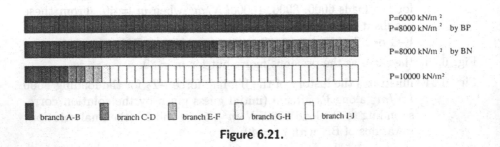

Figure 6.21.

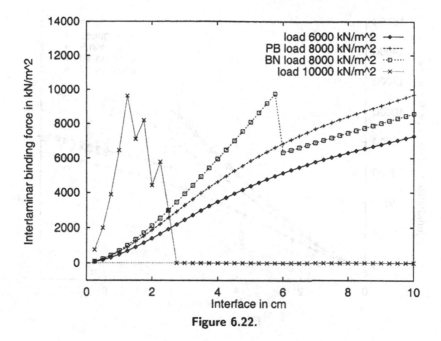

Figure 6.22.

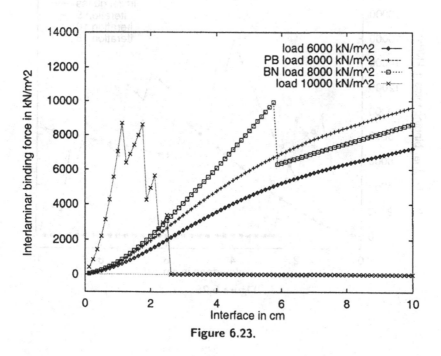

Figure 6.23.

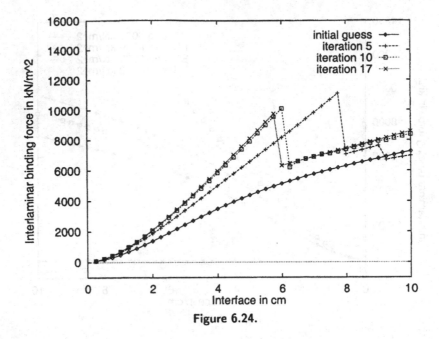

Figure 6.24.

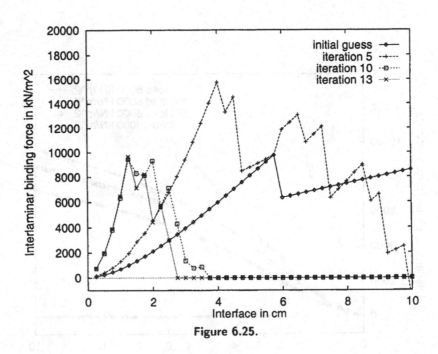

Figure 6.25.

Table 6.1. Discretization 10×2.

Code	$P=6000 \text{ kN/m}^2$		$P=8000 \text{ kN/m}^2$		$P=10000 \text{ kN/m}^2$	
	Ni	$\mathcal{L}$	Ni	$\mathcal{L}$	Ni	$\mathcal{L}$
BN	11	-0.134127	7	-0.243521	13	-1.506819
PB	171	-0.134127	216	-0.238448	392	-1.507079
PBM	1116	-0.134127	818	-0.243521	1698	-1.507082

Table 6.2. Discretization 20×4.

Code	$P=6000 \text{ kN/m}^2$		$P=8000 \text{ kN/m}^2$		$P=10000 \text{ kN/m}^2$	
	Ni	$\mathcal{L}$	Ni	$\mathcal{L}$	Ni	$\mathcal{L}$
BN	9	-0.145075	12	-0.272993	13	-4.195599
PB	246	-0.145075	254	-0.257911	902	-4.195599
PBM	6074	-0.145075	2429	-0.272992	3395	-4.195600

Table 6.3. Discretization 40×8.

Code	$P=6000 \text{ kN/m}^2$		$P=8000 \text{ kN/m}^2$		$P=10000 \text{ kN/m}^2$	
	Ni	$\mathcal{L}$	Ni	$\mathcal{L}$	Ni	$\mathcal{L}$
BN	10	-0.149333	16	-0.283673	13	-5.936222
PB	347	-0.149333	387	-0.265481	1544	-5.936187
PBM	2421	-0.149333	7586	-0.265481	5354	-5.936222

Table 6.4. Discretization 80×16.

Code	$P=6000 \text{ kN/m}^2$		$P=8000 \text{ kN/m}^2$		$P=10000 \text{ kN/m}^2$	
	Ni	$\mathcal{L}$	Ni	$\mathcal{L}$	Ni	$\mathcal{L}$
BN	8	-0.150651	16	-0.286784	13	-6.561239
PB	479	-0.150650	433	-0.267822	1605	-6.561173
PBM	3085	-0.150651	2605	-0.268245	8370	-6.561239

Table 6.5.   Average error of the normal displacements on Γ_c by BN.

Discretization	Load=6000 kN/m² Error in mm	Load=8000 kN/m² Error in mm	Load=10000 kN/m² Error in mm
10 × 2	0.019	0.064	2.4
20 × 4	0.0045	0.018	1.1
40 × 8	0.0011	0.0034	0.29
80 × 16	*	*	*

References

Dem'yanov, V. F., Stavroulakis, G. E., Polyakova, L. N., and Panagiotopoulos, P. D. (1996). *Quasidifferentiability and nonsmooth modelling in mechanics, engineering and economics.* Kluwer Academic, Dordrecht.

Lukšan, L. and Vlček, J. (1995). A bundle-newton method for nonsmooth unconstrained minimization. Technical report 654, Institute of Computer Science, Academy of Sciences of the Czech Republic.

Lukšan, L. and Vlček, J. (1998). A bundle-newton method for nonsmooth unconstrained minimization. *Math. Programming*, 83:373–391.

Mäkelä, M. M., Miettinen, M., Lukšan, L., and Vlček, J. (1999). Comparing nonsmooth nonconvex bundle methods in solving hemivariational inequalities. *J. Global Optim.*, 14:117–135.

Mäkelä, M. M. and Neittaanmäki, P. (1992). *Nonsmooth Optimization: Analysis and Algorithms with Applications to Optimal Control.* World Scientific Publishing Co., Singapore.

Miettinen, M. and Haslinger, J. (1995). Approximation of nonmonotone multivalued differential inclusions. *IMA J. Numer. Anal.*, 15:475–503.

Miettinen, M., Mäkelä, M. M., and Haslinger, J. (1995). On numerical solution of hemivariational inequalities by nonsmooth optimization methods. *J. Global Optim.*, 6:401–425.

Mistakidis, E. S. and Stavroulakis, G. E. (1998). *Nonconvex optimization in mechanics. Smooth and nonsmooth algorithms, heuristics and engineering applications by the F.E.M.* Kluwer Academic Publisher, Dordrecht, Boston, London.

Panagiotopoulos, P. D. (1993). *Hemivariational inequalities. Applications in mechanics and engineering.* Springer, Berlin, Heidelberg, New York.

Index